New Developments in the Analysis of Nonlocal Operators

CONTEMPORARY MATHEMATICS

723

New Developments in the Analysis of Nonlocal Operators

AMS Special Session
New Developments in the
Analysis of Nonlocal Operators
October 28–30, 2016
University of St. Thomas, Minneapolis, Minnesota

Donatella Danielli
Arshak Petrosyan
Camelia A. Pop
Editors

AMS AMERICAN MATHEMATICAL SOCIETY

2010 *Mathematics Subject Classification.* Primary 35R09, 35R11, 35R35;
Secondary 11M41, 26A33, 60G51, 91G80.

Library of Congress Cataloging-in-Publication Data

Names: AMS Special Session on New Developments in the Analysis of Nonlocal Operators (2016: Minneapolis, Minn.) | Danielli, Donatella, 1966– editor. | Petrosyan, Arshak, 1975– editor. | Pop, Camelia A., 1983– editor.

Title: New developments in the analysis of nonlocal operators: AMS Special Session on New Developments in the Analysis of Nonlocal Operators, October 28–30, 2016, University of St. Thomas, Minneapolis, Minnesota / Donatella Danielli, Arshak Petrosyan, Camelia A. Pop, editors.

Description: Providence, Rhode Island: American Mathematical Society, [2019] | Series: Contemporary mathematics; volume 723 | Includes bibliographical references.

Identifiers: LCCN 2018039664 | ISBN 9781470441104 (alk. paper)

Subjects: LCSH: Operator theory–Congresses. | Functional analysis–Congresses. | Differential equations–Congresses.

Classification: LCC QA329 .A49 2019 | DDC 515/.724–dc23

LC record available at https://lccn.loc.gov/2018039664

Contemporary Mathematics ISSN: 0271-4132 (print); ISSN: 1098-3627 (online)

DOI: https://doi.org/10.1090/conm/723

∞ The paper used in this book is acid-free and falls within the guidelines
established to ensure permanence and durability.
Visit the AMS home page at `https://www.ams.org/`

10 9 8 7 6 5 4 3 2 1 24 23 22 21 20 19

Contents

Preface

Over the last decade there has been a resurgence of interest in nonlocal operators. The distinctive feature of such operators, and of the associated equations, is that (unlike in the case of classical partial differential equations) the behavior of the solution at a point depends not only on the behavior of the function nearby but also on the values of the function far away. To a great extent, the study of nonlocal equations is motivated by applications. For instance, fully nonlinear integro-differential equations naturally arise in the study of certain problems in stochastic control. Another prime example of a nonlocal operator of elliptic type is the fractional Laplacian. In turn, the study of nonlocal operators has led to the development of a wide range of new mathematical tools and methods, and much progress has been made by researchers working in different areas.

This volume contains papers contributed by speakers and participants of the Special Session on New Developments in the Analysis of Nonlocal Operators, AMS Sectional Meeting at the University of St. Thomas, Minneapolis, Minnesota, October 28–30, 2016. The aim of the special session was to stimulate interaction on the latest developments of analytic, geometric, and probabilistic methods for problems involving nonlocal operators. The volume starts with the paper "Fractional thoughts" by N. Garofalo, which provides a comprehensive introduction to various aspects of the fractional Laplacian (also spelled Laplacean), with many historical remarks and an extensive and up-to-date list of references, suitable for the beginners and more seasoned researchers alike. M. Allen's paper proves uniqueness for weak solutions to abstract parabolic equations with the fractional Marchaud or Caputo time derivatives. H. Chang-Lara and O. Savin study the behavior of the one-phase Bernoulli free boundary problem near a fixed boundary by relating to a Signorini-type obstacle problem, which in turn is related to an obstacle problem for a half Laplacian. P. L. De Nápoli and P. R. Stinga reveal connections between fractional powers of the spherical Laplacian and functions from the analytic number theory and differential geometry such as Hurwitz zeta function and the Minakshisundaram zeta function. The volume concludes with a paper by the three of us on obstacle problems for a class of not stable-like nonlocal operators that include non-Gaussian asset price models widely used in mathematical finance, such as Variance Gamma Processes and Regular Lévy Processes of Exponential type.

We hope you will enjoy this volume!

Donatella Danielli
Arshak Petrosyan
Camelia A. Pop

Contemporary Mathematics
Volume **723**, 2019
https://doi.org/10.1090/conm/723/14569

Fractional thoughts

Nicola Garofalo

Dedicated to my family

ABSTRACT. In this note we present some of the most basic aspects of the operator $(-\Delta)^s$ with a self-contained and purely didactic intent, and with a somewhat different slant from the existing excellent references. Given the interest that nonlocal operators have generated since the extension paper of Caffarelli and Silvestre [Partial Differential Equations 32 (2007), no. 7–9, pp. 1245–1260], we feel it is appropriate offering to young researchers a quick additional guide to the subject which, we hope, will nicely complement the existing ones.

CONTENTS

2010 *Mathematics Subject Classification.* Primary 34A08, 34K37, 35R11.
This work was supported in part by a grant "Progetti d'Ateneo, 2013", University of Padova.

"...Beyond this there is nothing but prodigies and fictions, the only inhabitants are the poets and inventors of fables; there is no credit, or certainty any farther"

—Plutarch, Lives

1. Introduction

In his visionary papers [**R38**] and [**R49**] Marcel Riesz introduced the fractional powers of the Laplacean in Euclidean and Lorentzian space, developed the calculus of these nonlocal operators and studied the Dirichlet and Cauchy problems for respectively $(-\Delta)^s$ and $(\partial_{tt} - \Delta)^s$. The introduction of [**R38**] reads:... "On peut en particulier considérer certains procédés d'intégration de charactèr elliptique, hyperbolique et parabolique respectivement. Dans tout ces procédés l'intégrale d'ordre deux joue un rôle particulier, elle constitue l'inverse des opérations qui figurent respectivement dans l'équation de Laplace, celle des ondes et celle de la chaleur. Nous nous sommes occupé en particulier des deux premiers procédés et nous avons l'intention de rassembler nos recherches dans un mémoire élaboré. En attendant, nous donnons dans le présent travail un résumé assez détaillé de nos résultats concernant l'intégration elliptique et les potentiels qui y correspondant..."

Pseudo-differential operators such as $(-\Delta)^s$, $(\partial_{tt} - \Delta)^s$, $(\partial_t - \Delta)^s$, and the very different operators $\partial_{tt} + (-\Delta)^s$ and $\partial_t + (-\Delta)^s$, play an important role in many branches of the applied sciences ranging from fluid dynamics, to elasticity and to quantum mechanics. For instance, a main protagonist of geophysical fluid dynamics is the two-dimensional quasi-geostrophic equation (QGE)

$$\begin{cases} \theta_t + < u, \nabla\theta >= \kappa(-\Delta)^s\theta, \\ u = \nabla^\perp\psi, \qquad -\theta = (-\Delta)^{1/2}\psi, \end{cases}$$

where:

- ψ is the stream function;
- θ is the potential temperature,
- u is the velocity.

The parameter κ represents the viscosity, and $s \in (0,1)$. The QGE is one important instance in which the nonlocal operators $(-\Delta)^s$ and $\partial_t + (-\Delta)^s$ appear, see [**CC04**], [**CV10**], and the references therein. These nonlocal operators also present themselves in the convergence of nonlocal threshold dynamics approximations to front propagation, see [**CSo10**]. In [**FdlL86**] the authors study stability against collapse of a quantum mechanical system of N electrons and M nuclei interacting by pure Coulomb forces. For a single quantized electron attracted to a single nucleus having charge Z the relevant operator at study is the Hamiltonian

$$H = (-\Delta)^{1/2} - \frac{\alpha Z}{|x|},$$

where $\alpha > 0$ is the fine structure constant. Another example comes from elasticity, where the famous Signorini problem has been shown to be equivalent to the obstacle problem for $(-\Delta)^{1/2}$, see [**AC04**], [**CS07**], [**ACS08**], [**CSS08**], [**GP09**] and [**PSU12**]. Yet another instance is the phenomenon of osmosis, whose description can be converted into an obstacle problem for the fractional heat equation $(\partial_t - \Delta)^{1/2}$, see [**DL69**] and [**DGPT17**]. In the study of internal travelling solitary waves in a stable two-layer perfect fluid of infinite depth contained above a rigid horizontal bottom one, or in soliton theory, one has the Benjamin-Ono equation

$$(-\Delta)^{1/2}u + u - u^2 = 0$$

on the line $\mathbb{R}$. A basic question is the uniqueness of solutions, see [**AT91**], and also the more recent works [**FLe13**], [**FLeS16**] for important generalizations of the results in [**AT91**].

Besides these phenomena, the nonlocal operators listed above also arise prominently in other branches of mathematics, such as e.g. geometry, probability and financial mathematics. For some of these aspects we refer the reader to:

1. the classical volumes of E. Dynkin on Markov processes [**Dy65**];
2. the pioneering works of Silvestre [**Si07**], and Caffarelli and Silvestre [**CS07**];
3. the "obstacle" book by Petrosyan, Shahgholian and Uraltseva [**PSU12**];
4. the hitchiker's guide by Di Nezza, Palatucci and Valdinoci [**DPV12**];
5. the lecture notes of Bucur and Valdinoci [**BuV16**];
6. the survey paper [**Go16**] by M. del Mar González;
7. the variational book [**MRS16**] by Molica Bisci, Radulescu and Servadei;
8. the survey paper [**DS17**] by Danielli and Salsa;
9. the survey papers [**RO15**], [**RO17**] and [**RO17'**] by Ros-Oton;
10. the forthcoming volume edited by Kuusi and Palatucci [**KP17**];
11. the lecture notes [**DMV17**] by Dipierro, Medina and Valdinoci;
12. the "diffusion" lecture notes [**V17**] by J. L. Vazquez;
13. the recent lecture notes [**AV17**] by Abatangelo and Valdinoci.

For an introduction to the subject of fractional differentiation and integration from the point of view of analysis the essential references are:

14. M. Riesz' already cited original papers [**R38**] and [**R49**];
15. E. Stein's landmark book on singular integrals [**St70**];
16. Landkov's book on potential theory [**La72**];
17. the volume on fractional differentiation by Samko, Kilbas and Marichev [**SKM93**].

An interesting account of the fractional calculus, its applications and historical development can be found in:

18. the fractional book [**OS74**] by Oldham and Spanier;
19. the article [**Ro77**] by B. Ross.

Our objective in this note is presenting some of the most basic aspects of the operator $(-\Delta)^s$ with a self-contained and purely didactic intent, and with a somewhat different slant from the above cited references which of course reflects the taste of the author. Given the interest that nonlocal operators have generated since the extension paper of Caffarelli and Silvestre [**CS07**], we feel it is appropriate offering to young researchers a quick additional guide to the subject which, we hope, will nicely complement the (often more advanced) existing ones.

A list of the topics covered by this paper is provided by the table of content, but let us say something more in detail:

- In Section 2 we introduce the main pointwise definition of the nonlocal operator $(-\Delta)^s$, see (2.8) below. This is the starting point of the whole note as all the material presented here is, in one way or the other, derived from it. In Proposition 2.9 we show that the definition (2.8) implies a decay at infinity of the fractional Laplacean that plays an important role in its analysis.

- Section 3 contains a brief discussion of the maximum principle, the Harnack inequality and the theorem of Liouville in the fractional setting. We do not make any attempt at discussing these aspects extensively, but we simply confine ourselves to make the (unfamiliar) reader acquainted of the differences with their local counterparts, and then refer to the existing sources.
- Section 4 constitutes a brief interlude on two important protagonists of classical analysis which also play a central role in this note: the Fourier transform and Bessel functions. These two classical subjects are inextricably connected. One the one hand, the Bessel functions are eigenfunctions of the Laplacean. On the other, they also appear (the curvature of the unit sphere in $\mathbb{R}^n$ is lurking in the shadows here) as the Fourier transform of the measure carried by the unit sphere. In this connection, and since it is a recurrent ingredient in this note, we recall the classical Fourier-Bessel integral formula due to Bochner, see Theorem 4.4 below.
- Section 5 opens with the proof of Proposition 5.1, which describes the action of $(-\Delta)^s$ on the Fourier transform side. This result proves an important fact: the fractional Laplacean is a *pseudo-differential operator*, i.e., one of those nonlocal operators that can be written in the form

$$Tu(x) = \int_{\mathbb{R}^n} e^{2\pi i<x,\xi>} p(x,\xi)\hat{u}(\xi)d\xi,$$

where the function $p(x,\xi)$, known as the *symbol* of the operator, is required to belong to a certain class. A basic consequence of Proposition 5.1 is the semigroup property in Corollary 5.3 and the "integration by parts" Lemma 5.4, which shows that $(-\Delta)^s$ is a symmetric operator. We close the section with the computation in Proposition 5.6 of the normalization constant $\gamma(n,s)$ in the pointwise definition (2.8).
- Section 6 is devoted to discussing a basic question of interest in analysis and geometry which was asked by Strichartz in [**Str83**], and which has generated a considerable amount of work. We introduce the vector-valued Riesz transform $\mathcal{R} = \nabla(-\Delta)^{-1/2}$, and we show that being able to answer in the affirmative such question hinges upon the L^p mapping properties of $\mathcal{R}$. This is in turn intimately connected to the subject of Section 20 below.
- Similarly to the classical Laplacean, $(-\Delta)^s$ preserves spherical symmetry. In Section 7 we make this property more precise. Using Theorem 4.4 we provide an "explicit" formula for the fractional Laplacean of a spherically symmetric function.
- The purpose of Section 8 is multifold. Our declared intent is computing the fundamental solution of $(-\Delta)^s$, i.e., proving Theorem 8.4. This can be done in several ways. To the best of our knowledge, the approach we choose, although very classical, has not been tried before. We have introduced a regularization (8.6) of the fundamental solution, and in Lemma 8.5 we compute its Fourier transform. In Lemma 8.6 below we use this result to calculate the fractional Laplacean of the regularized fundamental solution, and with such result we finally prove Theorem 8.4.

- Using these results, in Section 9 we show that this approach leads in a natural way to the beautiful discovery of the functions (9.2). Remarkably, such functions are solutions of the nonlocal nonlinear equation

$$(-\Delta)^s u = u^{\frac{n+2s}{n-2s}},$$

which generalizes to the fractional setting the celebrated *Yamabe equation* from Riemannian geometry. The latter is obtained when $s = 1$.
- Section 10 presents in detail the central theme of the analysis of the fractional Laplacean: the extension problem of Caffarelli and Silvestre (10.1) below. We construct the Poisson kernel for the extension operator, and provide two proofs of (10.2), which characterizes $(-\Delta)^s$ as the weighted Dirichlet-to-Neumann map of the extension problem. The extension procedure is a very powerful tool which has been applied so far in many different directions, and it is hardly possible to accurately describe the impact of this paper in the field. A prominent one is the theory of *free boundaries*, which was in fact the main motivation behind the work [**CS07**] itself. Another remarkable application has been given to geometry in the work [**CG11**], where the authors used the extension procedure to characterize the fractional powers of the so-called *Paneitz operator*, a conformally covariant operator of order four, as the (weighted) Dirichlet-to-Neumann map on a conformally compact Einstein manifold. In the opening of Section 10 we also discuss briefly the beautiful 1965 paper [**MS65**] by Muchenhoupt and Stein which seems not known to the fractional community, but that deserves to be considered in connection with the extension procedure.
- In Section 11 we discuss one interesting aspect of the extension procedure which is perhaps not so well-known in the fractional community: the link between the nonlocal operator $(-\Delta)^s$ and the subelliptic operator $\mathcal{P}_\alpha$, which we define in (11.19) below, that was introduced by S. Baouendi in his 1967 Ph. D. Dissertation [**Ba67**]. Proposition 11.2 below shows that $(-\Delta)^s$ arises as the true Dirichlet-to-Neumann map of the so-called Baouendi-Grushin operator. Furthermore, it is possible to relate in a one-to-one onto fashion solutions of the extension operator L_a to those of $\mathcal{P}_\alpha$. In Proposition 11.4 we show that there is a direct link between the non-isotropic (sub-Riemannian) pseudo-balls naturally associated with $\mathcal{P}_\alpha$, and the Euclidean balls which are instead the natural ones for the extension operator L_a.
- In Section 12 we exploit this connection further to provide a proof of a fundamental property of the nonlocal operator $(-\Delta)^s$: its hypoellipticity. At first such property might appear surprising since now we are not dealing with solutions of a partial differential equation (pde). But, a moment's thought reveals that, in the end, what really makes harmonic functions infinitely smooth is the fact that they satisfy the crucial integral property (12.1) below, which then results into (12.2). The pde is only a vessel that takes harmonic functions into the blessed land of "averaging". Since this aspect is shared by solutions of the nonlocal equation $(-\Delta)^s$, we should expect solutions of the latter to be infinitely smooth. The approach we take to the hypoellipticity of $(-\Delta)^s$ is "elementary"

and runs much along the lines of the Caccioppoli-Cimmino-Weyl lemma for the classical Laplacean, but we make use of the extension procedure. With the intent of advertising the link, discussed in Section 11, with the theory of subelliptic equations, we start from Proposition 12.6, which is a representation formula involving the Baouendi operator $\mathcal{P}_\alpha$. We use it to establish a corresponding result for the extension operator L_a, see Proposition 12.7 below. With such result we prove Theorem 12.15 which provides an interesting mean-value formula for solutions to $(-\Delta)^s u = 0$. Finally, in Theorem 12.17 we establish our "elementary" version of the Caccioppoli-Cimmino-Weyl lemma for $(-\Delta)^s$. We do not discuss at all the real-analytic hypoellipticity of $(-\Delta)^s$.

- Section 13 is devoted to the question of the regularity at the boundary for solutions of the Dirichlet problem (13.4). Unlike the interior regularity, here the situation departs drastically from the local case, in the sense that there exist real-analytic domains and real-analytic "boundary values" for which the solution to (13.4) is not better then Hölder continuous at the boundary. One notable example of this negative phenomenon is the *torsion function* for the ball for $(-\Delta)^s$, i.e., the solution to the Dirichlet problem $(-\Delta)^s u = 1$ in $B(0,R)$, $u = 0$ in $\mathbb{R}^n \setminus B(0,R)$. The relevance of such function, which we construct in Proposition 13.1, is multi-faceted. Remaining within the framework of the subject of interest of this section, the torsion function shows that standard Schauder theory fails for $(-\Delta)^s$, or at least such theory needs to be suitably reinterpreted. This negative phenomenon is akin, and not by chance, to the failure of Schauder theory which occurs at the so-called characteristic points in the theory of subelliptic equations. For the fractional Laplacean the correct boundary regularity is provided by Theorems 13.4 and 13.5 below: the former, due to Ros-Oton and Serra, states that in the Dirichlet problem with zero "boundary data" in a $C^{1,1}$ domain Ω, the function $\frac{u}{\operatorname{dist}(\cdot,\partial\Omega)^s}$ is Hölder continuous up to the boundary. The latter, due to Grubb, states that in a C^∞ domain this same function is in fact C^∞ up to the boundary.
- In partial differential equations, the most fundamental property of interest is the so-called *strong unique continuation property*. It states that if a solution to a certain differential operator $P(x,\partial_x)$ vanishes to infinite order at a point of a connected open set, then it must vanish identically. This property is true when $P(x,\partial_x) = -\Delta$, but it is shared by large classes of second order partial differential equations, even with very rough coefficients. Section 14 is devoted to establishing the strong unique continuation property of the fractional Laplacean, see Theorem 14.2 below. We prove such result using monotonicity formulas of Almgren type. We resort again to the extension procedure, and use the monotonicity formula in Theorem 14.5 from [**CS07**]. The difficulty in using the extension procedure is that the information that a solution of $(-\Delta)^s u = 0$ vanishes to infinite order at a point does not transfer to the solution of the extension problem. One needs to further implement a delicate blowup analysis of a special family of rescalings first introduced in [**ACS08**] in the study of the Signorini problem.

- In Section 15 we discuss the nonlocal Poisson kernel for the ball and one of its direct consequences, the mean-value formula for $(-\Delta)^s$. These tools were introduced by M. Riesz in [**R38**] and are by now part of the fractional folklore. We consider the nonlocal mean-value operator $\mathscr{A}_r^{(s)}u(x)$ defined by (15.6), (15.7), and in Proposition 15.4 we show that

$$\lim_{s\to 1} \mathscr{A}_r^{(s)}u(x) = \mathscr{M}_r u(x),$$

where $\mathscr{M}_r u(x)$ is the spherical mean-value operator of the classical potential theory, see (2.3) below. In Proposition 15.6 we show that the nonlocal mean-value operator can be used to provide yet another expression of the fractional Laplacean, much like the Blaschke-Privalov Laplacean is used in classical potential theory to define the Laplacean on nonsmooth functions. In Corollary 15.8 we establish a nonlocal analogue of the classical theorem of Köebe.
- Section 16 departs from the previous ones in that we start discussing the heat flow associated with $(-\Delta)^s$. There is of course more than one nonlocal heat equation, but here we focus on $\partial_t u + (-\Delta)^s u = 0$. We introduce the fractional heat semigroup (16.7) below, and we spend most part of the section proving a basic property of the fractional heat kernel $G_s(x,t)$, namely its positivity, see Propositions 16.2 and 16.3.
- In Section 17 we use the principle of subordination introduced by Bochner to establish a pointwise representation of $(-\Delta)^s$ in terms of the classical heat semigroup, see Theorem 17.2 below.
- Section 18 contains yet another important instance of subordination. In (18.1) below we introduce Bochner's subordination function and in Theorem 18.3 show the important fact that the nonlocal heat semigroup is obtained through subordination with the standard heat semigroup.
- The chain rule is one of the most basic and useful tools in the theory of partial differential equations. In Section 19 we discuss a simple, yet quite remarkable nonlocal analogue of the chain rule which was first found in [**CC03**]. An important consequence of it is that when u is a solution to $(-\Delta)^s u = 0$, then u^2 is also a subsolution. More in general, if u is a nonnegative subsolution, then u^p is a nonnegative subsolution for every $p > 1$. We recall that, in the local case, such property is at the heart, for instance, of Moser's proof of the Harnack inequality for divergence form equations with bounded measurable coefficients.
- Over the recent years there has been an explosion of activity surrounding the so-called Gamma calculus of Bakry-Emery and various powerful generalizations of the latter. Section 20 is devoted to providing the reader with a bird's eye-view of the basics of such calculus. Our primary motivation is proposing the development of a nonlocal Gamma calculus. In this perspective the reader might consider this section just as a glimpse into a possibly rich theory to come. In Definition 20.1 we introduce the notion of nonlocal *carré du champ*. Such object defines a Dirichlet form whose associated energy $\mathscr{E}_{(s)}(u)$ is given in Definition 20.4. Proposition 20.5 shows that $(-\Delta)^s$ has a variational nature, since it arises as the Euler-Lagrange equation of the nonlocal energy $\mathscr{E}_{(s)}(u)$. The experienced fractional reader will immediately recognize familiar objects here. The section ends with

a discussion of the famous Bakry-Emery *curvature-dimension inequality*, at the heart of which there is the celebrated Bochner identity, and with a challenging open question.

- In Section 21 we continue the discussion from the previous one. Our intent is to provide the reader with an elementary motivation for undertaking a new bigger effort. Namely, understanding the beautiful Li-Yau theory. We introduce a special case of the celebrated Li-Yau inequality and in Theorem 21.3 we provide an elementary proof of such inequality for the classical heat semigroup in flat $\mathbb{R}^n$. We use this result to give a simple, yet elegant proof of the well-known scale invariant Harnack inequality for the standard heat equation independently proved by Pini and Hadamard in the 50's. We close the section with two equivalent interesting conjectures regarding the nonlocal heat semigroup.
- Section 22 is devoted to discussing a Li-Yau inequality for the Bessel semigroup (22.1) on the half-line. The Bessel process $\mathscr{B}_a$ is ubiquitous in the fractional world, especially in view of his role in the extension procedure. Since this topic is perhaps more frequented by workers in probability than analysts and geometers (with the exception of people in harmonic analysis), we provide a purely analytical construction of the fundamental solution with Neumann boundary conditions of the heat semigroup associated with $\mathscr{B}_a$, see Proposition 22.3 below. We close the section with Proposition 22.5, in which we show that the heat semigroup associated with the Bessel process satisfies an inequality of Li-Yau type.
- We end this note with the very brief Section 23 in which we discuss a fractional *nonlinear* operator which constitutes the nonlocal counterpart of the well-known *p-Laplacean* defined by $-\Delta_p u = \operatorname{div}(|\nabla u|^{p-2}\nabla u)$. Since its introduction in [**AMRT09**] and independently in [**IN10**], the nonlocal p-Laplacean $(-\Delta_p)^s$ has generated a great deal of interest in the fractional community and thus we could not close without a brief mention of the fundamental open question in the area: the optimal interior regularity of its variational solutions.

A notable omission in this note is the beautiful developing theory of nonlocal minimal surfaces. For this we refer the reader to the seminal works [**CRS10**], [**SV14**], [**CSV15**], [**CDS16**], [**DSV16**], [**DSV17'**], and the references therein, as well as to the survey paper [**BuV16**] which contains a nice introduction to the subject.

The reader understands that, for obvious considerations of space, it is not possible to formally introduce every definition or tool used in this paper. Thus, for instance, we will not discuss the Schwartz space $\mathscr{S}(\mathbb{R}^n)$ of rapidly decreasing functions in $\mathbb{R}^n$, and its topological dual, the space $\mathscr{S}'(\mathbb{R}^n)$ of tempered distributions. Similarly, we will not explicitly introduce the topology of the spaces $C^\infty(\mathbb{R}^n)$, or $C_0^\infty(\mathbb{R}^n)$, and their duals, the spaces $\mathscr{E}'(\mathbb{R}^n)$ of compactly supported distributions, and the larger space of all distributions $\mathscr{D}'(\mathbb{R}^n)$ on $\mathbb{R}^n$. Nor we will discuss in detail the Fourier transform in $\mathbb{R}^n$. For these topics there exist several excellent classical books, such as e.g.: [**BC49**], [**B59**], [**GS64**], [**Sch66**], [**T67**], [**St70**], [**SW71**] and [**Y78**]. One additional source that the reader is encouraged to peruse is the monograph [**La72**], in which the author provides an extensive discussion of the potential theoretic aspects of the nonlocal Laplacean, based on M. Riesz' paper [**R38**].

Finally, the present note has been written within the constraints imposed by timeliness. Many important and/or relevant references have been left out simply because it has been impossible, within the short amount of time available, to consult the ample existing literature on nonlocal equations. The author sincerely apologizes with all those people whose work is not properly acknowledged here.

2. The fractional Laplacean

In this section we introduce the protagonist of this note, M. Riesz' fractional Laplacean $(-\Delta)^s$, with $0 < s < 1$. At the onset we seek to define the action of such nonlocal operator on a suitable function in the pointwise sense. With this objective in mind, it will be convenient to work with the space $\mathscr{S}(\mathbb{R}^n)$ of L. Schwartz' rapidly decreasing functions (whose dual $\mathscr{S}'(\mathbb{R}^n)$ is the space of tempered distributions), although larger classes can be allowed, see Remark 2.5 and Proposition 2.15 below. We recall that $\mathscr{S}(\mathbb{R}^n)$ is the space $C^\infty(\mathbb{R}^n)$ endowed with the metric topology

$$d(f,g) = \sum_{p=0}^{\infty} 2^{-p} \frac{||f-g||_p}{1+||f-g||_p},$$

generated by the countable family of norms

$$||f||_p = \sup_{|\alpha|\le p} \sup_{x\in\mathbb{R}^n} (1+|x|^2)^{\frac{p}{2}} |\partial^\alpha f(x)|, \qquad p \in \mathbb{N}\cup\{0\}. \tag{2.1}$$

As it is customary, if $\alpha = (\alpha_1, ..., \alpha_n)$, then $|\alpha| = \alpha_1 + ... + \alpha_n$, and we have indicated with ∂^α the partial derivative $\frac{\partial^{|\alpha|}}{\partial x_1^{\alpha_1} ... \partial x_n^{\alpha_n}}$.

Our initial observation is the following simple calculus lemma which could be used to provide a probabilistic interpretation of the classical Laplacean on the real line.

LEMMA 2.1. *Let $f \in C^2(a,b)$, then for every $x \in (a,b)$ one has*

$$-f''(x) = \lim_{y\to 0} \frac{2f(x) - f(x+y) - f(x-y)}{y^2}.$$

The expression in the right-hand side in the equation in Lemma 2.1 is known as the symmetric difference quotient of order two. If we introduce the "spherical" surface and "solid" averaging operators

$$\mathscr{M}_y f(x) = \frac{f(x+y)+f(x-y)}{2}, \qquad \mathscr{A}_y f(x) = \frac{1}{2y}\int_{x-y}^{x+y} f(t)dt,$$

then we can reformulate the conclusion in Lemma 2.1 as follows:

$$-f''(x) = 2\lim_{y\to 0}\frac{f(x) - \mathscr{M}_y f(x)}{y^2} = 6\lim_{y\to 0}\frac{f(x)-\mathscr{A}_y f(x)}{y^2},$$

where it is easily seen that the second equality follows from the first one and L'Hopital's rule. The result that follows generalizes this observation to $n \geq 2$.

PROPOSITION 2.2. *Let $\Omega \subset \mathbb{R}^n$ be an open set. For any $f \in C^2(\Omega)$ and $x \in \Omega$ we have*

$$-\Delta f(x) = 2n \lim_{r\to 0}\frac{f(x)-\mathscr{M}_r f(x)}{r^2} = 2(n+2)\lim_{r\to 0}\frac{f(x)-\mathscr{A}_r f(x)}{r^2}, \tag{2.2}$$

where $\Delta f = \sum_{k=1}^n \frac{\partial^2 f}{\partial x_k^2}$ is the operator of Laplace.

In the equation (2.2) we have indicated with

$$\mathscr{M}_r u(x) = \frac{1}{\sigma_{n-1} r^{n-1}} \int_{S(x,r)} u(y) d\sigma(y), \quad \mathscr{A}_r u(x) = \frac{1}{\omega_n r^n} \int_{B(x,r)} u(y) dy, \tag{2.3}$$

the spherical surface and solid mean-value operators. Here, $B(x,r) = \{y \in \mathbb{R}^n \mid |y-x| < r\}$, $S(x,r) = \partial B(x,r)$, $d\sigma$ is the $(n-1)$-Lebesgue measure on $S(x,r)$, and the numbers σ_{n-1} and ω_n respectively represent the measure of the unit sphere and that of the unit ball in $\mathbb{R}^n$, see (4.6) below. Either one of the limits in the right-hand side of (2.2) is known as the *Blaschke-Privalov Laplacean*, and its relevance in potential theory is that, unlike the standard Laplacean, one can define such operator on functions which are not smooth, see for instance [**He69**] and [**DP70**]. For the probabilistic interpretation of the Laplacean one should see [**Dy65**].

Before proceeding, and in preparation for the central definition of this section, let us observe that using (4.6) it is easy to recognize that we can write the second identity in (2.2) in the more suggestive fashion:

$$-\Delta u(x) = \frac{(n+2)\Gamma(\frac{n}{2}+1)}{\pi^{\frac{n}{2}}} \lim_{r\to 0^+} \int_{\mathbb{R}^n} [2u(x) - u(x+y) - u(x-y)] \frac{1}{r^{n+2}} \mathbf{1}_{B(0,r)}(y) dy, \tag{2.4}$$

where we have denoted by $\mathbf{1}_E$ the indicator function of a set $E \subset \mathbb{R}^n$.

In the applied sciences it is of great importance to be able to consider fractional derivatives of functions. There exist many different definitions of such operations, see [**OS74**], [**SKM93**], and the recent paper [**BMST16**], but perhaps the most prominent one is based on the notion of (Marcel) Riesz' potential of a function. To motivate such operation let us assume that $n \geq 3$, and recall that in mathematical physics the *Newtonian potential* of a function $f \in \mathcal{S}(\mathbb{R}^n)$ is given by

$$I_2(f)(x) = \frac{1}{4\pi^{\frac{n}{2}}} \Gamma\left(\frac{n-2}{2}\right) \int_{\mathbb{R}^n} \frac{f(y)}{|x-y|^{n-2}} dy,$$

where we have denoted by $\Gamma(z)$ Euler's gamma function (for its definition and basic properties see Section 4 below). Now, using (4.6) and the properties of the gamma function, one recognizes that the convolution kernel $\frac{1}{4\pi^{\frac{n}{2}}}\Gamma\left(\frac{n-2}{2}\right)\frac{1}{|x|^{n-2}}$ in the definition of $I_2(f)$ is just the fundamental solution

$$E(x) = \frac{1}{(n-2)\sigma_{n-1}} \frac{1}{|x|^{n-2}}$$

of $-\Delta$. With this observation in mind, we recall the well-known identity of Gauss-Green that says that for any $f \in \mathcal{S}(\mathbb{R}^n)$ one has

$$I_2(-\Delta f) = f.$$

Recall M. Riesz' words in the opening of this note: "...l'intégrale d'ordre deux joue un rôle particulier, elle constitue l'inverse des opérations qui figurent respectivement dans l'équation de Laplace..." In other words, the Newtonian potential is the inverse of $-\Delta$, i.e., $I_2 = (-\Delta)^{-1}$. This important observation leads to the introduction of M. Riesz' generalization of the Newtonian potential.

DEFINITION 2.3 (Riesz' potentials). For any $n \in \mathbb{N}$, let $0 < \alpha < n$. The *Riesz potential* of order α is the operator whose action on a function $f \in \mathscr{S}(\mathbb{R}^n)$ is given by

$$I_\alpha(f)(x) = \frac{\Gamma\left(\frac{n-\alpha}{2}\right)}{\pi^{\frac{n}{2}} 2^\alpha \Gamma\left(\frac{\alpha}{2}\right)} \int_{\mathbb{R}^n} \frac{f(y)}{|x-y|^{n-\alpha}} dy.$$

It is not difficult to prove that $I_\alpha(f) \in C^\infty(\mathbb{R}^n)$ for any $f \in \mathcal{S}(\mathbb{R}^n)$. Concerning the definition of I_α, we note that the normalization constant in it matches that of I_2 when $\alpha = 2$. The important reason behind it is that such constant is chosen to guarantee the validity of the following crucial result, a kind of fractional fundamental theorem of calculus, stating that for any $f \in \mathscr{S}(\mathbb{R}^n)$ one has in $\mathscr{S}'(\mathbb{R}^n)$

$$I_\alpha(-\Delta)^{\alpha/2} f = (-\Delta)^{\alpha/2} I_\alpha f = f. \tag{2.5}$$

Of course (2.5) makes no sense unless we say what we mean by the fractional operator $(-\Delta)^{\alpha/2}$. The most natural way to introduce it (suggested in fact by the spectral theorem) is by defining the action of $(-\Delta)^{\alpha/2}$ on the Fourier transform side by the equation

$$\mathscr{F}((-\Delta)^{\alpha/2} u) = (2\pi|\cdot|)^\alpha \mathscr{F}(u), \quad u \in \mathscr{S}'(\mathbb{R}^n), \tag{2.6}$$

where $\mathscr{F}$ is defined as in (4.14) below. The equation (2.5) shows that I_α inverts the fractional powers of the Laplacean, i.e.,

$$I_\alpha = (-\Delta)^{-\alpha/2}, \qquad 0 < \alpha < n. \tag{2.7}$$

For this reason I_α is also called the *fractional integration operator* of order α, see [**R38**], but also [**St70**], [**La72**], [**SKM93**]. For a nice account of M. Riesz' work one should read the commemorative note [**Ga70**]. An interesting historical overview of the development of fractional calculus is provided by the book [**OS74**] and the article [**Ro77**].

Since our focus in this note is the fractional Laplacean $(-\Delta)^s$ in the range $0 < s < 1$, we will henceforth let $s = \alpha/2$ in the above formulas, or equivalently $\alpha = 2s$. Although we have formally introduced such operator in the equation (2.6) above, such definition has a major drawback: it is not easy to understand a given function (or a distribution) by prescribing its Fourier transform. It is for this reason that we begin our story by introducing a different pointwise definition of the fractional Laplacean that is more directly connected to the symmetric difference quotient of order two in the opening calculus Lemma 2.1, and with (2.4), and thus has the advantage of underscoring the probabilistic interpretation of the operator $(-\Delta)^s$ as a symmetric random process with jumps, see [**MO69**], [**Si07**], [**RO15**] and [**BuV16**]. Later in Proposition 5.1 we will reconcile the two definitions.

DEFINITION 2.4. Let $0 < s < 1$. The *fractional Laplacean* of a function $u \in \mathscr{S}(\mathbb{R}^n)$ is the nonlocal operator in $\mathbb{R}^n$ defined by the expression

$$(-\Delta)^s u(x) = \frac{\gamma(n,s)}{2} \int_{\mathbb{R}^n} \frac{2u(x) - u(x+y) - u(x-y)}{|y|^{n+2s}} dy, \tag{2.8}$$

where $\gamma(n,s) > 0$ is a suitable normalization constant that is given implicitly in (5.1), and explicitly in Proposition 5.6 below.

Before proceeding we remark that when dealing with the nonlocal operator $(-\Delta)^s$ one often needs to distinguish between the three possible cases:

- $0 < s < \frac{1}{2}$;
- $s = \frac{1}{2}$;
- $\frac{1}{2} < s < 1$.

Since as $s \to 1^-$ the fractional Laplacean tends (at least, formally right now) to $-\Delta$, one might surmise that in the regime $\frac{1}{2} < s < 1$ the operator $(-\Delta)^s$ should display properties closer to those of the classical Laplacean, whereas since

$(-\Delta)^s \to I$ as $s \to 0^+$, the stronger discrepancies might present themselves in the range $0 < s < \frac{1}{2}$.

Having said this, it will be good for the reader who is for the first time confronted with definition (2.8) above to have in mind the following quote from p. 51 in [**La72**]:...“*In the theory of M. Riesz kernels, the role of the Laplace operator, which has a local character, is taken...by a non-local integral operator...This circumstance often substantially complicates the theory...*”

It is obvious that (2.8) defines a linear operator since for any $u, v \in \mathscr{S}(\mathbb{R}^n)$ and $c \in \mathbb{R}$ one has

$$(-\Delta)^s(u+v) = (-\Delta)^s u + (-\Delta)^s v, \qquad (-\Delta)^s(cu) = c(-\Delta)^s u.$$

It is also important to observe that the integral in the right-hand side of (2.8) is convergent. To see this, it suffices to write

$$\int_{\mathbb{R}^n} \frac{2u(x) - u(x+y) - u(x-y)}{|y|^{n+2s}} dy = \int_{|y|\le 1} \frac{2u(x) - u(x+y) - u(x-y)}{|y|^{n+2s}} dy$$
$$+ \int_{|y|>1} \frac{2u(x) - u(x+y) - u(x-y)}{|y|^{n+2s}} dy.$$

Taylor's formula for C^2 functions gives for $|y| \le 1$

$$2u(x) - u(x+y) - u(x-y) = - < \nabla^2 u(x)y, y > + o(|y|^2),$$

where we have indicated with $\nabla^2 u$ the Hessian matrix of u. Therefore,

$$\left| \int_{|y|\le 1} \frac{2u(x) - u(x+y) - u(x-y)}{|y|^{n+2s}} dy \right| \le C \int_{|y|\le 1} \frac{dy}{|y|^{n-2(1-s)}} < \infty,$$

since $0 < s < 1$. On the other hand, keeping in mind that $u \in \mathscr{S}(\mathbb{R}^n)$ implies in particular that $u \in L^\infty(\mathbb{R}^n)$, we have

$$\left| \int_{|y|>1} \frac{2u(x) - u(x+y) - u(x-y)}{|y|^{n+2s}} dy \right| \le 4||u||_{L^\infty(\mathbb{R}^n)} \int_{|y|>1} \frac{dy}{|y|^{n+2s}} < \infty.$$

We have thus seen that for every $u \in \mathscr{S}(\mathbb{R}^n)$ definition (2.8) provides a well-defined function on $\mathbb{R}^n$.

REMARK 2.5. The reader should note that we have in fact just proved that $(-\Delta)^s u(x)$ is well-defined for every $u \in C^2(\mathbb{R}^n) \cap L^\infty(\mathbb{R}^n)$. For instance, one seemingly trivial, yet useful, situation to which this remark applies is when $u \equiv c \in \mathbb{R}$, for which we have

$$(-\Delta)^s c \equiv 0.$$

Note that such u is not in $\mathscr{S}(\mathbb{R}^n)$, unless $c = 0$, but of course for such u we have $u \in C^2(\mathbb{R}^n) \cap L^\infty(\mathbb{R}^n)$.

Two basic operations in analysis are the Euclidean translations and dilations

$$\tau_h f(x) = f(x+h), \quad h \in \mathbb{R}^n, \qquad \delta_\lambda f(x) = f(\lambda x), \quad \lambda > 0.$$

The next result clarifies the interplay of $(-\Delta)^s$ with them. Its simple proof based on (2.8) is left to the reader.

LEMMA 2.6. *For every function $u \in \mathscr{S}(\mathbb{R}^n)$ we have for every $h \in \mathbb{R}^n$*

$$(-\Delta)^s(\tau_h u) = \tau_h((-\Delta)^s u), \tag{2.9}$$

and every $\lambda > 0$

$$(-\Delta)^s(\delta_\lambda u) = \lambda^{2s}\delta_\lambda((-\Delta)^s u). \tag{2.10}$$

We note explicitly that the equation (2.10) says in particular that $(-\Delta)^s$ is a homogeneous operator of order $2s$. Since obviously $0 < 2s < 2$, one may surmise that this "decreased" homogeneity is bound to create problems near the boundary in the Dirichlet problem. This aspect will be discussed more precisely in Section 13 below.

A fundamental property of the Laplacean Δ is its invariance with respect to the action of the orthogonal group $\mathbb{O}(n)$ on $\mathbb{R}^n$. This means that if u is a function in $\mathbb{R}^n$, then for every $T \in \mathbb{O}(n)$ one has $\Delta(u \circ T) = \Delta u \circ T$. The following lemma shows that $(-\Delta)^s$ enjoys the same property.

LEMMA 2.7. *Let $u(x) = f(|x|)$ be a function with spherical symmetry in $C^2(\mathbb{R}^n) \cap L^\infty(\mathbb{R}^n)$. Then, also $(-\Delta)^s u$ has spherical symmetry.*

PROOF. This follows in a simple way from (2.8). In order to prove that $(-\Delta)^s u$ is spherically symmetric we need to show that for every $T \in \mathbb{O}(n)$ and every $x \in \mathbb{R}^n$ one has

$$(-\Delta)^s u(Tx) = (-\Delta)^s u(x).$$

We have

$$\begin{aligned}(-\Delta)^s u(Tx) &= \frac{\gamma(n,s)}{2}\int_{\mathbb{R}^n}\frac{2f(|Tx|) - f(|Tx+y|) - f(|Tx-y|)}{|y|^{n+2s}}dy \\ &= \frac{\gamma(n,s)}{2}\int_{\mathbb{R}^n}\frac{2f(|x|) - f(|x+T^ty|) - f(|x-T^ty|)}{|y|^{n+2s}}dy.\end{aligned}$$

If we make the change of variable $z = T^t y$, we conclude

$$\begin{aligned}(-\Delta)^s u(Tx) &= \frac{\gamma(n,s)}{2}\int_{\mathbb{R}^n}\frac{2f(|x|) - f(|x+z|) - f(|x-z|)}{|Tz|^{n+2s}}dz \\ &= \frac{\gamma(n,s)}{2}\int_{\mathbb{R}^n}\frac{2f(|x|) - f(|x+z|) - f(|x-z|)}{|z|^{n+2s}}dz = (-\Delta)^s u(x),\end{aligned}$$

and we are done.

□

Before proceeding we note the following alternative expression for $(-\Delta)^s$ that is at times quite useful in the computations.

PROPOSITION 2.8. *For any $u \in \mathscr{S}(\mathbb{R}^n)$ one has*

$$(-\Delta)^s u(x) = \gamma(n,s)\,\mathrm{PV}\int_{\mathbb{R}^n}\frac{u(x)-u(y)}{|x-y|^{n+2s}}dy, \tag{2.11}$$

where now the integral is taken according to Cauchy's principal value sense

$$\mathrm{PV}\int_{\mathbb{R}^n}\frac{u(x)-u(y)}{|x-y|^{n+2s}}dy = \lim_{\varepsilon\to 0^+}\int_{|y-x|>\varepsilon}\frac{u(x)-u(y)}{|x-y|^{n+2s}}dy.$$

PROOF. The expression (2.11) follows directly from (2.8) above as follows

$$\frac{1}{2}\int_{\mathbb{R}^n}\frac{2u(x)-u(x+y)-u(x-y)}{|y|^{n+2s}}dy=\frac{1}{2}\lim_{\varepsilon\to 0}\int_{|y|>\varepsilon}\frac{2u(x)-u(x+y)-u(x-y)}{|y|^{n+2s}}dy$$
$$=\frac{1}{2}\lim_{\varepsilon\to 0}\int_{|y|>\varepsilon}\frac{u(x)-u(x+y)}{|y|^{n+2s}}dy+\frac{1}{2}\lim_{\varepsilon\to 0}\int_{|y|>\varepsilon}\frac{u(x)-u(x-y)}{|y|^{n+2s}}dy$$
$$=\lim_{\varepsilon\to 0}\int_{|x-y|>\varepsilon}\frac{u(x)-u(y)}{|x-y|^{n+2s}}dy=\mathrm{PV}\int_{\mathbb{R}^n}\frac{u(x)-u(y)}{|x-y|^{n+2s}}dy.$$

However, it is now necessary to take the principal value of the integral since we have eliminated the cancellation of the linear terms in the symmetric difference of order two, and $u(x)-u(y)$ is only $O(|x-y|)$. Thus, the smoothness of u no longer guarantees the local integrability, unless we are in the regime $0<s<1/2$.

□

One can see from (5.2) in Proposition 5.1 below that for $u\in\mathscr{S}(\mathbb{R}^n)$ it is not true in general that $(-\Delta)^s u\in\mathscr{S}(\mathbb{R}^n)$. However, one can verify that $(-\Delta)^s u\in C^\infty(\mathbb{R}^n)$. But $(-\Delta)^s u$ is not only smooth, it also suitably decays at infinity according to the following result.

PROPOSITION 2.9. *Let $u\in\mathscr{S}(\mathbb{R}^n)$. Then, for every $x\in\mathbb{R}^n$ with $|x|>1$, we have*

$$|(-\Delta)^s u(x)|\le C_{u,n,s}|x|^{-(n+2s)},$$

where with $||u||_p$ as in (2.1), we have let

$$C_{u,n,s}=C_{n,s}\left(||u||_{n+2}+||u||_n+||u||_{L^1(\mathbb{R}^n)}\right).$$

PROOF. To see this we write

$$(\Delta)^s u(x)=\frac{\gamma(n,s)}{2}\Bigg(\int_{|y|<\frac{|x|}{2}}\frac{2u(x)-u(x+y)-u(x-y)}{|y|^{n+2s}}dy$$
$$+\int_{|y|\ge\frac{|x|}{2}}\frac{2u(x)-u(x+y)-u(x-y)}{|y|^{n+2s}}dy\Bigg).$$

Taylor's formula gives

$$2u(x)-u(x+y)-u(x-y)=-\frac{1}{2}<\nabla^2u(y^\star)y,y>-\frac{1}{2}<\nabla^2u(y^{\star\star})y,y>,$$

where $y^\star=x+t^\star y$, $y^{\star\star}=x-t^{\star\star}y$, for $t^\star,t^{\star\star}\in[0,1]$. We now observe that on the set where $|y|<\frac{|x|}{2}$ we have by the triangle inequality

$$|x|<2|y^\star|,\qquad |x|<2|y^{\star\star}|. \tag{2.12}$$

Using (2.12) and the definition (2.1) of the norm $||u||_{n+2}$ in $\mathscr{S}(\mathbb{R}^n)$, we find

$$\left|\int_{|y|<\frac{|x|}{2}}\frac{2u(x)-u(x+y)-u(x-y)}{|y|^{n+2s}}dy\right|\le\frac{1}{2}\int_{|y|<\frac{|x|}{2}}\frac{|\nabla^2u(y^\star)|+|\nabla^2u(y^{\star\star})|}{|y|^{n+2s}}|y|^2dy$$
$$\le C||u||_{n+2}\left(\int_{|y|<\frac{|x|}{2}}\frac{|y|^2}{(1+|y^\star|^2)^{\frac{n+2}{2}}|y|^{n+2s}}dy+\int_{|y|<\frac{|x|}{2}}\frac{|y|^2}{(1+|y^{\star\star}|^2)^{\frac{n+2}{2}}|y|^{n+2s}}dy\right)$$
$$\le C|x|^{-n-2}||u||_{n+2}\int_{|y|<\frac{|x|}{2}}\frac{dy}{|y|^{n+2s-2}}=C|x|^{-n-2}||u||_{n+2}|x|^{2-2s}$$
$$=C||u||_{n+2}|x|^{-(n+2s)},$$

where $C=C_{n,s}>0$.

Next, we estimate

$$\left|\int_{|y|\geq\frac{|x|}{2}} \frac{2u(x)-u(x+y)-u(x-y)}{|y|^{n+2s}}dy\right| \leq 2\int_{|y|\geq\frac{|x|}{2}} \frac{|u(x+y)-u(x)|}{|y|^{n+2s}}dy$$
$$\leq 2\int_{|y|\geq\frac{|x|}{2}} \frac{|u(x+y)|+|u(x)|}{|y|^{n+2s}}dy$$

We have

$$\int_{|y|\geq\frac{|x|}{2}} \frac{|u(x)|}{|y|^{n+2s}}dy \leq \sup_{x\in\mathbb{R}^n}\left((1+|x|^2)^{\frac{n}{2}}|u(x)|\right)\int_{|y|\geq\frac{|x|}{2}} \frac{dy}{(1+|x|^2)^{\frac{n}{2}}|y|^{n+2s}}$$
$$\leq \sup_{x\in\mathbb{R}^n}\left((1+|x|^2)^{\frac{n}{2}}|u(x)|\right)|x|^{-n}\int_{|y|\geq\frac{|x|}{2}} \frac{dy}{|y|^{n+2s}} = \frac{C||u||_n}{|x|^{n+2s}},$$

where $C = C_{n,s} > 0$. Finally, we have trivially

$$\int_{|y|\geq\frac{|x|}{2}} \frac{|u(x+y)|}{|y|^{n+2s}}dy \leq \frac{2^{n+2s}}{|x|^{n+2s}}\int_{|y|\geq\frac{|x|}{2}} |u(x+y)|dy = \frac{2^{n+2s}||u||_{L^1(\mathbb{R}^n)}}{|x|^{n+2s}}.$$

This completes the proof. □

Proposition 2.9 has the following nontrivial consequence.

COROLLARY 2.10. *Let* $u \in \mathscr{S}(\mathbb{R}^n)$. *Then,* $(-\Delta)^s u \subset C^\infty(\mathbb{R}^n) \cap L^1(\mathbb{R}^n)$.

The estimate in Proposition 2.9 can be written

$$-C_{u,n,s}|x|^{-(n+2s)} \leq -(-\Delta)^s u(x) \leq C_{u,n,s}|x|^{-(n+2s)}.$$

Let us notice that on a nonnegative bump function the estimate from below can be made stronger, a fact that reflects the nonlocal character of $(-\Delta)^s$. Suppose for instance that $u \in C_0^\infty(\mathbb{R}^n)$, with $0 \leq u \leq 1$, $u \equiv 1$ on $B(0,1)$ and supp $u \subset \overline{B}(0,2)$. Then, for $x \in \mathbb{R}^n \setminus B(0,3)$ one has from (2.8)

$$-(-\Delta)^s u(x) = \frac{\gamma(n,s)}{2}\int_{\mathbb{R}^n} \frac{u(x+y)+u(x-y)}{|y|^{n+2s}}dy$$
$$\geq \gamma(n,s)\int_{B(0,1)} \frac{dz}{|x-z|^{n+2s}}dz.$$

Since $|x-z| \geq 2$, for $|z| \leq 1$ we infer $|x| \geq |x-z| - |z| \geq |x-z| - 1 \geq |x-z|/2$. This gives for some $C(n,s) > 0$

$$-(-\Delta)^s u(x) \geq C(n,s)|x|^{-(n+2s)} > 0, \tag{2.13}$$

which shows that $(-\Delta)^s u$ needs not to vanish even far away from the support of u. This is clearly impossible for local operators $P(x,\partial_x)$, for which one has the obvious property supp $P(x,\partial_x)u \subset$ supp u.

In the next result we provide a useful expression of $(-\Delta)^s u(x)$ in terms of an integral involving the spherical mean-value operator $\mathscr{M}_r u(x)$. We will use this result later in the proof of Theorem 17.2.

PROPOSITION 2.11. *Let* $u \in \mathscr{S}(\mathbb{R}^n)$. *For every* $0 < s < 1$ *one has*

$$(-\Delta)^s u(x) = -\sigma_{n-1}\gamma(n,s)\int_0^\infty r^{-1-2s}[\mathscr{M}_r u(x) - u(x)]dr, \tag{2.14}$$

where $\gamma(n,s)$ is the constant in (2.8) *(and whose value is given in* (5.10) *in Proposition* 5.6*).*

PROOF. We observe that (2.2) in Proposition 2.2 shows that $\mathscr{M}_r u(x) - u(x) = O(r^2)$ as $r \to 0^+$. Therefore, the integrand in the right-hand side of (2.14) behaves like r^{1-2s} as $r \to 0^+$. Since at infinity it behaves like r^{-1-2s}, we conclude that the integral in the right-hand side of (2.14) is convergent. Next, we note that Cavalieri's principle allows to write (2.8) in the following way

$$\begin{aligned}(-\Delta)^s u(x) &= \gamma(n,s) \int_0^\infty \int_{S(x,r)} \frac{u(x)-u(y)}{|x-y|^{n+2s}} d\sigma(y) dr \\ &= \gamma(n,s) \int_0^\infty \frac{1}{r^{n+2s}} \int_{S(x,r)} [u(x)-u(y)] d\sigma(y) dr \\ &= -\sigma_{n-1}\gamma(n,s) \int_0^\infty \frac{r^{n-1}}{r^{n+2s}} [\mathscr{M}_r u(x) - u(x)] dr.\end{aligned}$$

This gives the desired conclusion.

□

We close this section by introducing another functional class that is relevant in connection with the nonlocal operator $(-\Delta)^s$ and whose motivation will become clearer later in this note. Up to now, in order to define $(-\Delta)^s u(x)$ pointwise as in formula (2.8) above we have considered functions u in $\mathscr{S}(\mathbb{R}^n)$ or in $C^2(\mathbb{R}^n) \cap L^\infty(\mathbb{R}^n)$. There exist however larger spaces in which it is still possible to define the nonlocal Laplacean either pointwise or as a tempered distribution.

DEFINITION 2.12. Let $0 < s < 1$. We denote by $\mathscr{L}_s(\mathbb{R}^n)$ the space of measurable functions $u : \mathbb{R}^n \to \overline{\mathbb{R}}$ for which the norm

$$||u||_{\mathscr{L}_s(\mathbb{R}^n)} = \int_{\mathbb{R}^n} \frac{|u(x)|}{1+|x|^{n+2s}} dx < \infty.$$

Notice that we trivially have $u \in \mathscr{L}_s(\mathbb{R}^n) \implies u \in L^1_{loc}(\mathbb{R}^n)$. This inclusion implies in particular that

$$\mathscr{S}(\mathbb{R}^n) \subset \mathscr{L}_s(\mathbb{R}^n) \subset \mathscr{S}'(\mathbb{R}^n).$$

We also note that for any $1 \le p \le \infty$, we have $L^p(\mathbb{R}^n) \subset \mathscr{L}_s(\mathbb{R}^n)$.

REMARK 2.13. Definition 2.12 can be found on p. 73 in [**Si07**], but the class $\mathscr{L}_s(\mathbb{R}^n)$ was implicitly first introduced in (1.6.4) in [**La72**] in connection with the nonlocal mean-value operator. For the latter one should see Definition 15.1 below.

From the definition (2.8) it should be clear to the reader that if $u \in C^2(\mathbb{R}^n) \cap \mathscr{L}_s(\mathbb{R}^n)$, then we can define $(-\Delta)^s u(x)$ at every $x \in \mathbb{R}^n$. This conclusion continues to be true provided that u has a minimal smoothness in terms of the following Hölder class.

DEFINITION 2.14. Given $0 < s < 1$ and $\varepsilon > 0$ sufficiently small, we define the class $C^{2s+\varepsilon}_{loc}$ according to the following convention:

$$C^{2s+\varepsilon}_{loc} = \begin{cases} C^{0,2s+\varepsilon}_{loc}, & \text{if } 0 < s < \frac{1}{2}, \\ C^{1,2s+\varepsilon-1}_{loc}, & \text{if } \frac{1}{2} \le s < 1. \end{cases}$$

The following result is a special case of Proposition 2.4 in [**Si07**]. For its simple proof we refer the reader to that source.

PROPOSITION 2.15. *Let $u \in \mathscr{L}_s(\mathbb{R}^n) \cap C^{2s+\varepsilon}_{loc}$, for some $\varepsilon \in (0,1)$. Then, for every $x \in \mathbb{R}^n$ the integral in (2.8) is convergent, and $(-\Delta)^s u \in C(\mathbb{R}^n)$.*

Although the next result follows directly from the inclusion $\mathscr{L}_s(\mathbb{R}^n) \subset \mathscr{S}'(\mathbb{R}^n)$, we nonetheless present a different simple proof.

COROLLARY 2.16. *Let $u \in \mathscr{L}_s(\mathbb{R}^n)$, then $(-\Delta)^s u \in \mathscr{S}'(\mathbb{R}^n)$.*

PROOF. We have in fact for every $\varphi \in \mathscr{S}(\mathbb{R}^n)$

$$
\begin{aligned}
&|< (-\Delta)^s u, \varphi >| = |< u, (-\Delta)^s \varphi >|\\
&= \left| \int_{\mathbb{R}^n} u(x)(-\Delta)^s \varphi(x) dx \right| \le C_{n,s} \int_{\mathbb{R}^n} \frac{|u(x)|}{1 + |x|^{n+2s}} dx < \infty,
\end{aligned}
$$

where in the second to the last inequality we have used Proposition 2.9 above.

□

3. Maximum principle, Harnack inequality and Liouville theorem

In this section we briefly discuss the three properties in the title with the intent of bringing to the reader's attention those changes that are imposed by the nonlocal nature of $(-\Delta)^s$. One of the most fundamental properties of the theory of second order elliptic and parabolic equations is the so called maximum principle. Let $\Omega \subset \mathbb{R}^n$ be an open set and let $u \in C^2(\Omega)$. A standard fact from calculus states that if u attains a local maximum at a point $x_0 \in \Omega$, then for the Hessian matrix of u we must have $-\nabla^2 u(x_0) \ge 0$. In particular, this gives $-\Delta u(x_0) = \text{trace}(-\nabla^2 u(x_0)) \ge 0$. The (weak) maximum principle for $-\Delta$ states that a subharmonic function in a bounded open set (or, for that matter, any convex function) must attain it supremum on the boundary. The strong maximum principle says much more. For a subharmonic function u in a connected open set we must have $u(x) < \sup_\Omega u$ for every $x \in \Omega$, unless $u \equiv \sup_\Omega u$.

What happens with s-subharmonic functions, i.e., solutions of $(-\Delta)^s u \le 0$? Suppose u has a *global maximum* at $x = x_0$, i.e., for instance $u(x) \le u(x_0)$ for every $x \in \mathbb{R}^n$. Then, a property analogous to the local case trivially holds. Under such assumption we have in fact from (2.8)

$$
(-\Delta)^s u(x_0) = \frac{\gamma(n,s)}{2} \int_{\mathbb{R}^n} \frac{2u(x_0) - u(x_0+y) - u(x_0-y)}{|y|^{n+2s}} dy \ge 0, \tag{3.1}
$$

with strict inequality if x_0 is a strict global maximum.

However, the inequality (3.1) fails to hold, in general, when x_0 is only a *local maximum* for u! Thus, in the nonlocal case the maximum principle, weak or strong, does not admit a formulation similar to the local one. This is caused by the nonlocal nature of $(-\Delta)^s$ which makes a s-harmonic (or, more in general, a s-subharmonic function) feel the effect of far away data. For an interesting example of this negative aspect one should see Theorem 2.2 in [**Ka11**], and also Theorem 3.3.1 in [**BuV16**]. The appropriate nonlocal maximum principle for $(-\Delta)^s$ is as follows.

PROPOSITION 3.1. *Let $u \in \mathscr{L}_s(\mathbb{R}^n) \cap C(\mathbb{R}^n)$. Let $\Omega \subset \mathbb{R}^n$ be a bounded open set and let $(-\Delta)^s u \ge 0$ in Ω. If $u \ge 0$ in $\mathbb{R}^n \setminus \Omega$, then $u \ge 0$ in Ω.*

We stress that in Proposition 3.1 the "boundary values" of the s-subharmonic function u are prescribed not just on $\partial\Omega$, but globally in $\mathbb{R}^n \setminus \Omega$. Interpreted in the sense of Proposition 3.1 also the strong maximum principle holds. For a proof of

Proposition 3.1 we refer the reader to [**Si07**] and [**BuV16**]. One should also see the discussion in the survey paper [**RO15**], which contains a version of the maximum principle for weak subsolutions of $(-\Delta)^s$ or more general nonlocal equations.

A direct consequence of the weak maximum principle for $-\Delta$ is the uniqueness in the Dirichlet problem: given $f \in C(\Omega)$ and $\varphi \in C(\partial\Omega)$, find a function $u \in C^2(\Omega) \cap C(\overline{\Omega})$ such that

$$\begin{cases} -\Delta u = f & \text{in } \Omega, \\ u = \varphi & \text{on } \partial\Omega. \end{cases}$$

Using Proposition 3.1 it is possible to obtain a similar result of uniqueness in the nonlocal Dirichlet problem which is formulated as follows: given a function $f \in C(\Omega)$ and $\varphi \in C(\mathbb{R}^n \setminus \Omega)$, find a function $u \in \mathscr{L}_s(\mathbb{R}^n) \cap C^{2s+\varepsilon}_{loc}$, for some $\varepsilon \in (0,1)$, such that

$$\begin{cases} (-\Delta)^s u = f & \text{in } \Omega, \\ u = \varphi & \text{in } \mathbb{R}^n \setminus \Omega. \end{cases}$$

Since in the nonlocal setting the strong maximum principle fails in its classical formulation, it is not surprising that also the Harnack inequality needs to be suitably stated.

For these important aspects of the theory we refer the reader to the paper [**Ka11'**] and the nice introductory presentations in [**BuV16**], [**RO15**], but also to the original paper by Riesz [**R38**] and the book [**La72**], as well as the works [**Bo97**], [**Bo99**], [**BB99**], [**BB99'**], [**Si07**] and [**FKV15**].

The Liouville theorem for harmonic functions states that there is no such function on the whole $\mathbb{R}^n$ which is one-side bounded other than the constants. A similar result holds in the nonlocal case.

THEOREM 3.2 (see Lemma 3.2 in [**BKN02**]). *Let $n \geq 2$ and $0 < s < 1$. Suppose that $u \geq 0$ satisfies $u \in \mathscr{L}_s(\mathbb{R}^n)$. If $(-\Delta)^s u = 0$ in $\mathscr{D}'(\mathbb{R}^n)$, then u must be constant.*

Interesting extensions of Theorem 3.2 have been recently independently obtained in [**CDL15**] and [**Fa16**]. In the former paper the condition $u \geq 0$ is relaxed to

$$\liminf_{|x|\to\infty} \frac{u(x)}{|x|^\gamma} \geq 0,$$

for some $\gamma \in [0,1]$ with $\gamma < 2s$. In the latter, the author shows that if $u \in \mathscr{L}_s(\mathbb{R}^n)$ and $(-\Delta)^s u = 0$ in $\mathscr{D}'(\mathbb{R}^n)$, then u must be affine, and constant if $0 < s \leq 1/2$. Liouville theorems for some anisotropic nonlocal operators are contained in [**FV17**], [**FV17'**].

The natural space of distributions for the operator $(-\Delta)^s$ is not the space $\mathscr{S}'(\mathbb{R}^n)$, but the smaller space $\mathscr{S}'_s(\mathbb{R}^n)$ introduced in the opening of Section 8 below. The motivation comes from Lemma 8.1 there. A natural question is whether the assumption $u \in \mathscr{L}_s(\mathbb{R}^n)$ in Theorem 3.2 above can be relaxed to $u \in \mathscr{S}'_s(\mathbb{R}^n)$. Nothing seems to be known about this intriguing aspect. In this connection we quote the interesting work [**DSV16'**] which develops an extension of the fractional Laplacean.

In closing, we mention the remarkable recent work [**DSV17**] in which the authors show the surprising result that, given any positive integer k, and a function

$f \in C^k(\overline{B}_1)$, there exists a solution of $(-\Delta)^s u = 0$ in $\mathbb{R}^n$, vanishing outside a suitably large ball B_R, which approximates f in C^k norm on B_1. This purely nonlocal result is in stunning contrast with what happens in the local case. It is not possible to approximate in C^k norm a function on a ball with harmonic functions since such functions are very rigid. For instance, if they have a local maximum in the ball, they must be constant.

4. A brief interlude about very classical stuff

To proceed with the analysis of the nonlocal operator $(-\Delta)^s$ we will need some basic properties of two important, and deeply interconnected, protagonists of classical analysis: the Fourier transform and Bessel functions. Since they both play a pervasive role in these notes, as a help to the reader in this section we recall their definition along with some elementary (and also not so elementary) facts. Before we do that, however, we introduce the ever present Euler's gamma function (see e.g. chapter 1 in [**Le72**]):

$$\Gamma(x) = \int_0^\infty t^{x-1} e^{-t}\, dt, \qquad x > 0.$$

The well-known identity $\Gamma(1/2) = \sqrt{\pi}$ is simply a reformulation of the famous integral

$$\int_{\mathbb{R}} e^{-x^2} dx = \sqrt{\pi}.$$

Of course, $\Gamma(z)$ can be equally defined as a holomorphic function for every $z \in \mathbb{C}$ with $\Re z > 0$. It is easy to check that for such z, one has

$$\Gamma(z+1) = z\Gamma(z). \tag{4.1}$$

This formula, and its iterations, can be used to meromorphically extend $\Gamma(z)$ to the whole complex plane having simple poles at $z = -k$, $k \in \mathbb{N} \cup \{0\}$, with residues $(-1)^k$. In particular, when $0 < s < 1$ one obtains from (4.1)

$$\Gamma(1-s) = -s\Gamma(-s). \tag{4.2}$$

Furthermore, one has the following basic relations:

$$\Gamma(z)\Gamma(1-z) = \frac{\pi}{\sin \pi z}. \tag{4.3}$$

and

$$2^{2z-1}\Gamma(z)\Gamma(z+\frac{1}{2}) = \sqrt{\pi}\Gamma(2z). \tag{4.4}$$

Stirling's formula provides the asymptotic behavior of the gamma function for large positive values of its argument

$$\Gamma(x) = \sqrt{2\pi}\, x^{x-\frac{1}{2}} e^{-x}\left(1 + O\left(\frac{1}{x}\right)\right), \qquad \text{as } x \to +\infty. \tag{4.5}$$

We close this brief prelude with a very classical formula which connects the gamma function to the $(n-1)$-dimensional Hausdorff measure of the unit sphere $\mathbb{S}^{n-1} \subset \mathbb{R}^n$, and the n-dimensional volume of the unit ball

$$\sigma_{n-1} = \frac{2\pi^{\frac{n}{2}}}{\Gamma(\frac{n}{2})}, \qquad \omega_n = \frac{\sigma_{n-1}}{n} = \frac{\pi^{\frac{n}{2}}}{\Gamma(\frac{n}{2}+1)}. \tag{4.6}$$

Here, Gaussians are lurking in the shadows! In fact, from Bonaventura Cavalieri's principle we easily see that for any $f \in L^1(\mathbb{R}^n)$ and spherically symmetric, i.e., such that $f(x) = f^\star(|x|)$, we have

$$\sigma_{n-1} = \frac{\int_{\mathbb{R}^n} f(x)dx}{\int_0^\infty f^\star(r)r^{n-1}dr}. \tag{4.7}$$

To compute σ_{n-1} it thus suffices to produce one spherically symmetric $f \in L^1(\mathbb{R}^n)$ for which we know how to compute both numerator and denominator in (4.7). As it turns out, Gaussians are the first prize winners. If in fact we take $f(x) = e^{-|x|^2}$, then we know that

$$\int_{\mathbb{R}^n} f(x)dx = \pi^{\frac{n}{2}},$$

whereas

$$\int_0^\infty f^\star(r)r^{n-1}dr = \int_0^\infty e^{-r^2} r^n \frac{dr}{r} = \frac{1}{2}\int_0^\infty e^{-t} t^{\frac{n}{2}} \frac{dt}{t} = \frac{1}{2}\Gamma(\frac{n}{2}).$$

Substituting the latter two formulas in (4.7) proves the first part of (4.6).

One identity that we will use is the following

$$\int_0^\infty u^{-s-1}(1-e^{-u})du = \frac{\Gamma(1-s)}{s}, \qquad 0 < s < 1. \tag{4.8}$$

It is easy to recognize that the integral converges absolutely. The proof of (4.8) then easily follows writing $u^{-s-1} = \frac{d}{du}(\frac{u^{-s}}{-s})$ and integrating by parts as follows

$$\int_0^\infty u^{-s-1}\left(1-e^{-u}\right)du = \frac{1}{s}\int_0^\infty u^{-s}e^{-u}du = \frac{\Gamma(1-s)}{s}.$$

Deeply connected with the gamma function is Euler's *beta function* which for $x, y > 0$ is defined as follows

$$B(x,y) = 2\int_0^{\frac{\pi}{2}} \left(\cos\vartheta\right)^{2x-1}\left(\sin\vartheta\right)^{2y-1} d\vartheta. \tag{4.9}$$

It is an easy exercise to recognize that

$$B(x,y) = 2\int_0^1 (1-\tau^2)^{x-1}\tau^{2y-1}d\tau = \int_0^1 (1-s)^{x-1}s^{y-1}ds. \tag{4.10}$$

The link between the beta and the gamma function is expressed by the following equation

$$B(x,y) = \frac{\Gamma(x)\Gamma(y)}{\Gamma(x+y)}, \tag{4.11}$$

see e.g. (1.5.6) on p. 14 in [**Le72**]. A useful integral which is expressed in terms of the beta, or gamma function is contained in the following proposition.

PROPOSITION 4.1. *Let $b > -n$ and $a > n+b$, then*

$$\int_{\mathbb{R}^n} \frac{|x|^b}{(1+|x|^2)^{\frac{a}{2}}}dx = \frac{\pi^{\frac{n}{2}}}{\Gamma\left(\frac{n}{2}\right)} \frac{\Gamma\left(\frac{b+n}{2}\right)\Gamma\left(\frac{a-b-n}{2}\right)}{\Gamma\left(\frac{a+b}{2}\right)}. \tag{4.12}$$

In particular, if $b = 0$ and $a = n+1$, then

$$\int_{\mathbb{R}^n} \frac{dx}{(1+|x|^2)^{\frac{n+1}{2}}} = \frac{\pi^{\frac{n+1}{2}}}{\Gamma\left(\frac{n+1}{2}\right)}. \tag{4.13}$$

PROOF. Let us observe preliminarily that the assumption $b > -n$ serves to guarantee that the integrand belongs to $L^1_{loc}(\mathbb{R}^n)$, whereas it is in $L^1(\mathbb{R}^n)$ if and only if $a - b > n$. Under these hypothesis we have

$$\int_{\mathbb{R}^n} \frac{|x|^b}{(1+|x|^2)^{\frac{a}{2}}} dx = \sigma_{n-1} \int_0^\infty \frac{r^{b+n-1}}{(1+r^2)^{\frac{a}{2}}} dr$$

(change of variable $r = \tan \xi$)

$$= \sigma_{n-1} \int_0^{\frac{\pi}{2}} \frac{(\tan \xi)^{b+n-1}}{(1+\tan^2 \xi)^{\frac{a-2}{2}}} d\xi = \sigma_{n-1} \int_0^{\frac{\pi}{2}} (\sin \xi)^{b+n-1} (\cos \xi)^{a-b-n-1} d\xi$$

$$= \frac{\sigma_{n-1}}{2} B\left(\frac{b+n}{2}, \frac{a-b-n}{2}\right),$$

where the last inequality follows by a comparison with (4.9). If we now apply formulas (4.6) and (4.11) we obtain

$$\frac{\sigma_{n-1}}{2} B\left(\frac{b+n}{2}, \frac{a-b-n}{2}\right) = \frac{\pi^{\frac{n}{2}}}{\Gamma\left(\frac{n}{2}\right)} \frac{\Gamma\left(\frac{b+n}{2}\right)\Gamma\left(\frac{a-b-n}{2}\right)}{\Gamma\left(\frac{a+b}{2}\right)},$$

which gives the desired conclusion (4.12). To obtain (4.13) it suffices to keep in mind that $\Gamma(1/2) = \sqrt{\pi}$.

□

We are ready to introduce the queen of classical analysis: given a function $u \in L^1(\mathbb{R}^n)$, we define its Fourier transform as

$$\mathscr{F}(u)(\xi) = \hat{u}(\xi) = \int_{\mathbb{R}^n} e^{-2\pi i <\xi, x>} u(x) dx. \tag{4.14}$$

We notice that the normalization that we have adopted in (4.14) is the one which makes $\mathscr{F}$ an isometry of $L^2(\mathbb{R}^n)$ onto itself, see [**SW71**]. We recall next some of the basic properties of $\mathscr{F}$. If $\tau_h u(x) = u(x+h)$ and $\delta_\lambda u(x) = u(\lambda x)$ are the translation and dilation operators in $\mathbb{R}^n$, then we have

$$\widehat{\tau_y u}(\xi) = e^{2\pi i <\xi, y>} \hat{u}(\xi), \tag{4.15}$$

and

$$\widehat{\delta_\lambda u}(\xi) = \lambda^{-n} \hat{u}\left(\frac{\xi}{\lambda}\right). \tag{4.16}$$

The Fourier transform is also invariant under the action of the orthogonal group $\mathbb{O}(n)$. We have in fact for every $T \in \mathbb{O}(n)$

$$\widehat{u \circ T} = \hat{u} \circ T. \tag{4.17}$$

Formula (4.17) says that the Fourier transform of a spherically symmetric function is spherically symmetric as well, see also Theorem 4.4 below for a deeper formulation of this fact.

Another crucial property is the Riemann-Lebesgue lemma:

$$u \in L^1(\mathbb{R}^n) \implies |\hat{u}(\xi)| \to 0 \text{ as } |\xi| \to \infty. \tag{4.18}$$

This result has important consequences when combined with the following two formulas. Let $u \in L^1(\mathbb{R}^n)$ be such that for $\alpha \in \mathbb{N}_0^n$ also $\partial^\alpha u \in L^1(\mathbb{R}^n)$. Then,

$$\widehat{(\partial^\alpha u)}(\xi) = (2\pi i)^{|\alpha|} \xi^\alpha \hat{u}(\xi). \tag{4.19}$$

In particular, (4.19) and (4.18) give: $|\xi^\alpha||\hat{u}(\xi)| \to 0$ as $|\xi| \to \infty$. Furthermore, if $u \in L^1(\mathbb{R}^n)$ is such that for $\alpha \in \mathbb{N}_0^n$ one has $x \to x^\alpha u(x) \in L^1(\mathbb{R}^n)$, then,

$$\partial^\alpha \hat{u}(\xi) = (-2\pi i)^{|\alpha|} \widehat{(\cdot)^\alpha u}(\xi). \tag{4.20}$$

In particular, (4.20) and (4.18) imply that: $\partial^\alpha \hat{u} \in C(\mathbb{R}^n)$ and $|\partial^\alpha \hat{u}(\xi)| \to 0$ as $|\xi| \to \infty$.

Combining these observations one derives one of the central properties of $\mathscr{F}$: it maps continuously $\mathscr{S}(\mathbb{R}^n)$ onto itself and is an isomorphism. Its inverse is also continuous, and is given by the Fourier inversion formula

$$\mathscr{F}^{-1}(u)(x) = \int_{\mathbb{R}^n} e^{2\pi i<\xi,x>} \hat{u}(\xi) d\xi.$$

We next introduce the second main character of this section: the Bessel functions. The book [**Le72**] provides a rewarding account of this beautiful classical subject. For a comprehensive study the reader can also consult G. N. Watson's classical treatise [**W62**], as well as the first two volumes of the Bateman manuscript project [**EMOT53**].

DEFINITION 4.2. For every $\nu \in \mathbb{C}$ such that $\Re\nu > -\frac{1}{2}$ we define the *Bessel function of the first kind* and of complex order ν by the formula

$$J_\nu(z) = \frac{1}{\Gamma(\frac{1}{2})\Gamma(\nu + \frac{1}{2})} \left(\frac{z}{2}\right)^\nu \int_{-1}^1 e^{izt}(1 - t^2)^{\frac{2\nu-1}{2}} dt, \tag{4.21}$$

where $\Gamma(x)$ denotes the Euler gamma function.

The function $J_\nu(z)$ in (4.21) derives its name from the fact that it solves the linear ordinary differential equation known as *Bessel equation of order* ν

$$z^2 \frac{d^2 J}{dz^2} + z\frac{dJ}{dz} + (z^2 - \nu^2)J = 0. \tag{4.22}$$

An expression of J_ν as a power series for an arbitrary value of $\nu \in \mathbb{C}$ is provided by

$$J_\nu(z) = \sum_{k=0}^\infty (-1)^k \frac{(z/2)^{\nu+2k}}{\Gamma(k+1)\Gamma(k+\nu+1)}, \qquad |z| < \infty, \quad |\arg(z)| < \pi, \tag{4.23}$$

see e.g. (5.3.2) on p. 102 in [**Le72**]. When $\nu \notin \mathbb{Z}$, another linearly independent solution of (4.22) is provided by the function $J_{-\nu}(z)$. When $\nu \in \mathbb{Z}$ the two functions J_ν and $J_{-\nu}$ are linearly dependent, and in order to find a second solution linearly independent from J_ν one has to proceed differently.

The observation that follows is very important in most concrete applications of the theory. Suppose that $\Phi(z)$ be a solution to the Bessel equation (4.22), and consider the function defined by the transformation

$$u(y) = y^\alpha \Phi(\beta y^\gamma). \tag{4.24}$$

Then, one easily verifies that $u(y)$ satisfies the *generalized Bessel equation*

$$y^2 u''(y) + (1 - 2\alpha) y u'(y) + \left[\beta^2\gamma^2 y^{2\gamma} + (\alpha^2 - \nu^2\gamma^2)\right] u(y) = 0. \tag{4.25}$$

We show with an example how this observation is applied. Consider the ball $B_R = \{x \in \mathbb{R}^n \mid |x| < R\}$ and the cylinder $\mathcal{C} = B_R \times \mathbb{R} \subset \mathbb{R}^{n+1}$. Denote the variable in $\mathcal{C}$ as (x,t), with $x \in \mathbb{R}^n, t \in \mathbb{R}$. Suppose we look for nontrivial functions $u(x,t)$ which are harmonic in $\mathcal{C}$. A classical tool is using separation of variables, i.e., look for u in the form $u(x,t)=\varphi(x)h(t)$. Imposing that $\Delta u=0$ in $\mathcal{C}$ leads to the equation

$$h(t)\Delta\varphi(x)+h''(t)\varphi(x)=0.$$

If we divide by $\varphi(x)h(t)$, we conclude that we must have

$$-\frac{\Delta\varphi(x)}{\varphi(x)}=\frac{h''(t)}{h(t)}.$$

This is possible only if there exists a number $\lambda\geq 0$ such that

$$-\Delta\varphi=\lambda\varphi,\qquad h''=\lambda h.$$

The second equation clearly gives $h(t)=Ae^{\sqrt{\lambda}t}+Be^{-\sqrt{\lambda}t}$, whereas for the first one we seek a solution which is spherically symmetric, i.e., $\varphi(x)=F(|x|)$. We easily see that $F(r)$ must satisfy the equation

$$r^2F''(r)+(n-1)rF'(r)+\lambda r^2F(r)=0. \tag{4.26}$$

It is not obvious at first glance that this is a Bessel equation. However, if we compare (4.26) with (4.25), we conclude that the former is a special instance of the latter if we take

$$1-2\alpha=n-1,\qquad \beta^2=\lambda,\qquad \gamma=1,\qquad \nu^2=\alpha^2.$$

We must thus have

$$\alpha=-\frac{n-2}{2},\qquad \beta=\sqrt{\lambda},\qquad \gamma=1,\qquad \nu=\pm\frac{n-2}{2}.$$

But then (4.24) gives us, at least for $n\neq 2k+2$, the two linearly independent solutions

$$\varphi_1(x)=|x|^{-\frac{n-2}{2}}J_{\frac{n-2}{2}}(\sqrt{\lambda}|x|),\qquad \varphi_2(x)=|x|^{-\frac{n-2}{2}}J_{-\frac{n-2}{2}}(\sqrt{\lambda}|x|).$$

Since from (4.27) below we see that $\varphi_2\notin C^2$, we must discard such solution and keep only φ_1. In conclusion, the function

$$u(x,t)=\left(Ae^{\sqrt{\lambda}t}+Be^{-\sqrt{\lambda}t}\right)|x|^{-\frac{n-2}{2}}J_{\frac{n-2}{2}}(\sqrt{\lambda}|x|),$$

provides a harmonic function in the cylinder $\mathcal{C}$, and the appearance of the Bessel function $J_{\frac{n-2}{2}}$ in such formula is the reason for which solutions of (4.22) are also called in the literature *cylinder functions*.

Returning to Definition 4.2, from (4.21) and (4.10) we immediately find

$$z^{-\nu}J_\nu(z)\underset{z\to 0}{\longrightarrow}\frac{2^{-\nu+1}}{\Gamma(\frac12)\Gamma(\nu+\frac12)}\int_0^1(1-s^2)^{\frac{2\nu-1}{2}}ds=\frac{2^{-\nu+1}}{\Gamma(\frac12)\Gamma(\nu+\frac12)}B(\nu+\frac12,\frac12).$$

From this asymptotic relation and (4.11) one obtains

$$J_\nu(z)\cong\frac{2^{-\nu}}{\Gamma(\nu+1)}z^\nu,\quad \text{as } z\to 0. \tag{4.27}$$

Unlike the simple expression of the asymptotic of $J_\nu(z)$ as $z\to 0$, the behavior at infinity of $J_\nu(z)$ is more delicate to come by. We have the following result, see for instance Lemma 3.11 in [**SW71**], or also (5.11.6) on p. 122 in [**Le72**].

LEMMA 4.3. *Let* $\Re\nu > -\dfrac{1}{2}$. *One has*

(4.28)
$$J_\nu(z) = \sqrt{\frac{2}{\pi z}}\cos\left(z - \frac{\pi\nu}{2} - \frac{\pi}{4}\right) + O(z^{-\frac{3}{2}})$$
$$\text{as } |z| \to \infty, \quad -\pi + \delta < \arg z < \pi - \delta.$$

In particular,

(4.29)
$$J_\nu(z) = O(z^{-1/2}), \qquad \text{as } z \to \infty,\ z \geq 0.$$

Along with the Bessel equation (4.22), in Sections 8 and 10 below we will need the *modified Bessel equation* of order $\nu \in \mathbb{C}$,

(4.30)
$$z^2\frac{d^2\Phi}{dz^2} + z\frac{d\Phi}{dz} - (z^2 + \nu^2)\Phi = 0.$$

We stress that the (substantial) difference between (4.30) and (4.22) is the sign of the coefficient z^2 in the zero order term.

Two linearly independent solutions of (4.30) are the modified Bessel function of the first kind,

(4.31)
$$I_\nu(z) = \sum_{k=0}^{\infty}\frac{(z/2)^{\nu+2k}}{\Gamma(k+1)\Gamma(k+\nu+1)}, \qquad |z| < \infty,\ \ |\arg(z)| < \pi,$$

and the modified Bessel function of the third kind, or Macdonald function, which for order $\nu \neq 0, \pm 1, \pm 2, ...$, is given by

(4.32)
$$K_\nu(z) = \frac{\pi}{2}\frac{I_{-\nu}(z) - I_\nu(z)}{\sin \pi\nu}, \qquad |\arg(z)| < \pi.$$

Notice that $K_\nu(z) = K_{-\nu}(z)$.

Similarly to what was observed for (4.33) above, it is easy to verify that if $\Phi(z)$ is a solution to the modified Bessel equation (4.30), then the function defined by the transformation (4.24) satisfies the *generalized modified Bessel equation*

(4.33)
$$y^2u''(y) + (1-2\alpha)yu'(y) + \left[(\alpha^2 - \nu^2\gamma^2) - \beta^2\gamma^2y^{2\gamma}\right]u(y) = 0.$$

As we have stated in the opening of this section the Fourier transform and the Bessel functions are deeply connected. One important instance of this link is the following result which provides a deeper meaning to the invariance of the Fourier transform with respect to the action of the orthogonal group $\mathbb{O}(n)$. We emphasize that the presence of Bessel functions in Theorem 4.4 below underscores the interplay between curvature (that of the unit sphere $\mathbb{S}^{n-1} \subset \mathbb{R}^n$) and Fourier analysis. The reader who wants to become familiar with this deep aspect can look at the beautiful expository paper of E. Stein [**St76**] and also [**SW78**]. For the following result we refer to Theorem 40 on p. 69 in [**BC49**].

THEOREM 4.4 (Fourier-Bessel representation). *Let* $u(x) = f(|x|)$, *and suppose that*
$$t \to t^{\frac{n}{2}}f(t)J_{\frac{n}{2}-1}(t) \in L^1(\mathbb{R}^+),$$
where we have denoted by $J_{\frac{n}{2}-1}$ *the Bessel function of order* $\nu = \frac{n}{2} - 1$ *defined by* (4.21). *Then,*
$$\hat{u}(\xi) = 2\pi|\xi|^{-\frac{n}{2}+1}\int_0^\infty t^{\frac{n}{2}}f(t)J_{\frac{n}{2}-1}(2\pi|\xi|t)dt.$$

To check the integrability assumption in Theorem 4.4 we can use the above given asymptotic (4.27) and (4.29) for the Bessel function J_ν.

One interesting application of Theorem 4.4 that will be needed in Section 13 is the following result about the *Bochner-Riesz kernels* in $\mathbb{R}^n$:

$$B_z(x) = \frac{\left(1-|x|^2\right)_+^z}{\Gamma(z+1)}, \qquad \Re z > -1, \tag{4.34}$$

where we have denoted $a_+ = \max\{a,0\}$. Notice that, thanks to the assumption $\Re z > -1$, we have $B_z \in L^1(\mathbb{R}^n)$.

LEMMA 4.5. *For every $z \in \mathbb{C}$, such that $\Re z > -1$, one has*

$$\hat{B}_z(\xi) = \pi^{-z}|\xi|^{-\left(\frac{n}{2}+z\right)} J_{\frac{n}{2}+z}(2\pi|\xi|), \qquad \xi \in \mathbb{R}^n. \tag{4.35}$$

PROOF. Since $B_z(x) = B^\star(|x|)$, with $B^\star(r) = \frac{\left(1-r^2\right)_+^z}{\Gamma(z+1)}$, applying Theorem 4.4 we find

$$\hat{B}_z(\xi) = \frac{2\pi|\xi|^{-\frac{n-2}{2}}}{\Gamma(z+1)} \int_0^1 \left(1-r^2\right)^z J_{\frac{n-2}{2}}(2\pi|\xi|r) r^{\frac{n}{2}} dr.$$

Next, we use the following formula 6.567 in [**GR80**]

$$\int_0^1 (1-r^2)^z s^{\nu+1} J_\nu(ar) dr = 2^z \Gamma(z+1) a^{-(z+1)} J_{\nu+z+1}(a), \tag{4.36}$$

which is valid for any $a > 0$, $\Re z > -1$, and $\Re \nu > -1$. Applying (4.36) with $\nu = \dfrac{n-2}{2}$, $a = 2\pi|\xi|$, we finally obtain (4.35).

□

Another important instance of Theorem 4.4 is the following. Consider the spherical mean-value operator

$$\mathscr{M}_r(u,x) = \frac{1}{\sigma_{n-1} r^{n-1}} \int_{S(x,r)} u(y) d\sigma(y),$$

introduced in (2.3). The normalized surface measure on the sphere $S(0,r)$, i.e.,

$$d\sigma_r = \frac{1}{\sigma_{n-1} r^{n-1}} d\sigma(y), \tag{4.37}$$

is the compactly supported distribution whose action on a test function φ is defined by

$$< d\sigma_r, \varphi >= \int_{S(0,r)} \varphi(y) d\sigma_r(y).$$

Clearly, supp $d\sigma_r = S(0,r)$. Since $d\sigma_r \in \mathscr{E}'(\mathbb{R}^n)$, by the Theorem of Paley-Wiener its Fourier transform $\mathscr{F}(d\sigma_r)(\xi)$ is not just in $\mathscr{S}'(\mathbb{R}^n)$, but it is in fact a C^∞ function in $\mathbb{R}^n$. An easy exchange of order of integration argument shows that

$$\mathscr{F}(d\sigma_r)(\xi) = \frac{1}{\sigma_{n-1} r^{n-1}} \int_{S(0,r)} e^{-2\pi i <\xi, x>} d\sigma(x).$$

The following beautiful classical result, which is a special case of Theorem 4.4, provides us with an explicit expression of such function, see p. 154 in [**SW71**].

PROPOSITION 4.6. *For any $r > 0$ one has*

$$\mathscr{F}(d\sigma_r)(\xi) = \frac{2\pi}{\sigma_{n-1}} \left(r|\xi|\right)^{-\frac{n}{2}+1} J_{\frac{n}{2}-1}(2\pi r|\xi|).$$

Now, the operator $\mathscr{M}_r u(\cdot)$ can be seen as the convolution of $d\sigma_r$ with the function u. In fact, for any $u \in \mathscr{S}(\mathbb{R}^n)$, if we let $\check{u}(x) = u(-x)$, then we have

$$
\begin{aligned}
< d\sigma_r \star u, \varphi > =< d\sigma_r, \check{u} \star \varphi >= \frac{1}{\sigma_{n-1} r^{n-1}} \int_{S(0,r)} \check{u} \star \varphi(y) d\sigma(y) \\
= \frac{1}{\sigma_{n-1} r^{n-1}} \int_{S(0,r)} \int_{\mathbb{R}^n} \check{u}(y-x)\varphi(x) dx d\sigma(y) \\
= \int_{\mathbb{R}^n} \left(\frac{1}{\sigma_{n-1} r^{n-1}} \int_{S(0,r)} u(x-y) d\sigma(y) \right) \varphi(x) dx \\
= \int_{\mathbb{R}^n} \mathscr{M}_r u(x) \varphi(x) dx =< \mathscr{M}_r u, \varphi > .
\end{aligned}
$$

This shows that in $\mathscr{S}'(\mathbb{R}^n)$ we have

$$
\mathscr{M}_r u = d\sigma_r \star u. \tag{4.38}
$$

If we take the Fourier transform of both sides of (4.38) we obtain

$$
\mathscr{F}(\mathscr{M}_r u)(\xi) = \mathscr{F}(d\sigma_r)(\xi) \mathscr{F}(u)(\xi). \tag{4.39}
$$

Combining (4.39) with Proposition 4.6, we conclude that the Fourier transform of the spherical mean-value operator is given by

$$
\mathscr{F}(\mathscr{M}_r u)(\xi) = \frac{2\pi}{\sigma_{n-1}} (r|\xi|)^{-\frac{n}{2}+1} J_{\frac{n}{2}-1}(2\pi r|\xi|) \mathscr{F}(u)(\xi), \tag{4.40}
$$

yet one more instance of the fascinating interplay between a classical operation of analysis, such as taking the spherical mean-value of a function, and the Bessel functions. Again, the presence of these special functions is a manifestation of curvature, see [**St76**] and [**SW78**].

Another family of special functions that will be needed in this paper are the so-called *hypergeometric functions*. In order to introduce them we recall the definition of the Pochammer's symbols

$$
\alpha_0 = 1, \qquad \alpha_k \overset{def}{=} \frac{\Gamma(\alpha+k)}{\Gamma(\alpha)} = \alpha(\alpha+1)...(\alpha+k-1), \qquad k \in \mathbb{N}.
$$

Notice that since, as we have said, the gamma function has a pole in $z = 0$, we have

$$
0_k = \begin{cases} 1 & \text{if } k = 0 \\ 0 & \text{for } k \geq 1. \end{cases}
$$

DEFINITION 4.7. Let $p, q \in \mathbb{N}_0$ be such that $p \leq q+1$, and let $\alpha_1, ..., \alpha_p$ and $\beta_1, ..., \beta_q$ be given parameters such that $-\beta_j \notin \mathbb{N}_0$ for $j = 1, ..., q$. Given a number $z \in \mathbb{C}$, the power series

$$
{}_pF_q(\alpha_1, ..., \alpha_p; \beta_1, ..., \beta_q; z) = \sum_{k=0}^{\infty} \frac{(\alpha_1)_k ... (\alpha_p)_k}{(\beta_1)_k ... (\beta_q)_k} \frac{z^k}{k!}
$$

is called the *generalized hypergeometric function*. When $p = 2$ and $q = 1$, then the function ${}_2F_1(\alpha_1, \alpha_2; \beta_1; z)$ is the *Gauss' hypergeometric function*, and it is usually denoted by $F(\alpha_1, \alpha_2; \beta_1; z)$.

Using the ratio test one easily verifies that the radius of convergence of the above hypergeometric series is ∞ when $p \leq q$, whereas it equals 1 when $p = q+1$.

For later reference we record the following facts that follow easily from Definition 4.7:

$$F(\alpha,0;\beta;z)=F(0,\alpha;\beta;z)=1, \tag{4.41}$$

and (see also p. 275 in [**Le72**])

$$F(\alpha,\beta;\beta;-z)={}_1F_0(\alpha;-z)=(1+z)^{-\alpha}. \tag{4.42}$$

It is also interesting to observe that the hypergeometric function ${}_0F_1$ is in essence a Bessel function, up to powers and rescaling. One has in fact from (4.31) and Definition 4.7,

$$I_\nu(z)=\frac{1}{\Gamma(\nu+1)}\left(\frac{z}{2}\right)^\nu {}_0F_1(\nu+1;(z/2)^2). \tag{4.43}$$

Formula (4.43), Definition 4.7 and a change of variable give

$$\begin{aligned}\int_0^z t^a I_\nu(t)dt &= \frac{2^{\frac{a-\nu-1}{2}}}{\Gamma(\nu+1)}\int_0^{(z/2)^2} t^{\frac{a+\nu-1}{2}}\ {}_0F_1(\nu+1;t)dt\\ &= \frac{2^{-2\nu}}{(a+\nu+1)\Gamma(\nu+1)}\ {}_1F_2(\frac{a+\nu+1}{2};\nu+1,\frac{a+\nu+1}{2}+1;(z/2)^2).\end{aligned}$$

We will also need the following beautiful result that connects Bessel functions to the Gauss' hypergeometric function $F(\alpha_1,\alpha_2;\beta_1;z)$. For its proof see p.51 and forward in Vol.2 of [**EMOT53**]. If one is interested only in the result, see 6.574, formula 3. on p.692 in [**GR80**].

LEMMA 4.8 (The discontinuous integral of Weber and Schafheitlin). *Let* $\Re(\nu+\mu-\lambda+1)>0$, $\Re\lambda>-1$, $0<b<a$. *Then,*

$$\begin{aligned}\int_0^\infty t^{-\lambda}J_\nu(at)J_\mu(bt)dt &= \frac{b^\mu\Gamma(\frac{\nu+\mu-\lambda+1}{2})}{2^\lambda a^{\mu-\lambda+1}\Gamma(\frac{\nu-\mu+\lambda+1}{2})\Gamma(\mu+1)}\\ &\quad\times F\left(\frac{\nu+\mu-\lambda+1}{2},\frac{-\nu+\mu-\lambda+1}{2};\mu+1,\frac{b^2}{a^2}\right).\end{aligned}$$

REMARK 4.9. Let us note explicitly that, thanks to (4.27), the assumption $\Re(\nu+\mu-\lambda+1)>0$ guarantees that $t\to t^{-\lambda}J_\nu(at)J_\mu(bt)\in L^1(0,\delta)$ for every $\delta>0$. On the other hand, the hypothesis $\Re\lambda>-1$ guarantees, in view of Lemma 4.3, that $t\to t^{-\lambda}J_\nu(at)J_\mu(bt)\in L^1(\delta,\infty))$. Therefore, under the given assumptions we do have $t\to t^{-\lambda}J_\nu(at)J_\mu(bt)\in L^1(\mathbb{R}^+)$.

5. Fourier transform, Bessel functions and $(-\Delta)^s$

After our brief interlude on the Fourier transform and Bessel functions, we now return to the main protagonist of these notes. The two main objectives of this section are:

(i) to establish an alternative way of computing $(-\Delta)^s$ based on the Fourier transform, see Proposition 5.1 below, and show that the fractional Laplacean is an elliptic pseudodifferential operator in the Kohn-Nirenberg's class $\Psi^{2s}_{1,0}$, see Proposition 5.2;

(ii) to compute explicitly the constant $\gamma(n,s)$ in (2.8), see Proposition 5.6 below.

PROPOSITION 5.1 (Pseudodifferential nature of $(-\Delta)^s$). *Let $\gamma(n,s)>0$ be the number identified by the following formula*

$$\gamma(n,s)\int_{\mathbb{R}^n}\frac{1-\cos(z_n)}{|z|^{n+2s}}dz=1. \tag{5.1}$$

Then, for any $u\in\mathscr{S}(\mathbb{R}^n)$ we have

$$\widehat{(-\Delta)^s u}(\xi)=(2\pi|\xi|)^{2s}\hat{u}(\xi). \tag{5.2}$$

PROOF. Let us observe that in view of Corollary 2.10 we know that $(-\Delta)^s u\in L^1(\mathbb{R}^n)$ and thus we can take its Fourier transform in the sense of L^1. Having said this, if we denote by $\tau_h u(x)=u(x+h)$ the translation operator in $\mathbb{R}^n$, we can rewrite (2.8) in the following way

$$(-\Delta)^s u(x)=\frac{\gamma(n,s)}{2}\int_{\mathbb{R}^n}\frac{2u(x)-\tau_y u(x)-\tau_{-y}u(x)}{|y|^{n+2s}}dy. \tag{5.3}$$

Using (4.15) we easily find

$$\widehat{(-\Delta)^s u}(\xi)=\gamma(n,s)\left(\int_{\mathbb{R}^n}\frac{1-\cos(2\pi<\xi,y>)}{|y|^{n+2s}}dy\right)\hat{u}(\xi)=J(\xi)\hat{u}(\xi), \tag{5.4}$$

where we have let

$$J(\xi)=\gamma(n,s)\int_{\mathbb{R}^n}\frac{1-\cos(2\pi<\xi,y>)}{|y|^{n+2s}}dy.$$

We notice that the integral defining $J(\xi)$ only depends on $|\xi|$. For every $T\in\mathbb{O}(n)$ one in fact easily verifies that $J(T\xi)=J(\xi)$. For $\xi\neq 0$ we can thus write

$$J(\xi)=\gamma(n,s)\int_{\mathbb{R}^n}\frac{1-\cos(<\frac{\xi}{|\xi|},2\pi|\xi|y>)}{|y|^{n+2s}}dy.$$

The change of variable $z=2\pi|\xi|y$ now gives

$$\begin{aligned}J(\xi)&=(2\pi|\xi|)^{2s}\gamma(n,s)\int_{\mathbb{R}^n}\frac{1-\cos(<\frac{\xi}{|\xi|},z>)}{|z|^{n+2s}}dz\\&=(2\pi|\xi|)^{2s}\gamma(n,s)\int_{\mathbb{R}^n}\frac{1-\cos(<e_n,z>)}{|z|^{n+2s}}dz\\&=(2\pi|\xi|)^{2s}\gamma(n,s)\int_{\mathbb{R}^n}\frac{1-\cos(z_n)}{|z|^{n+2s}}dz.\end{aligned} \tag{5.5}$$

Notice that the integrand in the right-hand side of the latter equation is nonnegative, and that the integral is convergent. We have in fact

$$\int_{\mathbb{R}^n}\frac{1-\cos(z_n)}{|z|^{n+2s}}dz=\int_{|z|\le 1}\frac{1-\cos(z_n)}{|z|^{n+2s}}dz+\int_{|z|>1}\frac{1-\cos(z_n)}{|z|^{n+2s}}dz$$
$$\le C\int_{|z|\le 1}\frac{dz}{|z|^{n-2(1-s)}}+2\int_{|z|>1}\frac{dz}{|z|^{n+2s}}<\infty.$$

Finally, if we substitute in (5.4) the expression given by (5.5), it becomes clear that if we choose $\gamma(n,s)>0$ as in (5.1), then (5.2) holds.

□

Formula (5.2) in Proposition 5.1 shows that the fractional Laplacean $(-\Delta)^s$ belongs to a class of operators known as *pseudodifferential operators*. Their action on functions is specified by the following formula

$$Pf(x) = \int_{\mathbb{R}^n} e^{2\pi i<\xi,x>} p(x,\xi)\hat{f}(\xi)d\xi, \tag{5.6}$$

where the function $p(x,\xi)$, known as a *symbol*, is requested to fulfill suitable hypothesis, see e.g. [**Ta81**]. For instance, if $p(x,\xi)$ is a C^∞ function on $\mathbb{R}^n \times \mathbb{R}^n$ with the property that there exist $m \in \mathbb{R}$ such that for every $\alpha, \beta \in \mathbb{N} \cup \{0\}$ and every $x, \xi \in \mathbb{R}^n$ one has

$$|\partial_\xi^\alpha \partial_x^\beta p(x,\xi)| \leq C_{\alpha,\beta} \left(1 + |\xi|\right)^{m-|\alpha|},$$

for some constant $C_{\alpha,\beta}$, then we say that $p(x,\xi)$ belongs to the symbol class $S^m_{1,0}$ introduced by Kohn and Nirenberg in their seminal works [**KN65**], [**KN65'**]. A more general class of symbols, denoted by $S^m_{\rho,\delta}$, was introduced by Hörmander, see [**Ho66**], and also [**Ta81**]. If $p(x,\xi) \in S^m_{1,0}$, then the corresponding operator P defined by (5.6) is called a pseudodifferential operator of order m and it is said to belong to the class $\Psi^m_{1,0}$. A pseudodifferential operator $P \in \Psi^m_{1,0}$ is called *elliptic* if there exists $r > 0$ such that its symbol $p(x,\xi)$ satisfies the following condition

$$|p(x,\xi)^{-1}| \leq \frac{C}{1 + |\xi|^m}, \qquad |x| \geq r.$$

From the equation (5.2) above, we see that the symbol of $(-\Delta)^s$ is $p(x,\xi) = (2\pi|\xi|)^{2s}$, and therefore one easily sees that $p \in S^{2s}_{1,0}$ and that $(-\Delta)^s$ is elliptic. We state this observation in a proposition since it is a basic aspect of $(-\Delta)^s$ which has important repercussions. For a notable one we refer the reader to Theorem 12.19 below.

PROPOSITION 5.2. *The operator $(-\Delta)^s$ is an elliptic pseudodifferential operator in the class $\Psi^{2s}_{1,0}$.*

The pseudodifferential character of the operator $(-\Delta)^s$ has been exploited in the recent work by Epstein and Pop [**EP16**] to study the regularity theory for the fractional Laplacean with a drift in the supercritical range $0 < s < 1/2$. More general pseudodifferential operators which include $(-\Delta)^s$ as a special case have been treated in the works of G. Grubb [**Gr14**], [**Gr15**], [**Gr16**].

Equation (5.2) in Proposition 5.1 has the following immediate consequence.

COROLLARY 5.3 (Semigroup property). *Let $0 < s, s' < 1$, with $s + s' \leq 1$. Then, for any $u \in \mathscr{S}(\mathbb{R}^n)$ we have*

$$(-\Delta)^{s+s'}u = (-\Delta)^s(-\Delta)^{s'}u = (-\Delta)^{s'}(-\Delta)^s u.$$

PROOF. It is enough to verify the desired inequality on the Fourier transform side. Using (5.2) we find

$$\begin{aligned}\mathscr{F}\left((-\Delta)^{s+s'}u\right) &= (2\pi|\xi|)^{2(s+s')}\hat{u} = (2\pi|\xi|)^{2s}(2\pi|\xi|)^{2s'}\hat{u} \\ &= \mathscr{F}((-\Delta)^s(-\Delta)^{s'}u) = \mathscr{F}((-\Delta)^{s'}(-\Delta)^s u).\end{aligned}$$

□

With Proposition 5.1 in hands we can now prove the following important "integration by parts" formula.

Lemma 5.4 (Symmetry). *Let $0 < s \leq 1$. Then, for any $u, v \in \mathscr{S}(\mathbb{R}^n)$ we have*

$$\int_{\mathbb{R}^n} u(x)(-\Delta)^s v(x)dx = \int_{\mathbb{R}^n} (-\Delta)^s u(x)v(x)dx. \tag{5.7}$$

Proof. The case $s = 1$ is well-known, and it is just integration by parts, so let us focus on $0 < s < 1$. Since by Corollary 2.10 we know $\widehat{(-\Delta)^s u}, \widehat{(-\Delta)^s v} \in L^1(\mathbb{R}^n)$, we can use the following formula, valid for any $f, g \in L^1(\mathbb{R}^n)$,

$$\int_{\mathbb{R}^n} \hat{f}(\xi)g(\xi)d\xi = \int_{\mathbb{R}^n} f(\xi)\hat{g}(\xi)d\xi. \tag{5.8}$$

Applying (5.8) and (5.2) in Proposition 5.1, we find

$$\int_{\mathbb{R}^n} (-\Delta)^s u(x)v(x)dx = \int_{\mathbb{R}^n} (-\Delta)^s u(x)\mathscr{F}(\mathscr{F}^{-1}v)(x)dx$$
$$= \int_{\mathbb{R}^n} \mathscr{F}((-\Delta)^s u)(\xi)\mathscr{F}^{-1}v(\xi)d\xi$$
$$= \int_{\mathbb{R}^n} (2\pi|\xi|)^{2s}\hat{u}(\xi)\mathscr{F}^{-1}v(\xi) = \int_{\mathbb{R}^n} \hat{u}(\xi)(2\pi|\xi|)^{2s}\mathscr{F}^{-1}v(\xi)d\xi.$$

Using (5.2) again we have

$$\mathscr{F}^{-1}((-\Delta)^s v)(\xi) = (2\pi|\xi|)^{2s}\mathscr{F}^{-1}v(\xi). \tag{5.9}$$

Inserting this information in the above equation, and applying (5.8) again, we find

$$\int_{\mathbb{R}^n} (-\Delta)^s u(x)v(x)dx = \int_{\mathbb{R}^n} \hat{u}(\xi)\mathscr{F}^{-1}((-\Delta)^s v)(\xi)d\xi$$
$$= \int_{\mathbb{R}^n} \mathscr{F}^{-1}(\hat{u})(x)(-\Delta)^s v(x)dx = \int_{\mathbb{R}^n} u(x)(-\Delta)^s v(x)dx.$$

□

Remark 5.5. For an extension of Lemma 5.4 one should see [**BB99'**].

We next turn to computing explicitly the constant $\gamma(n, s)$ in (5.1).

Proposition 5.6. *Let $0 < s < 1$. Then, we have*

$$\gamma(n, s) = \frac{s 2^{2s}\Gamma\left(\frac{n+2s}{2}\right)}{\pi^{\frac{n}{2}}\Gamma(1-s)}. \tag{5.10}$$

Proof. We use the beautiful idea of Bochner in his proof of Theorem 4.4 above. If we denote by $\theta \in [0, \pi]$ the angle that the vector $z \in \mathbb{R}^n \setminus \{0\}$ forms with the positive direction of the z_n-axis, then Cavalieri's principle, and Fubini's theorem, give

$$\int_{\mathbb{R}^n} \frac{1 - \cos z_n}{|z|^{n+2s}} dz = \int_0^\infty \int_{\mathbb{S}^{n-1}} \frac{1 - \cos(r\cos\theta)}{r^{n+2s}} d\sigma r^{n-1} dr$$
$$= \int_0^\infty \frac{1}{r^{1+2s}} \int_0^\pi [1 - \cos(r\cos\theta)] \int_{L_\theta} d\sigma' d\theta dr,$$

where we have indicated by $L_\theta = \{y \in \mathbb{S}^{n-1} \mid < y, e_n >= \cos\theta\}$ the $(n-2)$-dimensional sphere in $\mathbb{R}^n$ with radius $\sin\theta$ obtained by intersecting $\mathbb{S}^{n-1}$ with the hyperplane $y_n = \cos\theta$. Since with σ_{n-2} given by (4.6) above we have

$$\int_{L_\theta} d\sigma' = \sigma_{n-2}(\sin\theta)^{n-2},$$

we obtain
(5.11)

$$\int_{\mathbb{R}^n} \frac{1-\cos z_n}{|z|^{n+2s}} dz = \sigma_{n-2} \int_0^\infty \frac{1}{r^{1+2s}} \int_0^\pi [1-\cos(r\cos\theta)] (\sin\theta)^{n-2} d\theta dr$$

$$= \sigma_{n-2} \int_0^\infty \frac{1}{r^{1+2s}} \int_0^\pi [1-\cos(r\cos\theta)] (1-\cos^2\theta)^{\frac{n-3}{2}} \sin\theta d\theta dr \quad (\text{set } u = \cos\theta)$$

$$= \sigma_{n-2} \int_0^\infty \frac{1}{r^{1+2s}} \int_{-1}^1 [1-\cos(ru)] (1-u^2)^{\frac{n-3}{2}} du dr$$

$$= \sigma_{n-2} \int_0^\infty \frac{1}{r^{1+2s}} \left[\int_{-1}^1 (1-u^2)^{\frac{n-3}{2}} du - \int_{-1}^1 \cos(ru)(1-u^2)^{\frac{n-3}{2}} du\right] dr.$$

From (4.9) and (4.10) we thus find

$$\int_{-1}^1 (1-s^2)^{\frac{2\nu-1}{2}} ds = 2\int_0^1 (\cos\theta)^{2\nu} d\theta = B\left(\nu+\frac{1}{2},\frac{1}{2}\right) = \frac{\Gamma\left(\nu+\frac{1}{2}\right)\Gamma\left(\frac{1}{2}\right)}{\Gamma(\nu+1)}.$$

This gives

$$\int_{-1}^1 (1-u^2)^{\frac{n-3}{2}} du = \frac{\Gamma(\frac{n-1}{2})\Gamma(\frac{1}{2})}{\Gamma(\frac{n}{2})}.$$

On the other hand, we have

$$\int_{-1}^1 \cos(ru)(1-u^2)^{\frac{n-3}{2}} du = \int_{-1}^1 e^{iru}(1-u^2)^{\frac{n-3}{2}} du.$$

From this equation and (4.21) in Definition 4.2 we obtain with $\nu = \frac{n-2}{2}$ and $z = r$,

$$\int_{-1}^1 \cos(ru)(1-u^2)^{\frac{n-3}{2}} du = \Gamma(\frac{n-1}{2})\Gamma(\frac{1}{2})\left(\frac{2}{r}\right)^{\frac{n-2}{2}} J_{\frac{n-2}{2}}(r)$$

Substituting in (5.11) above, we find

$$\int_{\mathbb{R}^n} \frac{1-\cos(z_n)}{|z|^{n+2s}} dz = \sigma_{n-2} \frac{\Gamma(\frac{n-1}{2})\Gamma(\frac{1}{2})}{\Gamma(\frac{n}{2})} \int_0^\infty \frac{1}{r^{1+2s}} \left[1-\Gamma(\frac{n}{2})\left(\frac{2}{r}\right)^{\frac{n-2}{2}} J_{\frac{n-2}{2}}(r)\right] dr.$$

Keeping (4.6) in mind, which gives

$$\sigma_{n-2} = \frac{2\pi^{\frac{n-1}{2}}}{\Gamma(\frac{n-1}{2})},$$

and that $\sqrt{\pi} = \Gamma(1/2)$, we conclude that

$$\int_{\mathbb{R}^n} \frac{1-\cos(z_n)}{|z|^{n+2s}} dz = \sigma_{n-1} \int_0^\infty \frac{1}{r^{1+2s}} \left[1-\Gamma(\frac{n}{2})\left(\frac{2}{r}\right)^{\frac{n-2}{2}} J_{\frac{n-2}{2}}(r)\right] dr.$$

From this equation and (5.1) in Proposition 5.1 above, it is clear that the constant $\gamma(n,s)$ must be chosen so that

$$\gamma(n,s)\sigma_{n-1} \int_0^\infty \frac{1}{r^{1+2s}} \left[1-\Gamma(\frac{n}{2})\left(\frac{2}{r}\right)^{\frac{n-2}{2}} J_{\frac{n-2}{2}}(r)\right] dr = 1. \tag{5.12}$$

In order to complete the proof, we are thus left with computing explicitly the integral in the right-hand side of (5.12).

With $\nu = \frac{n}{2} - 1$, consider now the function

$$\Psi_\nu(r) = 1 - \Gamma(\nu+1)\left(\frac{2}{r}\right)^\nu J_\nu(r).$$

From the series expansion of $J_\nu(r)$, see (4.23) above, we have

$$J_\nu(r) = \frac{\left(\frac{r}{2}\right)^\nu}{\Gamma(\nu+1)} - \frac{\left(\frac{r}{2}\right)^{\nu+2}}{\Gamma(\nu+2)} + \frac{\left(\frac{r}{2}\right)^{\nu+4}}{\Gamma(\nu+3)} - \ldots$$

This expansion gives for some function $h(r) = O(r^2)$ as $r \to 0$,

$$\Psi_\nu(r) = (1 + h(r))\left(\frac{r}{2}\right)^2. \tag{5.13}$$

On the other hand, (4.29) implies that as $r \to \infty$

$$\Psi_\nu(r) = 1 + O(r^{-(\nu+\frac{1}{2})}), \tag{5.14}$$

and thus, in particular, $\Psi_\nu \in L^\infty[0,\infty)$. We thus find

$$\int_0^\infty \frac{1}{r^{1+2s}}\left[1 - \Gamma(\tfrac{n}{2})\left(\frac{2}{r}\right)^{\frac{n-2}{2}} J_{\frac{n-2}{2}}(r)\right] dr = \int_0^\infty \left(\frac{r^{-2s}}{-2s}\right)' \Psi_\nu(r)dr$$

$$= \lim_{R\to\infty,\varepsilon\to 0^+} \int_\varepsilon^R \left(\frac{r^{-2s}}{-2s}\right)' \Psi_\nu(r)dr = -\lim_{R\to\infty} \frac{R^{-2s}}{2s}\Psi_\nu(R) + \lim_{\varepsilon\to 0^+} \frac{\varepsilon^{-2s}}{2s}\Psi_\nu(\varepsilon)$$

$$+ \int_0^\infty \frac{r^{-2s}}{2s}\Psi_\nu'(r)dr.$$

Since as we have observed $\Psi_\nu \in L^\infty[0,\infty)$, we clearly have

$$\lim_{R\to\infty} \frac{R^{-2s}}{2s}\Psi_\nu(R) = 0.$$

From (5.13) and the fact that $0 < s < 1$, we obtain

$$\lim_{\varepsilon\to 0^+} \frac{\varepsilon^{-2s}}{2s}\Psi_\nu(\varepsilon) = 0.$$

We thus infer that

$$\int_0^\infty \frac{1}{r^{1+2s}}\left[1 - \Gamma(\tfrac{n}{2})\left(\frac{2}{r}\right)^{\frac{n-2}{2}} J_{\frac{n-2}{2}}(r)\right] dr = \frac{1}{2s}\int_0^\infty \frac{1}{r^{2s}}\Psi_\nu'(r)dr.$$

On the other hand, the recursion formula for J_ν, see e.g. (5.3.5) on p. 103 in [**Le72**],

$$(z^{-\nu}J_\nu(z))' = -z^{-\nu}J_{\nu+1}(z),$$

gives

$$\Psi'(r) = -2^\nu\Gamma(\nu+1)\left(r^{-\nu}J_\nu(r)\right)' = 2^\nu\Gamma(\nu+1)r^{-\nu}J_{\nu+1}(r).$$

We thus find

$$\int_0^\infty \frac{1}{r^{1+2s}}\left[1 - \Gamma(\tfrac{n}{2})\left(\frac{2}{r}\right)^{\frac{n-2}{2}} J_{\frac{n-2}{2}}(r)\right] dr = \frac{2^\nu\Gamma(\nu+1)}{2s}\int_0^\infty \frac{1}{r^{\frac{n}{2}-1+2s}} J_{\frac{n}{2}}(r)dr.$$

Recalling that $\nu = \frac{n}{2} - 1$ we can write the right-hand side as follows

$$\frac{2^\nu\Gamma(\nu+1)}{2s}\int_0^\infty \frac{1}{r^{\nu+2s}}J_{\nu+1}(r)dr = \frac{2^\nu\Gamma(\nu+1)}{2s}\int_0^\infty \frac{1}{r^{\mu-q}}J_\mu(r)dr,$$

where $\mu = \nu + 1 = \frac{n}{2}$, and $q = 1 - 2s$. We now invoke the following result, which is formula (17) on p. 684 in [**GR80**]:

$$\int_0^\infty \frac{1}{r^{\mu-q}} J_\mu(ar)dr = \frac{\Gamma(\frac{q+1}{2})}{2^{\mu-q}a^{q-\mu+1}\Gamma(\mu - \frac{q}{2} + \frac{1}{2})}, \tag{5.15}$$

provided that

$$-1 < \Re q < \Re\mu - \frac{1}{2}.$$

With the above values of the parameters μ and q this condition becomes

$$-1 < 1 - 2s < \frac{n}{2} - \frac{1}{2}.$$

Now, the former inequality is satisfied since it is equivalent to $s < 1$, and the second is also satisfied since it is equivalent to $s > \frac{1-n}{4}$, which is of course true since $s > 0$, whereas $\frac{1-n}{4} \le 0$. In conclusion, we obtain from (5.15)

$$\begin{aligned}\frac{2^\nu\Gamma(\nu+1)}{2s}\int_0^\infty \frac{1}{r^{\nu+2s}} J_{\nu+1}(r)dr &= \frac{2^{\frac{n}{2}-1}\Gamma(\frac{n}{2})}{2s}\frac{\Gamma(1-s)}{2^{\frac{n}{2}-1+2s}\Gamma(\frac{n}{2}+s)} \\ &= \frac{\Gamma(\frac{n}{2})}{2s}\frac{\Gamma(1-s)}{2^{2s}\Gamma(\frac{n}{2}+s)}.\end{aligned}$$

Returning to (5.12), and keeping the first identity in (4.6) in mind, we reach the conclusion that the constant $\gamma(n,s)$ is given by the equation

$$\gamma(n,s)\frac{2\pi^{\frac{n}{2}}}{\Gamma(\frac{n}{2})}\frac{\Gamma(\frac{n}{2})}{2s}\frac{\Gamma(1-s)}{2^{2s}\Gamma(\frac{n}{2}+s)} = 1,$$

which finally gives

$$\gamma(n,s) = \frac{s2^{2s}\Gamma(\frac{n}{2}+s)}{\pi^{\frac{n}{2}}\Gamma(1-s)}.$$

This proves (5.10), thus completing the lemma.

□

6. The fractional Laplacean and Riesz transforms

In this section we pause for discussing some interesting consequences of Corollary 5.3 and Lemma 5.4. In analysis and geometry one is interested in the following basic question. Consider a n-dimensional Riemannian manifold M, with gradient ∇ and Laplacean Δ. Given $1 < p < \infty$, when is it true that the two Sobolev spaces of order one obtained by completion of $C_0^\infty(M)$ with respect to the semi-norms $||\nabla f||_{L^p(M)}$ and $||\Delta^{1/2} f||_{L^p(M)}$ coincide? This question is important for the purpose of developing analysis on the manifold M and was raised in 1983 by R. Strichartz in [**Str83**].

In order to understand it, let us remain within the familiar surroundings of flat Euclidean space, i.e., when $M = \mathbb{R}^n$. If we denote by $(\cdot,\cdot)$ the inner product in $L^2(\mathbb{R}^n)$, a simple integration by parts shows that for any $u \in \mathscr{S}(\mathbb{R}^n)$ we have

$$(-\Delta u, u) = (\nabla u, \nabla u) = ||\nabla u||^2_{L^2(\mathbb{R}^n)}.$$

However, applying Corollary 5.3 and Lemma 5.4 we find

$$(-\Delta u, u) = ((-\Delta)^{1/2}(-\Delta)^{1/2}u, u) = ((-\Delta)^{1/2}u, (-\Delta)^{1/2}u) = ||(-\Delta)^{1/2}u||^2_{L^2(\mathbb{R}^n)}.$$

The reader should not underestimate the apparent simplicity of the latter conclusion. In a way, it is rather unintuitive that composing two nonlocal operators, such as $(-\Delta)^{1/2}$, we obtain a local operator.

Combining the latter two equations we obtain the remarkable conclusion

$$||\nabla u||^2_{L^2(\mathbb{R}^n)} = ||(-\Delta)^{1/2}u||^2_{L^2(\mathbb{R}^n)}. \tag{6.1}$$

We emphasize at this point that (6.1) allows to identify the first-order Sobolev subspaces of $L^2(\mathbb{R}^n)$ obtained by completion of $C_0^\infty(\mathbb{R}^n)$ with respect to the seminorms $||\nabla u||^2_{L^2(\mathbb{R}^n)}$ and $||(-\Delta)^{1/2}u||^2_{L^2(\mathbb{R}^n)}$. For the definition of the latter we refer the reader to (20.5) below.

However, when $1 < p < \infty$ and $p \neq 2$, a similar identification with respect to the seminorms $\|\nabla f\|_{L^p(\mathbb{R}^n)}$ and $\|(-\Delta)^{1/2} f\|_{L^p(\mathbb{R}^n)}$ is no longer such a simple matter. It is a easy to recognize that an estimate such as

(6.2)

$$A_p\|(-\Delta)^{1/2}f\|_{L^p(\mathbb{R}^n)} \le \|\nabla f\|_{L^p(\mathbb{R}^n)} \le B_p\|(-\Delta)^{1/2}f\|_{L^p(\mathbb{R}^n)}, \qquad f \in C_0^\infty(\mathbb{R}^n),$$

would suffice for such identification. It is also easy to see (by a duality argument) that the validity of the right-hand inequality in (6.2) for a certain $1 < p < \infty$ implies that of the left-hand inequality in $L^{p'}(\mathbb{R}^n)$, where $\frac{1}{p} + \frac{1}{p}' = 1$.

It is at this point that the Riesz transform enters the stage. One operator that occupies a central position in analysis is the k-th *Riesz transform* R_k, which, on the Fourier transform side, is defined by the formula

$$\widehat{R_k u}(\xi) = i\frac{\xi_k}{|\xi|}\hat{u}(\xi), \qquad k = 1, ..., n.$$

The vector-valued Riesz transform $\mathcal{R} = (R_1, ..., R_n)$ are the first basic examples of singular integrals, as they generalize to dimension $n \ge 2$ the classical Hilbert transform, see [**St70**]. Using the Fourier transform it is immediate to verify that

$$R_k = \frac{\partial}{\partial x_k}(-\Delta)^{-1/2}, \qquad k = 1, ..., n, \tag{6.3}$$

which in vector-valued form can be compactly written as $\mathcal{R} = \nabla(-\Delta)^{-1/2}$. If we now apply (6.1) to the function $u = (-\Delta)^{-1/2} f$ (this is fine, if $f \in \mathscr{S}(\mathbb{R}^n)$), we have proved the following result.

PROPOSITION 6.1. *The vector-valued Riesz transform R maps $L^2(\mathbb{R}^n)$ to $L^2(\mathbb{R}^n)$, and one has for any $f \in L^2(\mathbb{R}^n)$*

$$||\mathcal{R}f||^2_{L^2(\mathbb{R}^n)} = ||f||^2_{L^2(\mathbb{R}^n)}.$$

Thus (6.1) is equivalent to the $L^2(\mathbb{R}^n)$ continuity of the Riesz operator $\mathcal{R}$ (obviously, in $\mathbb{R}^n$ we could have proved Proposition 6.1 using the Fourier transform as well, but the above proof works as well in situations in which such tool is not available). In a similar way, the right-hand inequality in (6.2) is equivalent to the L^p continuity of the Riesz operator $\mathcal{R}$. In conclusion, the inequality (6.2) is true for all $1 < p < \infty$ if

$$||\mathcal{R}f||_{L^p(\mathbb{R}^n)} \le C_p||f||_{L^p(\mathbb{R}^n)}, \qquad f \in C_0^\infty(\mathbb{R}^n). \tag{6.4}$$

These arguments show that, at least when $M = \mathbb{R}^n$, the above question whether the two Sobolev spaces of order one obtained by completion of $C_0^\infty(M)$ with respect to the seminorms $\|\nabla f\|_{L^p(M)}$ and $\|(-\Delta)^{1/2} f\|_{L^p(M)}$ coincide can be answered affirmatively if one knows that (6.4) holds. On the other hand, one of the major accomplishments of the theorem of singular integrals is precisely their continuity in $L^p(\mathbb{R}^n)$, $1 < p < \infty$, and thus the opening question of this section admits an affirmative answer in $\mathbb{R}^n$.

It was because of the above considerations that in the above cited paper [**Str83**] R. Strichartz asked what hypothesis on a Riemannian manifold M would ensure the continuity of the Riesz operator $\mathcal{R} = \nabla(-\Delta)^{-1/2}$ in $L^p(M)$ for $1 < p < \infty$. An interesting answer was given in 1987 by D. Bakry who proved in [**Bak87**] that if the Ricci tensor of M is bounded from below by a non negative constant then (6.4), and therefore (6.2) hold for every $1 < p < \infty$. The reader should also consult the subsequent developments in the papers [**ACDH04**], [**CD99**], and the more recent generalization to sub-Riemannian geometry in [**BaG13**]. These developments are intimately connected to the framework introduced in Section 20 below.

7. The fractional Laplacean of a radial function

We have seen in Lemma 2.7 that when $u(x) = f(|x|)$, then $(-\Delta)^s u(x)$ also has spherical symmetry. The next result provides a useful recipe for actually computing such function. It constitutes the non-local replacement of the well-known formula $\Delta u(x) = f''(|x|) + \frac{n-1}{|x|} f'(|x|)$ of the Laplacean of a spherically symmetric function.

LEMMA 7.1. *Let $u(x) = f(|x|)$. Then,*

(7.1)
$$\begin{aligned}(-\Delta)^s u(x) &= \frac{(2\pi)^{2s+2}}{|x|^{\frac{n}{2}-1}} \int_0^\infty t^{2s+1} J_{\frac{n}{2}-1}(2\pi|x|t) \left(\int_0^\infty \tau^{\frac{n}{2}} f(\tau) J_{\frac{n}{2}-1}(2\pi t\tau) d\tau \right) dt. \\ &= |x|^{-\frac{n}{2}-2s-1} \int_0^\infty t^{2s+1} J_{\frac{n}{2}-1}(t) \left(\int_0^\infty \tau^{\frac{n}{2}} f(\tau) J_{\frac{n}{2}-1}(t|x|^{-1}\tau) d\tau \right) dt,\end{aligned}$$

provided that the integrals exist and are convergent.

PROOF. Let $U(x) = (-\Delta)^s u(x)$. We know from Lemma 2.7 that $U(x) = F(|x|)$. We also know by (5.2) in Proposition 5.1 that $\hat{U}(\xi) = (2\pi|\xi|)^{2s}\hat{u}(\xi)$. Combining this with Theorem 4.4, we find

$$\hat{U}(\xi) = (2\pi)^{2s+1}|\xi|^{-\frac{n}{2}+1+2s} \int_0^\infty \tau^{\frac{n}{2}} f(\tau) J_{\frac{n}{2}-1}(2\pi|\xi|\tau) d\tau.$$

Applying again Theorem 4.4 we obtain

$$\begin{aligned}U(x) = (2\pi)^{2s+2}|x|^{-\frac{n}{2}+1} \int_0^\infty & t^{\frac{n}{2}} J_{\frac{n}{2}-1}(2\pi|x|t) t^{-\frac{n}{2}+1+2s} \\ & \times \left(\int_0^\infty \tau^{\frac{n}{2}} f(\tau) J_{\frac{n}{2}-1}(2\pi t\tau) d\tau \right) dt.\end{aligned}$$

This gives the desired conclusion (7.1). □

For more elaborated representations related to Lemma 7.1 the reader should see the paper [**FV12**].

8. The fundamental solution of $(-\Delta)^s$

In this section we compute the fundamental solution of the fractional Laplacean operator. In most parts of this paper we will be implicitly assuming that the dimension of the ambient space is $n \geq 2$. Since $0 < s < 1$, this assumption obviously forces $s < \frac{n}{2}$. However, unlike the local case of the Laplacean, the situation when $n = 1$ has its own interest when dealing with $(-\Delta)^s$ and at times it needs to be discussed separately. Theorem 8.4 below is one instance of this situation. The main reason is that, when $n = 1$, then the case $s = \frac{n}{2}$ does occur when $s = \frac{1}{2}$. A remarkable study of the nondegeneracy and uniqueness for the nonlocal nonlinear equation

$$(-\Delta)^s u + u - |u|^\alpha u = 0, \qquad \alpha > 0,$$

entirely in the case $n = 1$ is [**FLe13**]. One should also see the sequel paper [**FLeS16**] in which the authors obtain a generalization to any dimension $n \geq 1$.

Before we turn to the proof of the main results we pause for a moment to recall that there exist spaces larger than $\mathscr{S}(\mathbb{R}^n)$, or $L^\infty(\mathbb{R}^n) \cap C^2(\mathbb{R}^n)$, in which it is still possible to define the nonlocal Laplacean either pointwise or as a tempered distribution. We have seen an instance of this in Proposition 2.15 above. Following Definition 2.3 in [**Si07**], given $0 < s < 1$ we can also consider the linear space of the functions $u \in C^\infty(\mathbb{R}^n)$ such that for every multi-index $\alpha \in \mathbb{N}_0^n$

$$[u]_\alpha = \sup_{x \in \mathbb{R}^n} \left(1 + |x|^{n+2s}\right) |\partial^\alpha u(x)| < \infty.$$

We denote by $\mathscr{S}_s(\mathbb{R}^n)$ the space $C^\infty(\mathbb{R}^n)$ endowed with the countable family of seminorms $[\cdot]_\alpha$, and by $\mathscr{S}_s'(\mathbb{R}^n)$ its topological dual. We clearly have the inclusions

$$C_0^\infty(\mathbb{R}^n) \hookrightarrow \mathscr{S}(\mathbb{R}^n) \hookrightarrow \mathscr{S}_s(\mathbb{R}^n) \hookrightarrow C^\infty(\mathbb{R}^n), \tag{8.1}$$

with the dual inclusions given by

$$\mathscr{E}'(\mathbb{R}^n) \hookrightarrow \mathscr{S}_s'(\mathbb{R}^n) \hookrightarrow \mathscr{S}'(\mathbb{R}^n) \hookrightarrow \mathscr{D}'(\mathbb{R}^n), \tag{8.2}$$

where we recall that $\mathscr{E}'(\mathbb{R}^n)$ indicates the space of distributions with compact support. The next lemma justifies the introduction of the space $\mathscr{S}_s(\mathbb{R}^n)$.

LEMMA 8.1. *Let $u \in \mathscr{S}(\mathbb{R}^n)$. Then, $(-\Delta)^s u \in \mathscr{S}_s(\mathbb{R}^n)$.*

PROOF. We have already observed that $(-\Delta)^s u \in C^\infty(\mathbb{R}^n)$, and that it is not true in general that $(-\Delta)^s u \in \mathscr{S}(\mathbb{R}^n)$. From Proposition 2.9 we know however that

$$[(-\Delta)^s u]_0 = \sup_{x \in \mathbb{R}^n} \left(1 + |x|^{n+2s}\right) |(-\Delta)^s u(x)| < \infty.$$

Suppose now that $\alpha \in \mathbb{N}_0^n$ and $|\alpha| = 1$. We can write $\alpha = e_k$, where e_k indicate one of the vectors of the standard basis of $\mathbb{R}^n$. Applying (5.2) in Proposition 5.1 and (4.20), we have

$$\begin{aligned}
\partial^\alpha (-\Delta)^s u(x) &= \partial_k \mathscr{F}^{-1}(\widehat{(-\Delta)^s u})(x) = (-2\pi i)\mathscr{F}^{-1}\left(\xi_k \widehat{(-\Delta)^s u}\right)(x) \\
&= (-2\pi i)\mathscr{F}^{-1}\left(\xi_k (2\pi|\xi|)^{2s} \hat{u}(\xi)\right) \quad \text{(by (4.19))} \\
&= \mathscr{F}^{-1}\left((2\pi|\xi|)^{2s} \widehat{\partial_k u}(\xi)\right) \quad \text{(by (5.2) again)} \\
&= \mathscr{F}^{-1}\mathscr{F}((-\Delta)^s \partial_k u) = (-\Delta)^s \partial_k u.
\end{aligned}$$

Since $\partial_k u \in \mathscr{S}(\mathbb{R}^n)$, again by Proposition 2.9 we conclude that

$$[u]_{e_k} = \sup_{x\in\mathbb{R}^n} \left(1+|x|^{n+2s}\right)|\partial_k u(x)| < \infty.$$

Proceeding by induction on $|\alpha|$, for all $\alpha \in \mathbb{N}_0^n$, we reach the desired conclusion. □

With Lemma 8.1 in hands we can now extend the notion of solution to distributional ones.

DEFINITION 8.2. Let $T \in \mathscr{S}'(\mathbb{R}^n)$. We say that a distribution $u \in \mathscr{S}'_s(\mathbb{R}^n)$ solves $(-\Delta)^s u = T$ if for every test function $\varphi \in \mathscr{S}(\mathbb{R}^n)$ one has

$$< u, (-\Delta)^s\varphi > = < T, \varphi > .$$

In the special case in which $T = \delta$, the Dirac delta, then Definition 8.2 leads to the following.

DEFINITION 8.3 (Fundamental solution). We say that a distribution $E_s \in \mathscr{S}'_s(\mathbb{R}^n)$ is a *fundamental solution* of $(-\Delta)^s$ if $(-\Delta)^s E_s = \delta$. This means that for every $\varphi \in \mathscr{S}(\mathbb{R}^n)$ one has

$$< E_s, (-\Delta)^s\varphi > = \varphi(0).$$

It is clear from Definition 8.3 that if $E_s \in \mathscr{S}'_s(\mathbb{R}^n)$ is a fundamental solution of $(-\Delta)^s$, then one has $(-\Delta)^s E_s = 0$ in $\mathscr{D}'(\mathbb{R}^n \setminus \{0\})$. The following result establishes the existence of an explicit fundamental solution $E_s \in C^\infty(\mathbb{R}^n \setminus \{0\})$ of $(-\Delta)^s$. As we will see in the important Theorem 12.19 below, the smoothness of such E_s in $\mathbb{R}^n \setminus \{0\}$ is in agreement with the above observed fact that $(-\Delta)^s E_s = 0$ in $\mathscr{D}'(\mathbb{R}^n \setminus \{0\})$.

THEOREM 8.4. *Let $n \geq 2$ and $0 < s < 1$. Denote by*

$$E_s(x) = \alpha(n,s)|x|^{-(n-2s)}, \tag{8.3}$$

where the normalizing constant in (8.3) *is given by*

$$\alpha(n,s) = \frac{\Gamma(\frac{n}{2}-s)}{2^{2s}\pi^{\frac{n}{2}}\Gamma(s)}. \tag{8.4}$$

Then, E_s is a fundamental solution for $(-\Delta)^s$.

The proof of Theorem 8.4 will be given after Lemma 8.6 below. For such proof we have chosen one approach that, although very classical, to the best of our knowledge has not been pursued elsewhere. We have done so since, in the course of proving Theorem 8.4, we establish some auxiliary results that have an interest in their own right (but also play an important role later in this note, see (10.15) in the proof of Theorem 10.1 below). In particular, we are led to discover in a natural way some remarkable solutions of the following semilinear nonlocal equation which generalizes the celebrated *Yamabe equation* from Riemannian geometry

$$(-\Delta)^s u = u^{\frac{n+2s}{n-2s}}. \tag{8.5}$$

Of course, there exist proofs of Theorem 8.4 different from the one presented here. Besides the original work of M. Riesz [**R38**], the reader should also consult Stein's landmark book [**St70**] and Landkof's cited monograph [**La72**]. We begin with the following preparatory result.

LEMMA 8.5. *Suppose that either $n \geq 2$, or $n = 1$ and $0 < s < 1/2$. For every $y > 0$ consider the regularized fundamental solution*

$$E_{s,y}(x) = \alpha(n,s)(y^2 + |x|^2)^{-\frac{n-2s}{2}}. \tag{8.6}$$

Then,

$$\widehat{E_{s,y}}(\xi) = \frac{y^s}{2^{2s-1}\pi^s\Gamma(s)}|\xi|^{-s}K_s(2\pi y|\xi|), \tag{8.7}$$

where we have denoted by K_ν the modified Bessel function of the third kind, see (4.32) *above. From* (8.7) *we obtain for every $\xi \neq 0$*

$$\widehat{E_s}(\xi) = \lim_{y\to 0^+} \widehat{E_{s,y}}(\xi) = (2\pi|\xi|)^{-2s}. \tag{8.8}$$

PROOF. To prove (8.7) it suffices to show that for every $f \in \mathscr{S}(\mathbb{R}^n)$ we have

$$< \widehat{E_{s,y}}, f > = \frac{y^s}{2^{2s-1}\pi^s\Gamma(s)} \int_{\mathbb{R}^n} |\xi|^{-s}K_s(2\pi y|\xi|)\, f(\xi)d\xi. \tag{8.9}$$

To establish (8.9) we use the heat semigroup and Bochner's subordination. The idea is to start from the observation that for every $L > 0$ and $\alpha > 0$ one has

$$\int_0^\infty e^{-tL}t^\alpha \frac{dt}{t} = \frac{\Gamma(\alpha)}{L^\alpha}. \tag{8.10}$$

Using Fubini and (8.10) with $L = |\xi|^2 + y^2$, we obtain for any $\alpha > 0$

$$\begin{aligned}
&\int_0^\infty t^\alpha \left(\int_{\mathbb{R}^n} e^{-t(|\xi|^2+y^2)}\hat{f}(\xi)d\xi\right)\frac{dt}{t} \\
&= \int_{\mathbb{R}^n} \hat{f}(\xi)\left(\int_0^\infty t^\alpha e^{-t(|\xi|^2+y^2)}\frac{dt}{t}\right)d\xi \\
&= \Gamma(\alpha)\int_{\mathbb{R}^n} \hat{f}(\xi)(|\xi|^2+y^2)^{-\alpha}d\xi.
\end{aligned}$$

The above assumptions $n \geq 2$, or $n = 1$ and $0 < s < 1/2$, imply that $\alpha = \frac{n}{2} - s > 0$. If we thus let $\alpha = \frac{n}{2} - s$ in the latter formula we find

$$\int_0^\infty t^{\frac{n}{2}-s}\left(\int_{\mathbb{R}^n} e^{-t(|\xi|^2+y^2)}\hat{f}(\xi)d\xi\right)\frac{dt}{t} = \Gamma\left(\frac{n-2s}{2}\right)\int_{\mathbb{R}^n} \hat{f}(\xi)(|\xi|^2+y^2)^{-(\frac{n-2s}{2})}d\xi. \tag{8.11}$$

On the other hand, (5.8) above gives for any $f \in \mathscr{S}(\mathbb{R}^n)$ and $y > 0$

$$\int_{\mathbb{R}^n} \mathscr{F}_{x\to\xi}\left(e^{-t(|x|^2+y^2)}\right)f(\xi)d\xi = \int_{\mathbb{R}^n} e^{-t(|\xi|^2+y^2)}\hat{f}(\xi)d\xi.$$

Multiplying both sides of this equation by $t^{\frac{n}{2}-s}$ and integrating between 0 and ∞ with respect to the dilation invariant measure $\frac{dt}{t}$ we obtain

$$\begin{aligned}
&\int_0^\infty t^{\frac{n}{2}-s}\int_{\mathbb{R}^n} \mathscr{F}_{x\to\xi}\left(e^{-t(|x|^2+y^2)}\right)f(\xi)d\xi\frac{dt}{t} \\
&= \int_0^\infty t^{\frac{n}{2}-s}e^{-y^2t}\int_{\mathbb{R}^n} \widehat{e^{-t|\cdot|^2}}(\xi)\; f(\xi)d\xi\frac{dt}{t}
\end{aligned}$$

We next recall the following notable Fourier transform in $\mathbb{R}^n$: for every $t > 0$, and every $\xi \in \mathbb{R}^n$, one has

$$(\widehat{e^{-t|\cdot|^2}})(\xi) = \frac{\pi^{\frac{n}{2}}}{t^{\frac{n}{2}}}\exp\left(-\pi^2\frac{|\xi|^2}{t}\right). \tag{8.12}$$

Substituting (8.12) in the preceeding formula, we find

$$\begin{aligned}&\int_0^\infty t^{\frac{n}{2}-s}\int_{\mathbb{R}^n}\mathscr{F}_{x\to\xi}\left(e^{-t(|x|^2+y^2)}\right)\ f(\xi)d\xi\frac{dt}{t}\\&=\pi^{\frac{n}{2}}\int_0^\infty t^{-s}e^{-y^2t}\int_{\mathbb{R}^n}\exp\left(-\pi^2\frac{|\xi|^2}{t}\right)f(\xi)d\xi\frac{dt}{t}\\&=\pi^{\frac{n}{2}}\int_{\mathbb{R}^n}f(\xi)\left(\int_0^\infty t^{-s}e^{-y^2t}\exp\left(-\pi^2\frac{|\xi|^2}{t}\right)\frac{dt}{t}\right)d\xi.\end{aligned}$$

We now use the following formula that can be found in 9. on p. 340 of [**GR80**]

$$\int_0^\infty t^{\nu-1}e^{-(\frac{\beta}{t}+\gamma t)}dt=2\left(\frac{\beta}{\gamma}\right)^{\frac{\nu}{2}}K_\nu(2\sqrt{\beta\gamma}),\tag{8.13}$$

provided $\Re\beta,\Re\gamma>0$. Applying (8.13) with

$$\nu=-s,\quad \beta=\pi^2|\xi|^2,\quad \gamma=y^2,$$

and keeping in mind that, as we have already observed, $K_\nu=K_{-\nu}$ (see 5.7.10 in [**Le72**]), we find

$$\int_0^\infty t^{-s}e^{-y^2t}\exp\left(-\pi^2\frac{|\xi|^2}{t}\right)\frac{dt}{t}=2\left(\frac{y}{\pi|\xi|}\right)^s K_s(2\pi y|\xi|).\tag{8.14}$$

Substituting (8.14) in the above integral, we conclude

(8.15)

$$\int_0^\infty t^{\frac{n}{2}-s}\int_{\mathbb{R}^n}\widehat{e^{-t(|\cdot|^2+y^2)}}(\xi)\ f(\xi)d\xi\frac{dt}{t}=2\pi^{\frac{n}{2}-s}y^s\int_{\mathbb{R}^n}|\xi|^{-s}K_s(2\pi y|\xi|)\ f(\xi)d\xi.$$

Since the integral in the left-hand side of (8.15) equals that in the left-hand side of (8.11), we finally have

(8.16)

$$\alpha(n,s)\int_{\mathbb{R}^n}\hat f(\xi)(|\xi|^2+y^2)^{-(\frac{n-2s}{2})}d\xi=\alpha(n,s)\frac{2\pi^{\frac{n}{2}-s}y^s}{\Gamma(\frac{n-2s}{2})}\int_{\mathbb{R}^n}|\xi|^{-s}K_s(2\pi y|\xi|)\ f(\xi)d\xi.$$

Recalling (8.4), which gives $\alpha(n,s)=\frac{\Gamma(\frac{n}{2}-s)}{2^{2s}\pi^{\frac{n}{2}}\Gamma(s)}$, we infer from (8.16) that

(8.17)

$$\alpha(n,s)\int_{\mathbb{R}^n}\hat f(\xi)(|\xi|^2+y^2)^{-(\frac{n-2s}{2})}d\xi=\frac{y^s}{2^{2s-1}\pi^s\Gamma(s)}\int_{\mathbb{R}^n}|\xi|^{-s}K_s(2\pi y|\xi|)\ f(\xi)d\xi.$$

Keeping (8.6) in mind, we can rewrite (8.17) as follows

$$<E_{s,y},\hat f>=\frac{y^s}{2^{2s-1}\pi^s\Gamma(s)}\int_{\mathbb{R}^n}|\xi|^{-s}K_s(2\pi y|\xi|)\ f(\xi)d\xi.$$

Since by definition $<\widehat{E_{s,y}},f>=<E_{s,y},\hat f>$, we conclude that (8.9) holds, thus completing the proof.

□

We next prove a remarkable result concerning the function $E_{s,y}$ defined by (8.3) and (8.4) above.

LEMMA 8.6. *For every $y>0$ the function $E_{s,y}$ satisfies the equation*

$$(-\Delta)^sE_{s,y}(x)=y^{2s}\frac{\Gamma(\frac{n}{2}+s)}{\pi^{\frac{n}{2}}\Gamma(s)}\left(y^2+|x|^2\right)^{-(\frac{n}{2}+s)}.\tag{8.18}$$

PROOF. In order to establish (8.18) we begin by computing the function

$$F_{s,y}(x) \overset{def}{=} (-\Delta)^s E_{s,y}(x).$$

With this objective in mind we appeal to (5.2), which gives

$$\widehat{F_{s,y}}(\xi) = \widehat{(-\Delta)^s E_{s,y}}(\xi) = (2\pi|\xi|)^{2s}\widehat{E_{s,y}}(\xi). \tag{8.19}$$

We now use (8.7) in Lemma 8.5. Inserting such equation in (8.19) we obtain

$$\widehat{F_{s,y}}(\xi) = (2\pi|\xi|)^{2s}\frac{y^s}{2^{2s-1}\pi^s\Gamma(s)}|\xi|^{-s}K_s(2\pi y|\xi|) = \frac{2y^s\pi^s}{\Gamma(s)}|\xi|^s K_s(2\pi y|\xi|). \tag{8.20}$$

Using Theorem 4.4 we find from (8.20)

$$F_{s,y}(x) = \frac{4y^s\pi^{s+1}}{\Gamma(s)}\frac{1}{|x|^{\frac{n}{2}-1}}\int_0^\infty t^{\frac{n}{2}+s}K_s(2\pi yt)J_{\frac{n}{2}-1}(2\pi|x|t)dt. \tag{8.21}$$

If we now let

$$\lambda = -\frac{n}{2} - s, \quad \mu = s, \quad \nu = \frac{n}{2} - 1,$$

then we can write the integral in the right-hand side of (8.21) in the form

$$\int_0^\infty t^{-\lambda}K_\mu(at)J_\nu(bt)dt,$$

with

$$a = 2\pi y, \quad b = 2\pi|x|.$$

Under the assumption $\nu - \lambda + 1 > |\mu|$, that is presently equivalent to $n + s > s$, which is obviously true, we can appeal to formula 3. in 6.576 on p. 693 in [**GR80**]. Such formula states that

$$\int_0^\infty t^{-\lambda}K_\mu(at)J_\nu(bt)dt = \frac{b^\nu\Gamma(\frac{\nu-\lambda+\mu+1}{2})\Gamma(\frac{\nu-\lambda-\mu+1}{2})}{2^{\lambda+1}a^{\nu-\lambda+1}\Gamma(1+\nu)} \tag{8.22}$$
$$\times F\left(\frac{\nu-\lambda+\mu+1}{2}, \frac{\nu-\lambda-\mu+1}{2}; \nu+1; -\frac{b^2}{a^2}\right),$$

where, we recall, $F(\alpha,\beta;\gamma;z)$ indicates the hypergeometric function ${}_2F_1(\alpha,\beta;\gamma;z)$, see Definition 4.7 above. Since

$$\frac{\nu-\lambda+\mu+1}{2} = \frac{n}{2} + s, \quad \frac{\nu-\lambda-\mu+1}{2} = \frac{n}{2},$$

from (8.21) and (8.22) we obtain

$$\int_0^\infty t^{\frac{n}{2}+s}K_s(2\pi yt)J_{\frac{n}{2}-1}(2\pi|x|t)dt = \frac{(2\pi|x|)^{\frac{n}{2}-1}\Gamma(\frac{n}{2}+s)}{2^{-\frac{n}{2}-s+1}(2\pi y)^{n+s}} \tag{8.23}$$
$$\times F\left(\frac{n}{2}+s, \frac{n}{2}; \frac{n}{2}; -\frac{|x|^2}{y^2}\right).$$

We now apply (4.42) to find

$$F\left(\frac{n}{2}+s, \frac{n}{2}; \frac{n}{2}; -\frac{|x|^2}{y^2}\right) = \left(1+\frac{|x|^2}{y^2}\right)^{-(\frac{n}{2}+s)}.$$

Inserting this information into (8.23) we have

(8.24)

$$\int_0^\infty t^{\frac{n}{2}+s}K_s(2\pi yt)J_{\frac{n}{2}-1}(2\pi|x|t)dt = \frac{(2\pi|x|)^{\frac{n}{2}-1}\Gamma(\frac{n}{2}+s)}{2^{-\frac{n}{2}-s+1}(2\pi y)^{n+s}}\left(1+\frac{|x|^2}{y^2}\right)^{-(\frac{n}{2}+s)}.$$

From (8.21) and (8.24) we finally conclude

$$F_{s,y}(x)=\frac{\Gamma(\frac{n}{2}+s)}{y^n\pi^{\frac{n}{2}}\Gamma(s)}\left(1+\frac{|x|^2}{y^2}\right)^{-(\frac{n}{2}+s)}=\frac{y^{2s}\Gamma(\frac{n}{2}+s)}{\pi^{\frac{n}{2}}\Gamma(s)}\left(y^2+|x|^2\right)^{-(\frac{n}{2}+s)}.$$

This establishes (8.18), thus completing the proof.

□

We are now ready to provide the

PROOF OF THEOREM 8.4. Our objective is establishing

$$\int_{\mathbb{R}^n}E_s(x)(-\Delta)^s\varphi(x)dx=\varphi(0). \tag{8.25}$$

for every test function $\varphi\in\mathscr{S}(\mathbb{R}^n)$. We begin by observing that, since we are assuming that $n\geq 2$, we automatically have that $0<s<\frac{n}{2}$. For $y>0$ we now consider the regularization $E_{s,y}$ of the distribution E_s defined by (8.3) and (8.4) above. Notice that $E_{s,y}\in C^\infty(\mathbb{R}^n)$ and that it decays at ∞ like $|x|^{-(n-2s)}$. Since for $\varphi\in\mathscr{S}(\mathbb{R}^n)$ we know from Lemma 8.1 that $(\Delta)^s\varphi\in\mathscr{S}_s(\mathbb{R}^n)$, it should be clear that Lebesgue dominated convergence theorem gives

$$\int_{\mathbb{R}^n}E_{s,y}(x)\;(-\Delta)^s\varphi(x)dx\;\longrightarrow\;\int_{\mathbb{R}^n}E_s(x)\;(-\Delta)^s\varphi(x)dx$$

as $y\to 0^+$. On the other hand, Lemma 5.4 (which continues to be valid in the present situation) gives

$$\int_{\mathbb{R}^n}E_{s,y}(x)\;(-\Delta)^s\varphi(x)dx=\int_{\mathbb{R}^n}(-\Delta)^sE_{s,y}(x)\;\varphi(x)dx. \tag{8.26}$$

Therefore, in view of (8.26), in order to complete the proof it will suffice to show that as $y\to 0^+$

$$\int_{\mathbb{R}^n}(-\Delta)^sE_{s,y}(x)\;\varphi(x)dx\;\longrightarrow\;\varphi(0). \tag{8.27}$$

To establish (8.27) we use (8.18) in Lemma 8.6 which gives

$$\begin{aligned}\int_{\mathbb{R}^n}(-\Delta)^sE_{s,y}(x)\;\varphi(x)dx&=\frac{\Gamma(\frac{n}{2}+s)}{y^n\pi^{\frac{n}{2}}\Gamma(s)}\int_{\mathbb{R}^n}\left(1+\frac{|x|^2}{y^2}\right)^{-(\frac{n}{2}+s)}\varphi(x)dx\\&=\frac{\Gamma(\frac{n}{2}+s)}{\pi^{\frac{n}{2}}\Gamma(s)}\int_{\mathbb{R}^n}\left(1+|x'|^2\right)^{-(\frac{n}{2}+s)}\varphi(yx')dx'\\&\longrightarrow\;\varphi(0)\,\frac{\Gamma(\frac{n}{2}+s)}{\pi^{\frac{n}{2}}\Gamma(s)}\int_{\mathbb{R}^n}\left(1+|x'|^2\right)^{-(\frac{n}{2}+s)}dx',\end{aligned}$$

where in the last equality we have used Lebesgue dominated convergence theorem. To complete the proof of (8.27) it would be sufficient to prove that

$$\frac{\Gamma(\frac{n}{2}+s)}{\pi^{\frac{n}{2}}\Gamma(s)}\int_{\mathbb{R}^n}\left(1+|x'|^2\right)^{-(\frac{n}{2}+s)}dx'=1. \tag{8.28}$$

Now, the validity of (8.28) follows from a straightforward application of Proposition 4.1 with the choice $a=n+2s,b=0$.

□

REMARK 8.7. Our proof of Theorem 8.4 uses the fact that $n-2s>0$. Therefore, besides the situation $n \geq 2$, for which this is automatically true, our proof continues to work when $n=1$ and $0<s<\frac{1}{2}$ since in such case we still have $n-2s>0$. It does not cover instead the following two cases:

(i) $n=1$ and $\frac{1}{2}<s<1$;
(ii) $n=1$ and $s=\frac{1}{2}$.

In case (i) formulas (8.3), (8.4) continue to be valid unchanged, whereas in the case (ii) one has to replace them with the following

$$E_s(x) = -\frac{1}{\pi}\log|x|.$$

The interested reader can find a discussion of such cases in the paper [**Bu16**].

9. The nonlocal Yamabe equation

In the previous section we have proved that the function

$$E_{s,y}(x) = \alpha(n,s)(y^2+|x|^2)^{-\frac{n-2s}{2}}.$$

satisfies the equation

$$(-\Delta)^s E_{s,y}(x) = y^{2s}\frac{\Gamma(\frac{n}{2}+s)}{\pi^{\frac{n}{2}}\Gamma(s)}\left(y^2+|x|^2\right)^{-(\frac{n}{2}+s)},$$

see (8.18). If we consider

$$v(x) = \lambda y^{\gamma} E_{s,y}, \tag{9.1}$$

where $\lambda>0$ and $\gamma\in\mathbb{R}$ are to be chosen in a moment, then it is clear that v solves the equation

$$\begin{aligned}
(-\Delta)^s v(x) &= \lambda y^{\gamma+2s}\frac{\Gamma(\frac{n}{2}+s)}{\pi^{\frac{n}{2}}\Gamma(s)}\left(y^2+|x|^2\right)^{-(\frac{n}{2}+s)}\\
&= \lambda y^{\gamma+2s}\frac{\Gamma(\frac{n}{2}+s)}{\pi^{\frac{n}{2}}\Gamma(s)}\left(\frac{1}{(y^2+|x|^2)^{\frac{n}{2}-s}}\right)^{\frac{\frac{n}{2}+s}{\frac{n}{2}-s}}\\
&= \lambda y^{\gamma+2s}\frac{\Gamma(\frac{n}{2}+s)}{\pi^{\frac{n}{2}}\Gamma(s)}\left(\lambda y^{\gamma}\alpha(n,s)\frac{1}{(y^2+|x|^2)^{\frac{n}{2}-s}}\right)^{\frac{\frac{n}{2}+s}{\frac{n}{2}-s}}\left(\lambda y^{\gamma}\alpha(n,s)\right)^{-\frac{\frac{n}{2}+s}{\frac{n}{2}-s}}\\
&= \lambda y^{\gamma+2s}\frac{\Gamma(\frac{n}{2}+s)}{\pi^{\frac{n}{2}}\Gamma(s)}\left(\lambda y^{\gamma}\alpha(n,s)\right)^{-\frac{\frac{n}{2}+s}{\frac{n}{2}-s}} v(x)^{\frac{\frac{n}{2}+s}{\frac{n}{2}-s}}.
\end{aligned}$$

We now choose $\gamma>0$ in such a way that the powers of y add up to 0, and we also choose λ so that

$$\lambda\frac{\Gamma(\frac{n}{2}+s)}{\pi^{\frac{n}{2}}\Gamma(s)}\left(\lambda\alpha(n,s)\right)^{-\frac{\frac{n}{2}+s}{\frac{n}{2}-s}} = 1.$$

For the first condition to be true, we must have

$$\gamma = \frac{n-2s}{2},$$

whereas the second condition will be valid if

$$\lambda^{\frac{4s}{n-2s}} = \frac{\frac{\Gamma(\frac{n}{2}+s)}{\pi^{\frac{n}{2}}\Gamma(s)}}{\alpha(n,s)^{\frac{n+2s}{n-2s}}}.$$

Inserting this choices of γ and λ in (9.1), after some elementary computations we obtain

$$v_y(x) = \kappa(n,s)\left(\frac{y}{y^2+|x|^2}\right)^{\frac{n-2s}{2}}, \qquad y>0, \tag{9.2}$$

with

$$\kappa(n,s) = 2^{\frac{n-2s}{2}}\left(\frac{\Gamma(\frac{n+2s}{2})}{\Gamma(\frac{n-2s}{2})}\right)^{\frac{n-2s}{4s}}. \tag{9.3}$$

In conclusion, we have proved the following remarkable fact.

THEOREM 9.1. *Let $n\geq 2$ and $0<s<1$. Then, for every $y>0$ the function $v=v_y$ defined by (9.2), with $\kappa(n,s)$ given by (9.3), solves the* nonlocal Yamabe equation

$$(-\Delta)^s v = v^{\frac{n-2s}{n+2s}}. \tag{9.4}$$

Every translation in x of such function is also a solution.

We close this section by recalling that in [**CLO06**] the authors proved that, given $n\geq 1$ and $0<s<\frac{n}{2}$, then every positive solution of the integral equation

$$u(x) = \int_{\mathbb{R}^n} \frac{u(y)^{\frac{n-2s}{n+2s}}}{|x-y|^{n-2s}}dy,$$

such that $u\in L_{loc}^{\frac{2n}{n-2s}}(\mathbb{R}^n)$, is a translation of one of the functions in (9.2). They also showed that, under the same hypothesis, an analogous conclusion holds for all positive solutions of the nonlocal Yamabe equation (9.4). Since when $n\geq 2$ and $0<s<1$ the condition $0<s<\frac{n}{2}$ is automatically fulfilled, Theorem 1.2 in [**CLO06**] provides a deep converse to Theorem 9.1 above. We also mention the paper [**CT04**] in which the authors compute the sharp constants $S(n,s)$ in the Sobolev embedding $H^{s,2}(\mathbb{R}^n)\hookrightarrow L^q(\mathbb{R}^n)$, i.e.,

$$||u||_{L^q(\mathbb{R}^n)} \leq S(n,s)||(-\Delta)^{s/2}u||_{L^2(\mathbb{R}^n)}, \qquad u\in H^{s,2}(\mathbb{R}^n),$$

where for any $0<s<n/2$, the Sobolev exponent q is determined by the equation

$$\frac{1}{2}-\frac{1}{q}=\frac{s}{n}, \qquad \text{or, equivalently,} \qquad q=\frac{2n}{n-2s}.$$

In their Theorem 1.1 they prove that the minimizers in the Sobolev embedding are of the form (9.2), or translations of it. Here, the Sobolev space is the standard one

$$\begin{aligned} H^{s,2}(\mathbb{R}^n) &= \{u\in L^2(\mathbb{R}^n) \mid (-\Delta)^{s/2}u\in L^2(\mathbb{R}^n)\} \\ &= \{u\in L^2(\mathbb{R}^n) \mid (1+|\xi|^2)^{s/2}\hat{u}\in L^2(\mathbb{R}^n)\}, \end{aligned} \tag{9.5}$$

see e.g. [**LM72**], and also [**DPV12**]. Finally, for a beautiful introduction to the role of nonlocal operators in geometry the reader should see the paper [**Go16**]. Also, for works on nonlocal equations and geometry one should see [**GZ03**], [**CG11**], [**GSS14**], [**BDS15**], [**FF15**], [**FGMT15**], [**CLZ16**], [**DDGW17**]. For related works on nonlocal nonlinear equation at interface of analysis and geometry one should see [**Tan11**], [**FLe13**], [**BCPS13**], [**CR13**], [**Sec13**], [**JW14**], [**CS14**], [**CS15**], [**SV15**], [**Ab15**], [**FLeS16**], [**DMPS16**], [**PS16**], [**Ab17**], [**JKS17**]. The recent paper [**CLL17**] provides an interesting account of the method of moving planes applied directly to nonlocal equations, rather than going through the extension (discussed in the next Section 10), as it was done for instance in [**BCPS13**] and in [**JW14**]. This is an important aspect since it allows to cover the full range of fractional powers $s \in (0,1)$.

10. Traces of Bessel processes: The extension problem

When dealing with nonlocal operators such as $(-\Delta)^s$ a major difficulty is represented by the fact that they do not act on functions like differential operators do, but instead through nonlocal integral formulas such as (2.8). As a consequence, the rules of differentiation are not readily available, and in these notes we have already seen several instances of this obstruction. In this perspective it would be highly desirable to have some kind of procedure that allows to connect nonlocal problems to ones for which the rules of differential calculus are available. Exploring this connection is the principal objective of this section.

During the past decade there has been an explosion of interest in the analysis of nonlocal operators such as (2.8) in connection with various problems from the applied sciences, analysis and geometry. The majority of these developments has been motivated by the remarkable 2007 "extension paper" [**CS07**] by Caffarelli and Silvestre. In that paper the authors introduced a method that allows to convert nonlocal problems in $\mathbb{R}^n$ into ones that involve a certain (degenerate) differential operator in $\mathbb{R}^{n+1}_+$. Precisely, it was shown in [**CS07**] that if for a given $0 < s < 1$ and $u \in \mathscr{S}(\mathbb{R}^n)$ one considers the function $U(x,y)$ that solves the following Dirichlet problem in the half-space $\mathbb{R}^{n+1}_+$:

$$\begin{cases} L_a U(x,y) = \operatorname{div}_{x,y}(y^a \nabla_{x,y} U) = 0 & x \in \mathbb{R}^n, y > 0, \\ U(x,0) = u(x), \end{cases} \tag{10.1}$$

where now $a = 1 - 2s$, then one can recover $(-\Delta)^s u(x)$ by the following "trace" relation

$$-\frac{2^{2s-1}\Gamma(s)}{\Gamma(1-s)} \lim_{y \to 0^+} y^{1-2s} \frac{\partial U}{\partial y}(x,y) = (-\Delta)^s u(x). \tag{10.2}$$

Thus, remarkably, (10.2) provides yet another way of characterizing $(-\Delta)^s u(x)$ as the weighted *Dirichlet-to-Neumann* map of the extension problem (10.1).

Before turning to solving (10.1) and proving (10.2), we mention that, in connection with the extension problem, there is one reference that should be cited since, with a somewhat different perspective, it contains closely related circle of ideas. The 1965 paper [**MS65**] by Muckenhoupt and Stein does not seem well-known to people in the fractional community, or to workers in geometry. In that paper the authors developed a detailed analysis of the equation

$$\operatorname{div}(y^{2\lambda} \nabla u) = y^{2\lambda} \left(\frac{\partial^2 u}{\partial x^2} + \frac{\partial^2 u}{\partial y^2} + \frac{2\lambda}{y} \frac{\partial u}{\partial y} \right) = 0, \qquad \lambda > 0, \tag{10.3}$$

in the upper half-plane $\mathbb{R}_x \times \mathbb{R}^+_y$ (notice that (10.3) is precisely the extension equation (10.1) above). They made substantial use of the conjugate equation

$$\frac{\partial^2 v}{\partial x^2} + \frac{\partial^2 v}{\partial y^2} - \frac{2\lambda}{y} \frac{\partial v}{\partial y} = 0,$$

and of the fact that if u solves (10.3), then $v = y^{2\lambda} u_y$ solves the conjugate equation. Remarkably, they also proved a strong maximum principle in regions across the singular line $\{y = 0\}$ for solutions to (10.3) such that $u(x,-y) = u(x,y)$, see Theorem 1 in [**MS65**]. Many of these aspects have presently become common knowledge to users of the extension procedure.

We also mention that in probability the extension procedure was introduced by Molchanov and Ostrovskii in [**MO69**], see also the earlier related work by Spitzer [**Sp58**] and the more recent paper by Kolsrud [**K89**]. Although it is fair to say that the contribution of [**MO69**] to the development of nonlocal operators in analysis and geometry is not nearly comparable to that of [**CS07**], it should be said that in the probabilistic literature there is a wealth of works that have developed thanks to [**MO69**].

We also want to emphasize another important aspect of the extension operator L_a. If we let $X = (x,y) \in \mathbb{R}^{n+1}$, then L_a is a special example of the class of differential equations

$$\operatorname{div}_X(A(X)\nabla_X f) = 0, \tag{10.4}$$

first studied by Fabes, Kenig and Serapioni in [**FKS81**]. For such equations the authors assumed that $X \to A(X)$ be a symmetric matrix-valued function with bounded measurable coefficients verifying the following degenerate ellipticity assumption for a.e. $X \in \mathbb{R}^{n+1}$ and every $\xi \in \mathbb{R}^{n+1}$:

$$\lambda\omega(X)|\xi|^2 \leq < A(X)\xi, \xi > \leq \lambda^{-1}\omega(X)|\xi|^2,$$

for some $\lambda > 0$. Here, $\omega(X)$ is a so-called Muckenhoupt A_2-weight. This means that there exists a constant $A > 0$ such that for any ball $B \subset \mathbb{R}^{n+1}$ one has

$$\fint_B \omega(X)dX \fint_B \omega(X)^{-1}dX \leq A.$$

Under such hypothesis they established a strong Harnack inequality, and the local Hölder continuity of the weak solutions of (10.4). Now, the extension equation in (10.1) is a special case of (10.4) since, given that $a = 1-2s \in (-1,1)$, the function $\omega(X) = \omega(X) = |y|^a$ is an A_2-weight in $\mathbb{R}^{n+1}$ (for a very nice introduction to Muckenhoupt A_p-weights the reader is referred to the classical paper [**CF74**]). As a consequence, one can obtain quantitative information on solutions of $(-\Delta)^s u = 0$, say, from corresponding properties of solutions of the extension problem (10.1).

Suppose for instance we want to establish the scale invariant Harnack inequality on balls in $\mathbb{R}^n$ for solutions of $(-\Delta)^s u = 0$ that are globally nonnegative (this is an important hypothesis when dealing with nonlocal operators). We extend such a u to a nonnegative function U in $\mathbb{R}^{n+1}_+$ that solves (10.1) above. In view of (10.2) and of the fact that $(-\Delta)^s u = 0$, we obtain for every $x \in \mathbb{R}^n$

$$\lim_{y\to 0^+} y^{1-2s}\frac{\partial U}{\partial y}(x,y) = 0.$$

This condition implies (after some work!) that if we reflect U evenly in $y > 0$, the resulting function is a nonnegative local solution in a ball in the thick space $\mathbb{R}^{n+1}$ of $L_a U = 0$. Therefore, the Harnack inequality established in [**FKS81**] holds in such ball for U. If in such inequality we set $y = 0$, using the fact that $U(x,0) = u(x)$, we obtain a corresponding Harnack inequality for u. This is one example of how the extension procedure is used to turn nonlocal problems into local ones, see Theorem 5.1 in [**CS07**]. For a different approach based on probability the reader should see the papers [**Bo97**] and [**BGR10**]. In connection with the extension procedure one should also see the works by Stinga and Torrea and by Nyström and Sande. In

[**ST10**] using the theory of semigroups this method has been generalized to define $(-L)^s$, where $L = \operatorname{div}(A(x)\nabla)$ is a variable coefficient elliptic operator in divergence form, whereas in [**NS16**] and [**ST17**] it has been generalized to the nonlocal heat operator $(\partial_t - \Delta)^s$. We also mention the work [**CS16**] in which the authors develop a Schauder type regularity theory, both interior and at the boundary, in the Dirichlet and Neumann problems for the nonlocal operator $(-L)^s$, with L as above.

We now turn to the task of actually solving the extension problem. One key observation is that the second order degenerate elliptic equation in (10.1) can also be written in nondivergence form in the following way

$$\begin{cases} -\Delta_x U = \mathscr{B}_a U, & (x,y) \in \mathbb{R}^{n+1}_+ \\ U(x,0) = u(x), & x \in \mathbb{R}^n, \\ U(x,y) \to 0, \text{ as } y \to \infty, & x \in \mathbb{R}^n, \end{cases} \tag{10.5}$$

where we have denoted by

$$\mathscr{B}_a = \frac{\partial^2}{\partial y^2} + \frac{a}{y}\frac{\partial}{\partial y} \tag{10.6}$$

the generator of the *Bessel semigroup* on $(\mathbb{R}^+, y^a dy)$. We will return to the discussion of this semigroup in Section 22 below.

THEOREM 10.1. *Let $u \in \mathscr{S}(\mathbb{R}^n)$. Then, the solution U to the extension problem (10.1) is given by*

$$U(x,y) = P_s(\cdot,y) \star u(x) = \int_{\mathbb{R}^n} P_s(x-z,y)u(z)dz, \tag{10.7}$$

where

$$P_s(x,y) = \frac{\Gamma(\frac{n}{2}+s)}{\pi^{\frac{n}{2}}\Gamma(s)} \frac{y^{2s}}{(y^2+|x|^2)^{\frac{n+2s}{2}}} \tag{10.8}$$

is the Poisson kernel *for the extension problem in the half-space $\mathbb{R}^{n+1}_+$. For U as in (10.7) one has*

$$(-\Delta)^s u(x) = -\frac{2^{2s-1}\Gamma(s)}{\Gamma(1-s)} \lim_{y\to 0^+} y^a \frac{\partial U}{\partial y}(x,y). \tag{10.9}$$

PROOF. Consider the extension problem (10.1), written in the form (10.5). If we take a partial Fourier transform of the latter with respect to the variable $x \in \mathbb{R}^n$, we find

$$\begin{cases} \frac{\partial^2 \hat U}{\partial y^2}(\xi,y) + \frac{a}{y}\frac{\partial \hat U}{\partial y}(\xi,y) - 4\pi^2|\xi|^2 \hat U(\xi,y) = 0 & \text{in } \mathbb{R}^{n+1}_+, \\ \hat U(\xi,0) = \hat u(\xi), \quad \hat U(\xi,y) \to 0, \text{ as } y \to \infty, & x \in \mathbb{R}^n, \end{cases} \tag{10.10}$$

where we have denoted

$$\hat U(\xi,y) = \int_{\mathbb{R}^n} e^{-2\pi i <\xi,x>} U(x,y)dx.$$

In order to solve (10.10) we fix $\xi \in \mathbb{R}^n \setminus \{0\}$, and with $Y(y) = Y_\xi(y) = \hat U(\xi,y)$, we write (10.10) as

$$\begin{cases} y^2 Y''(y) + ayY'(y) - 4\pi^2|\xi|^2 y^2 Y(y) = 0, & y \in \mathbb{R}^+, \\ Y(0) = \hat u(\xi), \\ Y(y) \to 0, \text{ as } y \to \infty. \end{cases} \tag{10.11}$$

Comparing (10.11) with the generalized modified Bessel equation in (4.33) above we see that the former fits into the general form of the latter provided that

$$\alpha = s, \qquad \gamma = 1, \qquad \nu = s, \qquad \beta = 2\pi|\xi|.$$

Thus, according to (4.24), two linearly independent solutions of (10.11) are given by

$$u_1(y) = y^s I_s(2\pi|\xi|y), \qquad u_2(y) = y^s K_s(2\pi|\xi|y).$$

It ensues that, for every $\xi \neq 0$, the general solution of (10.10) is given by

$$\hat{U}(\xi, y) = Ay^s I_s(2\pi|\xi|y) + By^s K_s(2\pi|\xi|y).$$

The condition $\hat{U}(\xi, y) \to 0$ as $y \to \infty$ forces $A = 0$ (see e.g. formulas (5.11.9) and (5.11.10) on p. 123 of [**Le72**] for the asymptotic behavior at ∞ of K_s and I_s), and thus

$$\hat{U}(\xi, y) = By^s K_s(2\pi|\xi|y). \tag{10.12}$$

Next, we use the condition $\hat{U}(\xi, 0) = \hat{u}(\xi)$ to fix the constant B. When $y \to 0^+$ we have

$$\begin{aligned} \hat{U}(\xi, y) &= By^s K_s(2\pi|\xi|y) \\ &= B\frac{\pi}{2}\frac{y^s I_{-s}(2\pi|\xi|y) - y^s I_s(2\pi|\xi|y)}{\sin \pi s} \quad \longrightarrow \quad \frac{B\pi}{2\Gamma(1-s)\sin \pi s}(2\pi|\xi|)^{-s}, \end{aligned}$$

Now from formula (5.7.1) on p. 108 of [**Le72**], we have as $z \to 0$

$$I_s(z) \cong \frac{1}{\Gamma(s+1)}\left(\frac{z}{2}\right)^s, \qquad I_{-s}(z) \cong \frac{1}{\Gamma(1-s)}\left(\frac{z}{2}\right)^{-s}.$$

Using this asymptotic, along with the formula (4.3) above, we find that as $y \to 0^+$,

$$By^s K_s(2\pi|\xi|y) \longrightarrow \frac{B\pi 2^{s-1}}{\Gamma(1-s)\sin \pi s}(2\pi|\xi|)^{-s} = B2^{s-1}\Gamma(s)(2\pi|\xi|)^{-s}.$$

In order to fulfill the condition $\hat{U}(\xi, 0) = \hat{u}(\xi)$ we impose that the right-hand side of the latter equation equal $\hat{u}(\xi)$. For this to happen we must have

$$B = \frac{(2\pi|\xi|)^s \hat{u}(\xi)}{2^{s-1}\Gamma(s)}.$$

Substituting such value of B in (10.12), we finally obtain

$$\hat{U}(\xi, y) = \frac{(2\pi|\xi|)^s \hat{u}(\xi)}{2^{s-1}\Gamma(s)} y^s K_s(2\pi|\xi|y). \tag{10.13}$$

At this point we want to invert the Fourier transform in (10.13). In fact, it is clear from the latter equation that the function $U(x, y)$ will be given by (10.7), with $P_s(x, y)$ as in (10.8), if we can show that

$$\mathscr{F}^{-1}_{\xi \to x}\left(\frac{(2\pi|\xi|)^s}{2^{s-1}\Gamma(s)} y^s K_s(2\pi|\xi|y)\right) = \frac{\Gamma(\frac{n}{2}+s)}{\pi^{\frac{n}{2}}\Gamma(s)}\frac{y^{2s}}{(y^2+|x|^2)^{\frac{n+2s}{2}}}. \tag{10.14}$$

Since the function between parenthesis in the left-hand side of (10.14) is spherically symmetric, proving (10.14) is equivalent to establishing the following identity

$$\mathscr{F}_{\xi \to x}\left(2\pi^s|\xi|^s y^s K_s(2\pi|\xi|y)\right) = \frac{\Gamma(\frac{n}{2}+s)}{\pi^{\frac{n}{2}}}\frac{y^{2s}}{(y^2+|x|^2)^{\frac{n+2s}{2}}}.$$

In view of Theorem 4.4, the latter identity is equivalent to

$$\text{(10.15)}\quad \frac{2^2\pi^{s+1}y^s}{|x|^{\frac{n}{2}-1}}\int_0^\infty t^{\frac{n}{2}+s}K_s(2\pi yt)J_{\frac{n}{2}-1}(2\pi|x|t)dt = \frac{\Gamma(\frac{n}{2}+s)}{\pi^{\frac{n}{2}}}\frac{y^{2s}}{(y^2+|x|^2)^{\frac{n+2s}{2}}}.$$

We are thus left with proving (10.15). Remarkably, this identity has already been established in (8.24) above. Therefore, (10.15) does hold and, with it, (10.7) and (10.8) as well.

In order to complete the proof of the theorem we are thus left with establishing (10.9). With this objective in mind we note that in view of (5.2) in Proposition 5.1, proving (10.9) is equivalent to showing

$$\text{(10.16)}\qquad (2\pi|\xi|)^{2s}\hat{u}(\xi) = -\frac{2^{2s-1}\Gamma(s)}{\Gamma(1-s)}\lim_{y\to 0^+} y^a\frac{\partial\hat{U}}{\partial y}(\xi,y).$$

Keeping in mind that $a = 1-2s$, and using the formula

$$K_s'(z) = \frac{s}{z}K_s(z) - K_{s+1}(z)$$

(see (5.7.9) on p. 110 of [**Le72**]), we obtain

$$y^a\frac{\partial\hat{U}}{\partial y}(\xi,y) = \frac{(2\pi|\xi|)^{s+1}\hat{u}(\xi)}{2^{s-1}\Gamma(s)}y^{1-s}\left\{\frac{2s}{(2\pi|\xi|)y}K_s(2\pi|\xi|y) - K_{s+1}(2\pi|\xi|y)\right\}.$$

Since

$$\frac{2s}{z}K_s(z) - K_{s+1}(z) = -K_{s-1}(z) = -K_{1-s}(z)$$

(again, by (5.7.9) on p. 110 of [**Le72**]), we finally have

$$y^a\frac{\partial\hat{U}}{\partial y}(\xi,y) = -\frac{(2\pi|\xi|)^{s+1}\hat{u}(\xi)}{2^{s-1}\Gamma(s)}y^{1-s}K_{1-s}(2\pi|\xi|y).$$

Now, as before, we have as $y\to 0^+$,

$$y^{1-s}K_{1-s}(2\pi|\xi|y) \longrightarrow 2^{-s}\Gamma(1-s)(2\pi|\xi|)^{s-1}.$$

We finally reach the conclusion that

$$y^a\frac{\partial\hat{U}}{\partial y}(\xi,y) \underset{y\to 0^+}{\longrightarrow} -\frac{\Gamma(1-s)}{2^{2s-1}\Gamma(s)}(2\pi|\xi|)^{2s}\hat{u}(\xi).$$

This proves (10.16), thus completing the proof. For an alternative proof of (10.9) see Remark 10.5 below.

□

REMARK 10.2. Using Proposition 4.1 with the choice $b=0$ and $a = n+2s$, it is easy to recognize from (10.8) that

$$\text{(10.17)}\qquad ||P_s(\cdot,y)||_{L^1(\mathbb{R}^n)} = \int_{\mathbb{R}^n} P_s(x,y)dx = 1, \qquad \text{for every } y>0.$$

REMARK 10.3. Notice that when $s=\frac{1}{2}$ we have $a = 1-2s = 0$, and the extension operator L_a becomes the standard Laplacean $L_a = \Delta_x + \frac{\partial^2}{\partial y^2}$ in $\mathbb{R}^{n+1}$. From formula (10.8) we obtain in such case

$$P_{1/2}(x,y) = \frac{\Gamma(\frac{n+1}{2})}{\pi^{\frac{n+1}{2}}}\frac{y}{(y^2+|x|^2)^{\frac{n+1}{2}}},$$

which is in fact the standard Poisson kernel for the upper half-space $\mathbb{R}^{n+1}_+$, see e.g. [**SW71**].

REMARK 10.4. If we compare the expression of the Poisson kernel in (10.8) with (8.18) in Lemma 8.6, we conclude that, remarkably, we have shown that

$$P_s(x,y) = (-\Delta)^s E_{s,y}(x), \tag{10.18}$$

where for $y>0$ the function $E_{s,y} = c(n,s)(y^2+|x|^2)^{-\frac{n-2s}{2}}$ is the y-regularization of the fundamental solution of $(-\Delta)^s$. If we combine (10.18) with (8.27) above, we see that we can reformulate (8.27) as follows

$$\lim_{y\to 0^+} P_s(\cdot,y) = \delta \qquad \text{in } \mathscr{S}'(\mathbb{R}^n),$$

or, equivalently, for any $\varphi \in \mathscr{S}(\mathbb{R}^n)$

$$\lim_{y\to 0^+} \int_{\mathbb{R}^n} P_s(x,y)\varphi(x)dx = \varphi(0).$$

If we let $\check{\varphi}(x) = \varphi(-x)$, then we obtain from the latter limit relation

$$P_s(\cdot,y)\star\varphi(x) = \int_{\mathbb{R}^n} P_s(z,y)\tau_{-x}\check{\varphi}(z)dz \longrightarrow \tau_{-x}\check{\varphi}(0) = \varphi(x). \tag{10.19}$$

REMARK 10.5 (Alternative proof of (10.9)). Using the property (10.19) of the Poisson kernel $P_s(x,y)$ we can provide another "short" proof of (10.9) along the following lines, see Section 3.1 in [**CS07**]. Let $u \in \mathscr{S}(\mathbb{R}^n)$ and consider the solution $U(x,y) = P_s(\cdot,y)\star u(x)$ to the extension problem (10.1), see (10.7). Using (10.17) we can write

$$U(x,y) = \frac{\Gamma(\frac{n}{2}+s)}{\pi^{\frac{n}{2}}\Gamma(s)} \int_{\mathbb{R}^n} \frac{y^{2s}}{(y^2+|x-z|^2)^{\frac{n+2s}{2}}}(u(z)-u(x))dz + u(x).$$

Differentiating both sides of this formula with respect to y and keeping in mind that $a = 1-2s$, we obtain that as $y\to 0^+$

$$y^a\frac{\partial U}{\partial y}(x,y) = 2s\frac{\Gamma(\frac{n}{2}+s)}{\pi^{\frac{n}{2}}\Gamma(s)} \int_{\mathbb{R}^n} \frac{u(z)-u(x)}{(y^2+|z-x|^2)^{\frac{n+2s}{2}}}dz + O(y^2).$$

Letting $y\to 0^+$ and using Lebesgue dominated convergence theorem, we thus find

$$\begin{aligned}\lim_{y\to 0^+} y^a\frac{\partial U}{\partial y}(x,y) &= 2s\frac{\Gamma(\frac{n}{2}+s)}{\pi^{\frac{n}{2}}\Gamma(s)}\,\mathrm{PV}\int_{\mathbb{R}^n}\frac{u(z)-u(x)}{|z-x|^{n+2s}}dz\\ &= -2s\frac{\Gamma(\frac{n}{2}+s)}{\pi^{\frac{n}{2}}\Gamma(s)}\gamma(n,s)^{-1}(-\Delta)^s u(x),\end{aligned}$$

where in the second equality we have used (2.11) above. If in the latter equation we now replace the expression (5.10) of the constant $\gamma(n,s)$, we reach the conclusion that (10.9) is valid.

The Poisson kernel $P_s(x,y)$ is of course a solution of $L_a P_s = 0$ in $\mathbb{R}^{n+1}_+$. What is instead not obvious is that the y-regularization $E_{s,y}$ of the fundamental solution E_s of $(-\Delta)^s$ introduced in (8.6) in Lemma 8.5 is also a solution of the extension operator L_a. It was shown in [**CS07**] that, up to a constant, such function is in fact the fundamental solution of L_a. The heuristic motivation behind this is that, with $x \in \mathbb{R}^n$, and $\eta \in \mathbb{R}^{a+1}$, if $y = |\eta|$ then the operator

$$y^{-a} L_a = \Delta_x + \frac{\partial^2}{\partial y^2} + \frac{a}{y} \frac{\partial}{\partial y} \tag{10.20}$$

can be thought of as the Laplacean in the fractional dimension $N = n+a+1$ acting on functions $U(x, |\eta|)$. Such heuristic is confirmed by the following result.

PROPOSITION 10.6. *For $y \in \mathbb{R}$ consider the function $G(x, y) = (|x|^2 + y^2)^{-\frac{n-2s}{2}}$, see (8.6). Then, for every $(x, y) \in \mathbb{R}^{n+1}_+$, with $a = 1 - 2s$ we have*

$$L_a G(x, y) = 0.$$

PROOF. It is convenient to use the expression of (10.20) on functions depending on $r = |x|$ and y

$$y^{-a} L_a = \frac{\partial^2}{\partial r^2} + \frac{n-1}{r} \frac{\partial}{\partial r} + \frac{\partial^2}{\partial y^2} + \frac{a}{y} \frac{\partial}{\partial y}.$$

Then, the proof becomes a simple computation. Abusing the notation we write $G(x, y) = G(r, y) = (r^2 + y^2)^{-\frac{n-2s}{2}}$. We have

$$G_r = -(n + a - 1)(r^2 + y^2)^{-\frac{n+a-1}{2} - 1} r,$$

$$G_{rr} = (n + a - 1)(r^2 + y^2)^{-\frac{n+a-1}{2} - 2}((n + a)r^2 - y^2).$$

This gives

$$G_{rr} + \frac{n-1}{r} G_r = (n + a - 1)(r^2 + y^2)^{-\frac{n+a-1}{2} - 2}((1 + a)r^2 - ny^2).$$

On the other hand, a similar computation gives

$$G_{yy} + \frac{a}{y} G_y = -(n + a - 1)(r^2 + y^2)^{-\frac{n+a-1}{2} - 2}((1 + a)r^2 - ny^2).$$

Adding the latter two equations gives the desired conclusion $L_a G = 0$.

□

11. Fractional Laplacean and subelliptic equations

In Section 10 we have analyzed the important fact (10.2) that s-harmonic functions arise as weighted Dirichlet-to-Neumann traces of the solutions of the extension problem (10.1). This aspect underscores the deep connection between the fractional Laplacean and the class of second order partial differential equations of degenerate type introduced in [**FKS81**].

In this section we want to advertise another aspect of nonlocal equations, namely the link between the nonlocal operator $(-\Delta)^s$ and the theory of the so-called *subelliptic equations*. This name comes from the fact that, although the relevant differential operator L fails to satisfy the a priori estimates of the elliptic theory, it does satisfy the following replacement estimate *below* the elliptic index, hence the name subelliptic:

$$||u||_{H^{2\varepsilon}} \leq C\left(||u||_{L^2} + ||Lu||_{L^2}\right),$$

for all C_0^∞ functions u, and for some $0 < \varepsilon < 1$. Subelliptic operators typically display loss of control of derivatives in a set of directions.

The aspect that we have in mind originates with the following particular subelliptic operator

$$\mathcal{P}_\alpha = \frac{\partial^2}{\partial z^2} + |z|^{2\alpha}\frac{\partial^2}{\partial x^2}, \qquad \alpha > 0,$$

that was first introduced by S. Baouendi in 1967 in his Ph. D. Dissertation under the supervision of B. Malgrange, see [**Ba67**]. At that time M. Vishik was visiting Malgrange, who discussed with him the thesis project of Baouendi. Vishik subsequently asked Malgrange permission to suggest to his own Ph. D. student, Grushin, to work on some questions related to the hypoellipticity of $\mathcal{P}_\alpha$ when $\alpha \in \mathbb{N}$, see [**Gru70**] and [**Gru71**]. This is how the operator $\mathcal{P}_\alpha$ became known as the *Baouendi-Grushin operator.* A decade later, in the early 80's, Franchi and Lanconelli introduced a class of operators which include $\mathcal{P}_\alpha$, and they pioneered the study of the fine properties of their weak solutions, such as the Harnack type inequality and the Hölder continuity, by studying a control distance associated with the relevant operators, see [**FL82**]-[**FL85**], and also the subsequent work [**FS87**]. As we will see in this and the subsequent section, there is an underlying strong connection between these works, the paper [**FKS81**] of Fabes, Kenig and Serapioni mentioned in the previous section, and the fractional Laplacean $(-\Delta)^s$. We will further the discussion of the interconnection between these operators in Sections 12 and 14 below.

On one hand, we will see from Proposition 11.2 below that, at least in the range $0 < s \leq 1/2$, the fractional Laplacean arises as the true Dirichlet-to-Neumann map of the Baouendi-Grushin operator $\mathcal{P}_\alpha$ defined in (11.8). On the other hand, the recent work of Koch, Petrosyan and Shi [**KPS15**] has underscored an even deeper link between nonlocal and subelliptic equations. The central tool in their study of the real-analytic smoothness of the regular free boundary in the obstacle problem for $(-\Delta)^{1/2}$ is a partial hodograph transformation. After such change of variables, they obtain a fully nonlinear partial differential equation which has a subelliptic structure. In the sense that the linearization of such fully nonlinear equation is precisely a Baouendi operator such as (11.12) below with $\alpha = 1$, see Section 5 in [**KPS15**]. Before proceeding we mention that the C^∞ smoothness of the regular free boundary in the obstacle problem for $(-\Delta)^{1/2}$ has also been proved with a completely different approach by De Silva and Savin in [**DS16**], see also [**DS15**] for a related result in the Bernoulli problem. Their approach has been subsequently generalized to all $s \in (0,1)$ in [**JN17**].

These facts represent an interesting opportunity for interaction between two seemingly disjoint communities: that of workers in subelliptic equations and the closely connected field of sub-Riemannian geometry, and that of workers in nonlocal equations. We hope that the present discussion, as well as the content of Sections 12 and 14, will encourage such exchange. Most of the material that follows is borrowed from the papers [**G93**], [**CS07**] and [**GRO17**].

To introduce our discussion let us consider the fractional Laplacean $(-\Delta)^s$ with $0 < s < 1$. With $a = 1 - 2s$, we have $-1 < a < 1$, and we have seen in (10.20) that the extension operator L_a can be written in the form

$$L_a = y^a\left(\Delta_x + \frac{\partial^2}{\partial y^2} + \frac{a}{y}\frac{\partial}{\partial y}\right). \tag{11.1}$$

We now want to connect the operator L_a to another degenerate elliptic operator. Following [**CS07**] we introduce the change of variable $\Phi : \mathbb{R}^{n+1}_+ \to \mathbb{R}^{n+1}_+$

$$(x, z) = \Phi(x, y) = (x, h(y)), \tag{11.2}$$

where the function $h(y)$ is chosen so to eliminate the drift term in (11.1). To do this, given a function $U(x, z)$ defined for $(x, z) \in \mathbb{R}^{n+1}_+$, we define a function $\tilde{U}(x, y)$, with $(x, y) \in \mathbb{R}^{n+1}_+$, by the formula

$$\tilde{U}(x, y) := U(\Phi(x, y)) = U(x, h(y)). \tag{11.3}$$

A simple computation gives

$$\begin{aligned} &L_a\tilde{U}(x, y) \\ &= y^a \left[\Delta_x U(x, h(y)) + \left(h''(y) + \frac{a}{y}h'(y)\right) D_z U(x, h(y)) + h'(y)^2 D_{zz} U(x, h(y))\right]. \end{aligned}$$

From this equation it is clear that if $h(y)$ satisfies the differential equation

$$h''(y) + \frac{a}{y}h'(y) \equiv 0, \tag{11.4}$$

then we obtain

$$L_a\tilde{U}(x, y) = (h^{-1}(z))^a \left[\Delta_x U(x, z) + h'(h^{-1}(z))^2 D_{zz} U(x, z)\right]. \tag{11.5}$$

Solving (11.4) we find $h(y) = Ay^{1-a}$ for some $A \in \mathbb{R} \setminus \{0\}$. We now choose A so that $h'(h^{-1}(z)) = z^{-\frac{a}{1-a}}$, which gives $A = (1 - a)^{-(1-a)}$. Summarizing, $h : (0, \infty) \to (0, \infty)$ is the strictly increasing function given by

$$z = h(y) = \left(\frac{y}{1-a}\right)^{1-a}, \quad \text{with inverse} \quad y = h^{-1}(z) = (1 - a)z^{\frac{1}{1-a}}. \tag{11.6}$$

With this choice we have $h'(y) = (1 - a)^a y^{-a}$, and thus we conclude from (11.5) that

$$L_a\tilde{U}(x, y) = (1 - a)^a z^{\frac{a}{1-a}} \left[\Delta_x U(x, z) + z^{-\frac{2a}{1-a}} D_{zz} U(x, z)\right]. \tag{11.7}$$

The next proposition summarizes the content of (11.7).

PROPOSITION 11.1. *Let $-1 < a < 1$, and $\alpha = \frac{a}{1-a} \in (-1/2, \infty)$. The mapping $U \leftrightarrow \tilde{U}$ defined by (11.3), with $h(y)$ given by (11.6), converts in a one-to-one, onto fashion, solutions of the equation $L_a\tilde{U} = 0$ with respect to the variables $(x, y) \in \mathbb{R}^{n+1}_+$ into solutions with respect to the variables $(x, z) \in \mathbb{R}^{n+1}_+$ of the equation*

$$\mathcal{P}_\alpha U = D_{zz} U(x, z) + z^{2\alpha} \Delta_x U(x, z) = 0. \tag{11.8}$$

We note explicitly that, in the correspondence $U \leftrightarrow \tilde{U}$, the equation (11.7) can be more suggestively expressed as

$$y^a L_a\tilde{U}(x, y) = \mathcal{P}_\alpha u(x, z). \tag{11.9}$$

Using Proposition 11.1 one obtains the following interesting result.

PROPOSITION 11.2. *Given any $s \in (0, 1)$, the fractional Laplacean $(-\Delta)^s$ in $\mathbb{R}^n$ can be interpreted as the Dirichlet-to-Neumann map of the operator $\mathcal{P}_\alpha$ in $\mathbb{R}^{n+1}_+$*

defined in (11.8)*, where* $\alpha = \frac{1}{2s} - 1$*. By this we mean that if for any* $u \in \mathscr{S}(\mathbb{R}^n)$ *one considers the solution* $U(x,z)$ *to the Dirichlet problem*

$$\begin{cases} \mathcal{P}_\alpha U(x,z) = D_{zz}U(x,z) + z^{2\alpha}\Delta_x U(x,z) = 0, & (x,z) \in \mathbb{R}^{n+1}_+, \\ U(x,0) = u(x), \end{cases} \tag{11.10}$$

then one has

$$(-\Delta)^s u(x) = -\frac{\Gamma(1+s)}{\Gamma(1-s)} \lim_{z\to 0^+} \frac{\partial U}{\partial z}(x,z). \tag{11.11}$$

PROOF. Consider the solution to the Dirichlet problem (11.10). Denote by $\tilde{U}(x,y)$ the function associated to U in the variables $(x,y) \in \mathbb{R}^{n+1}_+$ by the correspondence (11.3), where $h(y)$ is given by (11.6), and a and α are related by the equation $\alpha = \frac{a}{1-a}$, or equivalently $a = \frac{\alpha}{\alpha+1}$. Since $\alpha = \frac{1}{2s} - 1$, we see that $a = 1-2s$. By Proposition 11.1 we know that $\tilde{U}$ satisfies $L_a\tilde{U} = 0$ in $\mathbb{R}^{n+1}_+$. Furthermore, we have $\tilde{U}(x,0) = U(x,0) = u(x)$. By (10.2) we have

$$(-\Delta)^s u(x) = -\frac{2^{2s-1}\Gamma(s)}{\Gamma(1-s)} \lim_{y\to 0^+} y^{1-2s}\frac{\partial \tilde{U}}{\partial y}(x,y).$$

On the other, (11.3) and the chain rule give

$$y^{1-2s}\frac{\partial \tilde{U}}{\partial y}(x,y) - y^{1-2s}\frac{\partial U}{\partial z}(x,h(y))h'(y) = (2s)^{-2s+1}\frac{\partial U}{\partial z}(x,(2s)^{-2s}y^{2s}).$$

From the latter two equations we obtain

$$\begin{aligned} (-\Delta)^s u(x) &= -\frac{2^{2s-1}\Gamma(s)}{\Gamma(1-s)} \lim_{y\to 0^+} y^{1-2s}\frac{\partial \tilde{U}}{\partial y}(x,y) \\ &= -\frac{2^{2s-1}\Gamma(s)}{\Gamma(1-s)}(2s)^{-2s+1} \lim_{z\to 0^+} \frac{\partial U}{\partial z}(x,z). \end{aligned}$$

Keeping (4.1) in mind, we have thus proved (11.11).

□

We pause for a moment to emphasize that Proposition 11.2 shows that the true Dirichlet-to-Neumann map that defines the nonlocal operator $(-\Delta)^s$ is the one associated with the degenerate elliptic operator $\mathcal{P}_\alpha$ in (11.8). If we now restrict the attention to the regime $0 < s \le 1/2$, then $\alpha = \frac{1}{2s} - 1 \ge 0$ and the operator in (11.8) is a model of the *Baouendi-Grushin operators* in $\mathbb{R}^n_x \times \mathbb{R}^m_z$ given by

$$\mathcal{P}_\alpha = \Delta_z + |z|^{2\alpha}\Delta_x, \qquad \alpha \ge 0. \tag{11.12}$$

When $\alpha > 0$ such operators are degenerate elliptic along the n-dimensional subspace $M = \mathbb{R}^n \times \{0\}_{\mathbb{R}^m}$. They are the prototype of a class of equations that continues to be much studied nowadays. One of the reasons for such continuing interest is that, as we next illustrate, (11.12) is closely connected with an object of fundamental relevance in harmonic analysis, partial differential equations and geometry, the Heisenberg group $\mathbb{H}^n$. This is the stratified nilpotent Lie group whose underlying manifold is $\mathbb{C}^n \times \mathbb{R} \cong \mathbb{R}^{2n+1}$ with noncommutative group law given in real coordinates by

$$(x,y,t)\circ(x',y',t') = (x+x', y+y', t+t' + \frac{1}{2}(<x,y'> - <x',y>)). \tag{11.13}$$

If we let $p = (x, y, t), p' = (x', y', t') \in \mathbb{H}^n$, and define the operator of left-translation by $L_p(p') = p \circ p'$, then denoting by dL_p the differential of the map (11.13), a basis for the real Lie algebra of left-invariant differential operators is given by applying dL_p to the standard basis in $\mathbb{R}^{2n+1}$. We thus find

$$\begin{aligned} &X_j = X_j(p) = dL_p(e_j) = \partial_{x_j} - \frac{y_j}{2}\partial_t, \quad j = 1, ..., n, \\ &X_{n+j} = X_{n+j}(p) = dL_p(e_{n+j}) = \partial_{y_j} + \frac{x_j}{2}\partial_t, \quad j = 1, ..., n, \\ &T = T(p) = dL_p(e_{2n+1}) = \partial_t. \end{aligned}$$

These vector fields satisfy the commutation relations

$$[X_j, X_{n+k}] = \delta_{jk} T, \quad j, k = 1, ..., n, \tag{11.14}$$

all other commutators being trivial (remember that the commutator of two vector fields X and Y is defined by $[X, Y] = XY - YX$). The name *Heisenberg group* comes from the fact that, when $n = 1$, then in $\mathbb{H}^1 = (\mathbb{R}^3, \circ)$ the resulting equation (11.14) represents an abstract version of Heisenberg's canonical commutation relations in quantum mechanics for position and momentum of a relativistic particle, see [**Fo89**]. In fact, much before than mathematicians christened it with such name, the three-dimensional Heisenberg group had long been known to physicist as the *Weyl's group*, and it was identified with the following group of 3×3 matrices:

$$\begin{pmatrix} 1 & x & z \\ 0 & 1 & y \\ 0 & 0 & 1 \end{pmatrix}, \quad x, y, z \in \mathbb{R}.$$

The Lie algebra is clearly spanned by the matrices

$$X = \begin{pmatrix} 0 & 1 & 0 \\ 0 & 0 & 0 \\ 0 & 0 & 0 \end{pmatrix}, \ Y = \begin{pmatrix} 0 & 0 & 0 \\ 0 & 0 & 1 \\ 0 & 0 & 0 \end{pmatrix}, \ Z = \begin{pmatrix} 0 & 0 & 1 \\ 0 & 0 & 0 \\ 0 & 0 & 0 \end{pmatrix},$$

for which the following commutation relations hold

$$[X, Y] = Z, \ [X, Z] = [Y, Z] = 0.$$

Now, much like the Laplacean in $\mathbb{R}^n$, in the Heisenberg group there is a second order partial differential operator which plays a fundamental role in the analysis of such group. Since according to (11.14) the vector fields $X_1, ..., X_{2n}$ generate the whole Lie algebra, it is natural to consider the following operator

$$\Delta_H = \sum_{j=1}^{2n} X_j^2, \tag{11.15}$$

which is known as the real part of the *Kohn-Spencer sub-Laplacean*. We will simply call it the *sub-Laplacean* on $\mathbb{H}^n$. Although it plays in the analysis of $\mathbb{H}^n$ a role quite similar to that played by the standard Laplacean in classical analysis, the differences between these two objects are stunning since the geometry of $\mathbb{H}^n$ is not Riemannian and it is not easy to grasp.

In the real coordinates $p = (x, y, t) \in \mathbb{H}^n$, if we indicate $z = (x, y) \in \mathbb{R}^{2n}$, then the operator (11.15) takes the form

$$\Delta_H = \Delta_z + \frac{|z|^2}{4}\partial_{tt} + \partial_t \sum_{j=1}^{n} (x_j \partial_{y_j} - y_j \partial_{x_j}). \tag{11.16}$$

One remarkable feature of (11.16) is that this operator fails to be elliptic at every point $p \in \mathbb{H}^n$. It is in fact an easy exercise to verify that the matrix of the quadratic form associated with (11.16) has a vanishing eigenvalue. However, since by (11.14) we know that the vector fields $\{X_1, ..., X_n, X_{n+1}, ..., X_{2n}\}$ generate the whole Lie algebra of left-invariant vector fields, then thanks to a celebrated theorem of Hörmander we know that solutions of $\Delta_H u = 0$ are C^∞, see [**Ho67**] (in fact, they are real-analytic, but that does not follow from Hörmander's theorem) and also the lecture notes [**G16**]. For an introduction to the Heisenberg group one should see [**CDPT**]. Second order partial differential equations such as the sub-Laplacean (11.15) on $\mathbb{H}^n$ and the Baouendi operator (11.12) are called *subelliptic.* The reason for this name is that, despite the fact that they may fail to be *elliptic* (at every point, in the case of (11.15), along a submanifold for (11.12)), they satisfy a so-called a priori subelliptic estimate. But the discussion of this deep aspect would take us too far, and thus we must leave it to the interested reader to possibly further it on his/her own.

Returning to the Baouendi operator (11.12), suppose that $n = 1$ and $\alpha = 1$, in which case (11.12) becomes

$$\mathcal{P} = \Delta_z + |z|^2 \partial_{xx}. \tag{11.17}$$

If we now consider (11.16), we see that in $\mathbb{H}^n$ every solution of $\Delta_H u = 0$ which is invariant under the action of the vector field

$$\Theta = \sum_{j=1}^{n} (x_j \partial_{y_j} - y_j \partial_{x_j}),$$

is a solution of the equation

$$\Delta_z u + \frac{|z|^2}{4} \partial_{tt} u = 0, \tag{11.18}$$

and, up to a rescaling factor, this is precisely (11.17). For instance, in the three-dimensional Heisenberg group every function $u : \mathbb{H}^1 \to \mathbb{R}$ that has cylindrical symmetry, i.e., $u(z,t) = u^\star(|z|, t)$, is such that $\Theta u = 0$, and vice-versa. Therefore, it solves $\Delta_H u = 0$ if and only if it solves (11.18). This explains the connection of the operator of Baouendi with the sub-Laplacean on the Heisenberg group $\mathbb{H}^n$.

We next explore further the connection between to Baouendi operator (11.12) and $(-\Delta)^s$. With this objective in mind we assume that $m = 1$ in (11.12), so that the resulting operator in $\mathbb{R}^n_x \times \mathbb{R}_z$ is

$$\mathcal{P}_\alpha U = \frac{\partial^2 U}{\partial z^2} + |z|^{2\alpha} \Delta_x U. \tag{11.19}$$

As stated in (11.8) we want $\alpha = \frac{1}{2s} - 1$, and since $0 < s < 1$, this means that in (11.19) we must have $-1/2 < \alpha < \infty$. We have three possibilities:

- $0 < s < 1/2 \implies \alpha > 0$ ($\mathcal{P}_\alpha$ is of Baouendi type);
- $s = 1/2 \implies \alpha = 0$ ($\mathcal{P}_\alpha$ is the standard Laplacean in $\mathbb{R}^n_x \times \mathbb{R}_z$);
- $1/2 < s < 1 \implies -1/2 < \alpha < 0$ ($\mathcal{P}_\alpha$ is not of Baouendi type).

REMARK 11.3. Although $\mathcal{P}_\alpha$ is not a Baouendi operator when $-1/2 < \alpha < 0$, all the subsequent discussion covers such case as well. This is true in particular of the results from [**G93**] that we are going to use, and which were in that paper obtained under the hypothesis that $\alpha \geq 0$.

First, we equip $\mathbb{R}^n_x \times \mathbb{R}_z$ with the following non-isotropic dilations

$$\delta_\lambda(x,z) = (\lambda^{\alpha+1}x, \lambda z), \qquad \lambda > 0. \tag{11.20}$$

A function u is said δ_λ-homogeneous of degree κ if

$$u(\delta_\lambda(x,z)) = \lambda^\kappa u(x,z), \qquad \lambda > 0.$$

It is straightforward to verify that the partial differential operator $\mathcal{P}_\alpha$ is δ_λ-homogeneous of degree two, i.e.,

$$\mathcal{P}_\alpha(\delta_\lambda \circ u) = \lambda^2 \delta_\lambda \circ (\mathcal{P}_\alpha u).$$

We note that Lebesgue measure in $\mathbb{R}^n_x \times \mathbb{R}_z$ changes according to the equation

$$d(\delta_\lambda(x,z)) = \lambda^{(\alpha+1)n+1} dx dz, \tag{11.21}$$

which motivates the definition of the *homogeneous dimension* for the number

$$Q = Q_\alpha = (\alpha+1)n + 1. \tag{11.22}$$

In the analysis of (11.19) the following pseudo-gauge introduced in [**G93**] plays an important role

$$\rho_\alpha(x,z) = \left((\alpha+1)^2|x|^2 + |z|^{2(\alpha+1)}\right)^{\frac{1}{2(\alpha+1)}}. \tag{11.23}$$

We clearly have

$$\rho_\alpha(\delta_\lambda(x,z)) = \lambda \rho_\alpha(x,z), \tag{11.24}$$

i.e., the pseudo-gauge is homogeneous of degree one. The pseudo-ball and sphere centered at the origin with radius $r > 0$ are respectively defined as

$$B_{\rho_\alpha}(r) = \{(x,z) \in \mathbb{R}^n_x \times \mathbb{R}_z \mid \rho_\alpha(x,z) < r\}, \qquad S_{\rho_\alpha}(r) = \partial B_{\rho_\alpha}(r). \tag{11.25}$$

In [**G93**] it was proved that, with Q as in (11.22), and $C_\alpha > 0$ given by

$$C_\alpha^{-1} = (Q+2\alpha)(Q-2)\int_{\mathbb{R}^n_x \times \mathbb{R}_z} \frac{|z|^\alpha dx dz}{[((\alpha+1)^2|x|^2 + |z|^{\alpha+1}+1)]^{1+\frac{Q+2\alpha}{2(\alpha+1)}}},$$

the function

$$\Gamma_\alpha(x,z) = \frac{C_\alpha}{\rho_\alpha(x,z)^{Q-2}} \tag{11.26}$$

is a fundamental solution for $-\mathcal{P}_\alpha$ with singularity at $(0,0)$. Since the operator is invariant with respect to translations along $M = \mathbb{R}^n \times \{0\}$, from (11.26) we immediately obtain the fundamental solution for $\mathcal{P}_\alpha$ with singularity at any point of the subspace M.

In what follows we indicate with $d_e(x,y) = (|x|^2+y^2)^{1/2}$ the standard Euclidean distance in $\mathbb{R}^{n+1}$, and to emphasize certain differences we will indicate with $B_e(r) = \{(x,y) \in \mathbb{R}^{n+1} \mid d_e(x,y) < r\}$ the Euclidean ball in $\mathbb{R}^{n+1}$ of radius r centered at the origin. Balls centered at a different point $X = (x,y)$ will be indicated with $B_e(X,r)$. When $X = (x,0)$, with a slight abuse of notation we will write $B_e(x,r)$ instead of $B_e((x,0),r)$. We will use analogous notations for the spheres $S_e(r), S_e(X,r), S_e(x,r)$ in $\mathbb{R}^{n+1}$.

The function $h(y)$ is that given by (11.6) above. We have the following simple yet important fact.

Proposition 11.4. *Given $a \in (-1,1)$, let $\alpha = \frac{a}{1-a}$. Then, for any $(x,y) \in \mathbb{R}^{n+1}_+$ we have*

$$\rho_\alpha(x, h(|y|)) = h(d_e(x,y)). \tag{11.27}$$

The equation (11.27) *implies in particular that*

$$B_{\rho_\alpha}(h(r)) = \Phi(B_e(r)), \qquad r > 0, \tag{11.28}$$

where $\Phi(x,y) = (x, h(|y|))$.

In view of (11.7), or Proposition 11.1, it is clear that if we consider the function in $\mathbb{R}^{n+1}_+$ given by

$$\tilde{\Gamma}(x,y) = (1-a)^a \Gamma_\alpha(x, h(y)), \tag{11.29}$$

then we have $L_a \tilde{\Gamma} = 0$ in $\mathbb{R}^{n+1}_+$. Notice that from (11.27), (11.26) we have

$$\tilde{\Gamma}(x,y) = \frac{(1-a)^a C_\alpha}{\rho_\alpha(x,h(y))^{Q-2}} = \frac{(1-a)^a C_\alpha}{h(d_e(x,y))^{Q-2}} = \frac{(1-a)^{n+2a-1} C_\alpha}{d_e(x,y)^{n+a-1}} = \frac{\tilde{C}_a}{d_e(x,y)^{n+a-1}}, \tag{11.30}$$

where in the secondo to the last equality we have used the above expression (11.22) of the homogeneous dimension associated with the dilations (11.20) in $\mathbb{R}^{n+1}$. Now, the function $\tilde{\Gamma}(x,y)$ is precisely the fundamental solution of the Laplacean

$$\Delta = \Delta_x + D_{yy} + \frac{a}{y} D_y$$

in the fractional dimension

$$\tilde{Q} = n + a + 1, \tag{11.31}$$

found by Caffarelli and Silvestre in formula (2.1) in [**CS07**]. Furthermore, if we keep (11.31) in mind, we see that the exponent $n + a - 1$ in (11.30) is nothing but $\tilde{Q} - 2$, whereas $\tilde{C}_a = C_\alpha (1-a)^{n+2a-1}$.

We close this section with the following result from [**GRO17**] that allows to connect integrals on the pseudo-balls and spheres $B_{\rho_\alpha}(r)$ and $S_{\rho_\alpha}(r)$ in the space of the variables (x,z), to corresponding integrals on the Euclidean balls and spheres in the variables (x,y).

Proposition 11.5. *Let U be a continuous function in the space $\mathbb{R}^{n+1}$ with the variables (x,z), even in z, and let $\tilde{U}(x,y) = U(x, h(|y|))$. Then, we have for every $r > 0$*

$$\int_{B_{\rho_\alpha}(h(r))} U(x,z)\,dx dz = (1-a)^a \int_{B_e(r)} \tilde{U}(x,y)|y|^{-a} dx dy, \tag{11.32}$$

and also

$$h'(r) \int_{S_{\rho_\alpha}(h(r))} \frac{U(x,z)}{|\nabla \rho_\alpha(x,z)|} dH_n(x,z) = (1-a)^a \int_{S_e(r)} \tilde{U}(x,y)|y|^{-a} dH_n(x,y). \tag{11.33}$$

12. Hypoellipticity of $(-\Delta)^s$

One of the most important properties of Laplace equation is its *hypoellipticity*. This means that distributional solutions of $\Delta u = f$ are C^∞ wherever f is. Here is the formal definition, see e.g. [**T75**].

DEFINITION 12.1. A linear partial differential operator $P(x,\partial_x)$ in an open set $\Omega \subset \mathbb{R}^n$ is said to be *hypoelliptic* if, given any open subset $U \subset \Omega$ and any distribution u in U, u is a C^∞ function in U if this is true of $P(x,\partial_x)u$.

To understand this aspect let us recall a classical fact. Let $\Omega \subset \mathbb{R}^n$ be an open set and for a function $u \in C(\Omega)$ consider the spherical mean-value $\mathscr{M}_r u(x)$ defined in (2.3).

PROPOSITION 12.2. *Let $\Omega \subset \mathbb{R}^n$ be an open set. For $x \in \Omega$ let $R > 0$ be such that $\overline{B}(x,R) \subset \Omega$. One has:*

(i) *if $u \in C^1(\Omega)$, then for every $0 < r < R$ one has*
$$\frac{\partial \mathscr{M}_r u}{\partial r}(x) = \frac{1}{\sigma_{n-1} r^{n-1}} \int_{S(x,r)} \frac{\partial u}{\partial \nu}(y) d\sigma(y);$$

(ii) *if $u \in C^2(\Omega)$, then for every $0 < r < R$ one has*
$$\frac{\partial \mathscr{M}_r u}{\partial r}(x) = \frac{1}{\sigma_{n-1} r^{n-1}} \int_{B(x,r)} \Delta u(y) dy.$$

A basic consequence of (ii) is that if u is harmonic in Ω, i.e., $u \in C^2(\Omega)$ and $\Delta u = 0$ in Ω, then for any $x \in \Omega$ and any $0 < r < \operatorname{dist}(x,\partial\Omega)$, one has

$$u(x) = \mathscr{M}_r u(x). \tag{12.1}$$

Now, the fact that a function satisfies the mean-value formula (12.1) has truly remarkable consequences. One of them, is the following well-known converse to Gauss' mean value theorem.

THEOREM 12.3 (of Koëbe). *Let $u \in C(\Omega)$ and suppose that for every $x \in \Omega$ and $0 < r < \operatorname{dist}(x,\partial\Omega)$ formula (12.1) hold. Then, $u \in C^\infty(\Omega)$ and in fact $\Delta u = 0$ in Ω.*

PROOF. To prove that $u \in C^\infty(\Omega)$ it is obviously enough to show that $u \in C^\infty(\Omega_\varepsilon)$ for every $\varepsilon > 0$, where $\Omega_\varepsilon = \{x \in \Omega \mid \operatorname{dist}(x,\partial\Omega) > \varepsilon\}$. With this objective in mind let K be a spherically symmetric Friedrichs' mollifier, i.e., $K \in C_0^\infty(\mathbb{R}^n)$, with $\int_{\mathbb{R}^n} K(y)dy = 1$, supp $K \subset \overline{B}(0,1)$, and $K(y) = K^\star(|y|)$, and denote by $K_\varepsilon(y) = \varepsilon^{-n} K(y/\varepsilon)$ the corresponding approximation to the identity. We claim that as consequence of (12.1) the following must be true in Ω_ε:

$$u = K_\varepsilon \star u. \tag{12.2}$$

To verify this claim we use Cavalieri's principle to write for every $x \in \Omega_\varepsilon$

$$\begin{aligned} K_\varepsilon \star u(x) &= \int_0^\infty \int_{|y|=r} K(y) u(x-y) d\sigma(y) dr = \int_0^\infty K^\star(r) \int_{|y|=r} u(x-y) d\sigma(y) dr \\ &= \sigma_{n-1} \int_0^\infty K^\star(r) r^{n-1} \mathscr{M}_r u(x) dr = u(x) \sigma_{n-1} \int_0^\infty K^\star(r) r^{n-1} dr \\ &= u(x) \int_{\mathbb{R}^n} K(y) dy = u(x), \end{aligned}$$

where in the second to the last equality we have used the spherical symmetry of K and Cavalieri's principle again.

From (12.2) and a well-known property of the convolution, we conclude that (12.1) implies that $u \in C^\infty(\Omega_\varepsilon)$, and therefore $u \in C^\infty(\Omega)$. We pause for a moment to emphasize this remarkable conclusion: *a continuous function which locally satisfies the mean value property* (12.1) *must in fact be infinitely smooth!* Once we know this we can appeal to (ii) in Proposition 12.2 above to infer that for every $x \in \Omega$ and $0 < r < \operatorname{dist}(x, \partial\Omega)$ we can differentiate at that point r the function $t \to \mathscr{M}_t u(x)$, and we have

$$\frac{\partial \mathscr{M}_r u}{\partial r}(x) = \frac{1}{\sigma_{n-1} r^{n-1}} \int_{B(x,r)} \Delta u(y) dy.$$

On the other hand, the constancy of $r \to \mathscr{M}_r u(x, r)$ that follows from by (12.1) implies that $\frac{\partial \mathscr{M}_r u}{\partial r}(x) = 0$ for every $0 < r < \operatorname{dist}(x, \partial\Omega)$, and thus in particular we have for every such value of r

$$\frac{1}{\omega_n r^n} \int_{B(x,r)} \Delta u(y) dy = \frac{n}{\sigma_{n-1} r^n} \int_{B(x,r)} \Delta u(y) dy = 0,$$

where in the first equality we have used the second identity in (4.6) above. Since $\Delta u \in C(\Omega)$, letting $r \to 0^+$ in the above equation we infer that it must be $\Delta u(x) = 0$. By the arbitrariness of $x \in \Omega$, we conclude $\Delta u = 0$ in Ω.

□

From Theorem 12.3 and the fact that harmonic functions satisfy (12.1), we immediately obtain the following important result.

COROLLARY 12.4 (Smoothing property of Δ). *Let $u \in C^2(\Omega)$ be such that $\Delta u = 0$ in Ω. Then, $u \in C^\infty(\Omega)$.*

Hypoellipticity means that the conclusion of Corollary 12.4 continues to be true if we replace the hypothesis that $u \in C^2(\Omega)$ and $\Delta u = 0$, with the much weaker assumption that u be harmonic in the distributional sense, i.e., $u \in \mathscr{D}'(\Omega)$ and $\Delta u = 0$ in $\mathscr{D}'(\Omega)$. Historically, this result is known as *Weyl's lemma*, after the famous 1940 paper by H. Weyl [**W40**]. Although less known, R. Caccioppoli in [**C37**] had already established such result for $n = 2$ in a more general form in 1937, and subsequently G. Cimmino extended Caccioppoli's theorem to all elliptic operators with smooth coefficients in the plane, see [**Ci38**], [**Ci38'**]. Since their results preceded Weyl's paper, it should be called the *Caccioppoli-Cimmino-Weyl lemma*.

After this prelude on the Laplacean, we return to the main focus on this note and ask the natural question: *how smooth are solutions of* $(-\Delta)^s u = 0$? The answer to this question is that the nonlocal operator $(-\Delta)^s$ behaves much like the standard Laplacean, and thus *distributional solutions of* $(-\Delta)^s u = 0$ *are* C^∞. This is contained in Theorem 12.19 below, whose proof however relies on important results from the theory of pseudo-differential operators. Since the declared intent of these notes is didactic and being self-contained, we next present a somewhat less general result whose proof has the advantage of being conceptually much simpler, and closer in spirit to the opening discussion on the Laplacean. It also keeps up with the spirit of Section 11 of connecting $(-\Delta)^s$ to the class of degenerate elliptic operators such as $\mathcal{P}_\alpha$ and L_a via the extension procedure.

With this comment in mind we return to the Baouendi operator $\mathcal{P}_\alpha$ in (11.19) in the space $\mathbb{R}^n_x \times \mathbb{R}_z$, and recall some results from [**G93**]. For any $\alpha \geq 0$ we define the α-gradient of a function U that lives in an open set of such space as follows

$$\nabla_\alpha U = (|z|^\alpha \nabla_x U, D_z U).$$

Given two functions U and V we set

$$\langle \nabla_\alpha U, \nabla_\alpha V\rangle = |z|^{2\alpha}\langle \nabla_x U, \nabla_x V\rangle + D_z U D_z V.$$

The square of the length of $\nabla_\alpha U$ is

$$|\nabla_\alpha U|^2 = |z|^{2\alpha}|\nabla_x U|^2 + (D_z U)^2. \tag{12.3}$$

The following lemma, collects the identities (2.12)-(2.14) in [**G93**].

LEMMA 12.5. *Let ρ_α be the pseudo-gauge in* (11.23) *above. One has in $\mathbb{R}^{n+1}\setminus\{0\}$,*

$$\psi_\alpha \overset{def}{=} |\nabla_\alpha \rho_\alpha|^2 = \frac{|z|^{2\alpha}}{\rho_\alpha^{2\alpha}}. \tag{12.4}$$

Moreover, given a function u one has

$$\langle \nabla_\alpha u, \nabla_\alpha \rho_\alpha\rangle = \frac{Z_\alpha u}{\rho_\alpha}\psi_\alpha, \tag{12.5}$$

where Z_α is the infinitesimal generator of the dilations (11.20)*, i.e.,*

$$Z_\alpha = (\alpha+1)\sum_{i=1}^n x_i\partial_{x_i} + z\partial_z. \tag{12.6}$$

The next result provides a generalization to the Baouendi operator $\mathcal{P}_\alpha$ of classical representation formulas. It combines Theorem 2.1 and Corollary 2.1 in [**G93**]. The pseudo-ball $B_{\rho_\alpha}(r)$ and the pseudo-sphere $S_{\rho_\alpha}(r)$ centered at the origin with radius $r>0$ are those defined in (11.25) above.

PROPOSITION 12.6. *Let $\alpha \ge 0$ and consider a sufficiently smooth function U in $\mathbb{R}^n_x \times \mathbb{R}_z$. For every $r>0$ one has*

$$\frac{1}{|S_{\rho_\alpha}(r)|_\alpha}\int_{S_{\rho_\alpha}(r)} U(x,z)\frac{\psi_\alpha(x,z)}{|\nabla \rho_\alpha(x,z)|}dH_n(x,z) = U(0,0) \tag{12.7}$$
$$+ \int_{B_{\rho_\alpha}(r)} \mathcal{P}_\alpha U(x,z)\left[\Gamma_\alpha(x,z) - \frac{C_\alpha}{r^{Q-2}}\right] dxdz,$$

where Γ_α and C_α are as in (11.26)*. In particular, if $\mathcal{P}_\alpha U = 0$, then we have for every $r>0$*

$$U(0,0) = \frac{1}{|S_{\rho_\alpha}(r)|_\alpha}\int_{S_{\rho_\alpha}(r)} U(x,z)\frac{\psi_\alpha(x,z)}{|\nabla \rho_\alpha(x,z)|}dH_n(x,z). \tag{12.8}$$

In the statement of Proposition 12.6 by slightly abusing the notation we have indicated with

$$|S_{\rho_\alpha}(r)|_\alpha = \int_{S_{\rho_\alpha}(r)} \frac{\psi_\alpha(x,z)}{|\nabla \rho_\alpha(x,z)|}dH_n(x,z).$$

If, by a similar abuse of notation, we set

$$|B_{\rho_\alpha}(r)|_\alpha = \int_{B_{\rho_\alpha}(r)} \psi_\alpha(x,z)dxdz,$$

then keeping in mind that from (12.4) we easily see that ψ_α is homogeneous of degree zero with respect to the anisotropic dilations (11.20), by a rescaling and (11.21) we obtain

$$|B_{\rho_\alpha}(r)|_\alpha = \omega_\alpha r^Q,$$

where

$$\omega_\alpha = |B_{\rho_\alpha}(1)|_\alpha = \int_{B_{\rho_\alpha}(1)} \psi_\alpha(x,z)dxdz,$$

and Q is as in (11.22). Differentiating this formula and using Federer's coarea formula (aka Bonaventura Cavalieri's principle), see for instance [**EG15**], we find (12.9)

$$Q\omega_\alpha r^{Q-1} = \frac{d}{dr}\int_{B_\alpha(r)} \psi_\alpha(x,z)dxdz = \int_{S_{\rho_\alpha}(r)} \frac{\psi_\alpha(x,z)}{|\nabla\rho_\alpha(x,z)|}dH_n(x,z) = |S_{\rho_\alpha}(r)|_\alpha.$$

We now apply the formula (11.33) in Proposition 11.5 to the function $U\psi_\alpha$, instead of just U, obtaining

$$h'(r)\int_{S_{\rho_\alpha}(h(r))} \frac{U(x,z)\psi_\alpha(x,z)}{|\nabla\rho_\alpha(x,z)|}dH_n(x,z)$$
$$= (1-a)^a \int_{S_e(r)} \tilde{U}(x,y)\tilde{\psi}_\alpha(x,y)|y|^{-a}dH_n(x,y).$$

If we observe that (11.6) and (11.27) give

$$\tilde{\psi}_\alpha(x,y) = \frac{h(|y|)^{2\alpha}}{\rho_\alpha^{2\alpha}(x,h(|y|))} = \frac{h(|y|)^{2\alpha}}{h(d_e(x,y))^{2\alpha}} = \frac{h(|y|)^{\frac{2a}{1-a}}}{h(d_e(x,y))^{\frac{2a}{1-a}}} = \frac{|y|^{2a}}{d_e(x,y)^{2a}},$$

then keeping in mind that $h'(r) - \frac{(1-a)^a}{r^a}$, we have

$$\int_{S_{\rho_\alpha}(h(r))} \frac{U(x,z)\psi_\alpha(x,z)}{|\nabla\rho_\alpha(x,z)|}dH_n(x,z) = \frac{1}{r^a}\int_{S_e(r)} \tilde{U}(x,y)|y|^a dH_n(x,y). \tag{12.10}$$

Since by (12.9) we have

$$|S_{\rho_\alpha}(h(r))|_\alpha = \frac{Q\omega_\alpha}{(1-a)^n}r^n,$$

we conclude from (12.10) that

$$\frac{1}{|S_{\rho_\alpha}(h(r))|}\int_{S_{\rho_\alpha}(h(r))} \frac{U(x,z)\psi_\alpha(x,z)}{|\nabla\rho_\alpha(x,z)|}dH_n(x,z) \tag{12.11}$$
$$= \frac{(1-a)^n}{Q\omega_\alpha r^{n+a}}\int_{S_e(r)} \tilde{U}(x,y)|y|^a dH_n(x,y).$$

Since when $U \equiv 1$ the left-hand side of (12.11) is 1, we obtain

$$|S_e(r)|_a \overset{def}{=} \int_{S_e(r)} |y|^a dH_n(x,y) = \frac{Q\omega_\alpha}{(1-a)^n}r^{n+a}. \tag{12.12}$$

Furthermore, if we combine (11.32) above with (11.9), (11.29) and (11.30), we find

$$\int_{B_{\rho_\alpha}(h(r))} \mathcal{P}_\alpha U(x,z)\left[\Gamma_\alpha(x,z) - \frac{C_\alpha}{h(r)^{Q-2}}\right]dxdz \tag{12.13}$$
$$= (1-a)^a\int_{B_e(r)} L_a\tilde{U}(x,y)\left[\Gamma_\alpha(x,h(y)) - \frac{(1-a)^{n+a-1}C_\alpha}{r^{n+a-1}}\right]dxdy,$$

Combining (12.10), (12.13) with (12.8) above, and using (12.9) and the translation invariance of either L_a or $\mathcal{P}_\alpha$ in the x-variable, we obtain the following.

PROPOSITION 12.7. *Let $\tilde{U}$ be a sufficiently regular function in $\mathbb{R}^n_x \times \mathbb{R}_y$. Then, for every $x \in \mathbb{R}^n$ and $r > 0$ we have*

$$\tilde{U}(x,0) + \int_{B_e(r)} L_a\tilde{U}(x-x',y)\left[\tilde{\Gamma}(x',y) - \frac{\tilde{C}_a}{r^{n+a-1}}\right]dx'dy \tag{12.14}$$
$$= \frac{(1-a)^n}{Q\omega_\alpha r^{n+a}}\int_{S_e(r)} \tilde{U}(x-x',y)|y|^a dH_n(x',y).$$

REMARK 12.8. We note that, since we have used Proposition 12.6, and since $a = \frac{\alpha}{\alpha+1}$, strictly speaking we have proved Proposition 12.7 only for the range $0 \le a < 1$. However, it is easy to recognize that (12.14) does in fact hold also in the range $-1 < a < 0$.

If we now define the L_a-spherical average of a function $\tilde{U}$ as

$$\mathscr{M}_{\tilde{U},a}(x,r) = \frac{1}{|S_e(r)|_a}\int_{S_e(r)} \tilde{U}(x-x',y)|y|^a dH_n(x',y), \tag{12.15}$$

then we can reformulate (12.14) as follows

$$\mathscr{M}_{\tilde{U},a}(x,r) = \tilde{U}(x,0) + \int_{B_e(r)} L_a\tilde{U}(x-x',y)\left[\tilde{\Gamma}(x',y) - \frac{\tilde{C}_a}{r^{n+a-1}}\right]dx'dy. \tag{12.16}$$

Differentiating (12.16) with respect to r produces the following interesting consequence.

PROPOSITION 12.9. *Let $\tilde{U}$ be a sufficiently regular function in $\mathbb{R}^n_x \times \mathbb{R}_y$. Then, for every $x \in \mathbb{R}^n$ and $r > 0$ we have*

$$\frac{\partial \mathscr{M}_{\tilde{U},a}}{\partial r}(x,r) = \frac{(n+a-1)\tilde{C}_a}{r^{n+a}}\int_{B_e(r)} L_a\tilde{U}(x-x',y)dx'dy. \tag{12.17}$$

We note explicitly that, since $-1 < a < 1$ and $n \ge 2$, we have $n + a - 1 > 0$.

COROLLARY 12.10. *Let $\tilde{U}$ be as in Proposition 12.9. If either $L_a\tilde{U} = 0$, or $\tilde{U} \ge 0$ and $L_a\tilde{U} \ge 0$, one has for every $x \in \mathbb{R}^n$ and $r > 0$*

$$\frac{\partial M_{\tilde{U}^2,a}}{\partial r}(x,r) \ge \frac{(n+a-1)\tilde{C}_a}{r^{n+a}}\int_{B_e(x,r)} |\nabla\tilde{U}(x',y)|^2|y|^a dx'dy. \tag{12.18}$$

Equality holds in (12.18) when $L_a\tilde{U} = 0$, and in either case the function $r \to \mathscr{M}_{\tilde{U}^2,a}(x,r)$ is monotonically increasing.

PROOF. It suffices to apply (12.17) to the function $\tilde{U}^2$ and observe that

$$L_a\tilde{U}^2 = 2\tilde{U}L_a\tilde{U} + 2|y|^a|\nabla\tilde{U}|^2.$$

If either $L_a\tilde{U} = 0$, or $\tilde{U} \ge 0$ and $L_a\tilde{U} \ge 0$, one has

$$L_a\tilde{U}^2 \ge 2|y|^a|\nabla\tilde{U}|^2,$$

with equality when $L_a\tilde{U} = 0$.

□

We will use the following inequality in the proof of Theorem 14.2 below.

PROPOSITION 12.11 (Caccioppoli inequality). *Let $\Omega \subset \mathbb{R}^n$ be open, and suppose that either $\tilde{U}$ be a solution to $L_a\tilde{U} = 0$ in the open set $\Omega \times (-d, d) \subset \mathbb{R}^n \times \mathbb{R}$, where $d = \mathrm{diam}(\Omega)$, or $\tilde{U} \geq 0$ and $L_a\tilde{U} \geq 0$ there. Then, there exists $C(n, a) > 0$ such that for every $x \in \Omega$ and every $0 < s < t < \mathrm{dist}(x, \partial\Omega)$, one has*

(12.19)

$$\int_{B_e(x,s)} |\nabla\tilde{U}(x', y)|^2 |y|^a dx' dy \leq \frac{C(n,a)}{t-s} \int_{S_e(x,t)} \tilde{U}(x', y)^2 |y|^a dH_n(x', y).$$

PROOF. Integrating (12.17) in r on the interval (s, t) we find

$$(n+a-1)\tilde{C}_a \int_s^t \frac{1}{r^{n+a+1}} \left(\int_{B_e(x,r)} |\nabla\tilde{U}(x', y)|^2 |y|^a dx' dy \right) dr$$
$$\leq \int_s^t \frac{1}{r} \frac{\partial \mathscr{M}_{\tilde{U}^2,a}}{\partial r}(x, r) dr.$$

Since we trivially have

$$\int_s^t \frac{1}{r^{n+a+1}} \left(\int_{B_e(x,r)} |\nabla\tilde{U}(x', y)|^2 |y|^a dx' dy \right) dr$$
$$\geq \int_s^t \frac{1}{r^{n+a+1}} dr \left(\int_{B_e(x,s)} |\nabla\tilde{U}(x', y)|^2 |y|^a dx' dy \right),$$

and since integrating and applying the mean value theorem we find

$$\int_s^t \frac{1}{r^{n+a+1}} dr \geq \frac{t-s}{st^{n+a}},$$

we obtain

$$(n+a-1)\tilde{C}_a \frac{t-s}{st^{n+a}} \int_{B_e(x,s)} |\nabla\tilde{U}(x', y)|^2 |y|^a dx' dy \leq \int_s^t \frac{1}{r} \frac{\partial \mathscr{M}_{\tilde{U}^2,a}}{\partial r}(x, r) dr.$$

On the other hand, an integration by parts gives

$$\int_s^t \frac{1}{r} \frac{\partial \mathscr{M}_{\tilde{U}^2,a}}{\partial r}(x, r) dr = \frac{1}{t}\mathscr{M}_{\tilde{U}^2,a}(x, t) - \frac{1}{s}\mathscr{M}_{\tilde{U}^2,a}(x, s) + \int_s^t \frac{1}{r^2}\mathscr{M}_{\tilde{U}^2,a}(x, r) dr$$
$$\leq \frac{1}{t}\mathscr{M}_{\tilde{U}^2,a}(x, t) - \frac{1}{s}\mathscr{M}_{\tilde{U}^2,a}(x, s) + \mathscr{M}_{\tilde{U}^2,a}(x, t)\left(\frac{1}{s} - \frac{1}{t}\right)$$
$$= \frac{1}{s}\left(\mathscr{M}_{\tilde{U}^2,a}(x, t) - \mathscr{M}_{\tilde{U}^2,a}(x, s)\right),$$

where the inequality is justified by the monotonicity of the function $r \to \mathscr{M}_{\tilde{U}^2,a}(x, r)$, see the second part of Corollary 12.10. In conclusion, we have found

$$(n+a-1)\tilde{C}_a \frac{t-s}{st^{n+a}} \int_{B_e(x,s)} |\nabla\tilde{U}(x', y)|^2 |y|^a dx' dy \leq \frac{1}{s}\left(\mathscr{M}_{\tilde{U}^2,a}(x, t) - \mathscr{M}_{\tilde{U}^2,a}(x, s)\right).$$

Keeping (12.12) and (12.15) in mind, it is clear that this inequality trivially implies the desired conclusion (12.19).

□

REMARK 12.12. The above proof of Proposition 12.11 is self-contained and provides a slightly stronger version of the usual Caccioppoli inequality since in the right-hand side one has the spherical L^2 norm of the function, instead of the solid one. Such proof needs to be modified when dealing with equations with rough coefficients. The reader can find the Caccioppoli inequality for more general degenerate elliptic equations in the paper [**FKS81**].

Suppose now that $\varphi \in C_0^\infty(\mathbb{R}^n_x \times \mathbb{R}_y)$ is a spherically symmetric bump function, i.e., $\varphi(X) = \varphi^\star(|X|)$, such that, say, supp $\varphi^\star \subset [1/4, 3/4]$, and

$$\int_{\mathbb{R}^n_x \times \mathbb{R}_y} \varphi(X)|y|^a dX = 1.$$

Using the coarea formula and (12.12) we can reformulate this latter assumption as follows

$$1 = \int_0^\infty \varphi^\star(r) \int_{S_e(r)} |y|^a dH_n(x,y)dr = \frac{Q\omega_\alpha}{(1-a)^n} \int_0^\infty \varphi^\star(r) r^{n+a} dr.$$

Now, if we multiply both sides of (12.14) by $\varphi^\star(r)r^{n+a}$, and we integrate with respect to $r \in (0,\infty)$, keeping in mind that

$$\int_{\mathbb{R}^n \times \mathbb{R}} \tilde{U}(x-x',y)|y|^a dx'dy = \int_0^\infty \int_{S_e(r)} \tilde{U}(x-x',y)|y|^a dH_n(x',y)dr,$$

we obtain the following result.

PROPOSITION 12.13. *Let $\tilde{U}$ be a solution to $L_a\tilde{U} = 0$ in $\mathbb{R}^n \times \mathbb{R}$. Then, for every $x \in \mathbb{R}^n$ we have*

$$\tilde{U}(x,0) = \int_{\mathbb{R}^n \times \mathbb{R}} \tilde{U}(x',y)\varphi(x-x',y)|y|^a dx'dy. \tag{12.20}$$

If instead $\Omega \subset \mathbb{R}^n$ is an open set, suppose that $\tilde{U}$ be a solution to $L_a\tilde{U} = 0$ in the open set $\Omega \times (-d,d) \subset \mathbb{R}^n \times \mathbb{R}$, where $d = \operatorname{diam}(\Omega)$. Then, for every $x \in \Omega$ and every $0 < r < \operatorname{dist}(x,\partial\Omega)$, one has

$$\tilde{U}(x,0) = \int_{\mathbb{R}^n \times \mathbb{R}} \tilde{U}(x',y)\varphi_r(x-x',y)|y|^a dx'dy, \tag{12.21}$$

where we have let $\varphi_r(X) = \frac{1}{r^{n+a+1}}\varphi(\frac{X}{r})$.

The reader should be aware that, while in keeping up with the spirit of the present section and of the previous one we have derived Proposition 12.13 from Proposition 12.6, the mean value formulas (12.16), (12.21) were independently obtained by a direct computation in Theorem 1 of the paper [**ABG15**]. In Theorem 2 of the same paper, combining (12.14) with Theorem 10.1 above, the authors established an interesting representation formula for solutions of $(-\Delta)^s$ that can be seen as the nonlocal counterpart of formula (12.2) in Theorem 12.3 above. Before we state such formula in Theorem 12.15 below, we introduce a useful lemma.

LEMMA 12.14. *Let $\varphi(X) = \varphi^\star(|X|)$ be a $C_0^\infty(\mathbb{R}^{n+1})$ function as in Proposition 12.13, and, with P_s as in (10.8) above, and $a = 1-2s$, define*

$$\Phi(x) = \int_{\mathbb{R}^n \times \mathbb{R}} \varphi(x-x',y)P_s(x',|y|)|y|^a dx'dy. \tag{12.22}$$

Then, one has

(i) *Φ is spherically symmetric, i.e., there exists a function $\Phi^\star : [0,\infty) \to \mathbb{R}$ such that $\Phi(x) = \Phi^\star(|x|)$.*

(ii) $\Phi \in C^\infty(\mathbb{R}^n)$, *and for every multi-index* α *such that* $|\alpha| = k$, *there exists* $C(n,s,k) > 0$ *such that one has for every* $x \in \mathbb{R}^n$

$$|\partial^\alpha \Phi(x)| \le \frac{C(n,s,k)}{(1+|x|)^{n+1-a}}.$$

PROOF. (i). Since $z \to \varphi(z,y)$ and $z \to P_s(z,y)$ are spherically symmetric functions, and since the convolution of two spherically symmetric functions is spherically symmetric, it is clear that there exists a function $\Phi^\star : [0,\infty) \to \mathbb{R}$ such that $\Phi(x) = \Phi^\star(|x|)$.

(ii) Observe that, from the definition (10.8) and the fact that $a = 1-2s$, we can alternatively write the Poisson kernel for L_a as

$$P_a(x,y) = \frac{\Gamma(\frac{n+1-a}{2})}{\pi^{\frac{n}{2}}\Gamma(\frac{1-a}{2})} \frac{|y|^{1-a}}{(y^2+|x|^2)^{\frac{n+1-a}{2}}} = c(n,a)\frac{|y|^{1-a}}{(y^2+|x|^2)^{\frac{n+1-a}{2}}}.$$

We thus have

$$\Phi(x) = c(n,a)\int_{\mathbb{R}}\int_{\mathbb{R}^n} \varphi(x-x',y)\frac{|y|}{(y^2+|x'|^2)^{\frac{n+1-a}{2}}}dx'dy.$$

Differentiating under the integral sign in the definition of $\Phi(x)$, and keeping in mind that $\operatorname{supp}\varphi \subset \overline{B}_e(0,1)$, for every $\alpha \in (\mathbb{N}\cup\{0\})^n$ such that $|\alpha| = k$ we have

$$|\partial^\alpha \Phi(x)| \le C(n,a,k)\int_{|y|\le 1}\int_{|x'-x|\le 1} |\partial^\alpha\varphi(x-x',y)|\frac{|y|}{(y^2+|x'|^2)^{\frac{n+1-a}{2}}}dx'dy.$$

If now $|x| > 2$, then when $|x'-x| \le 1$ we have $|x'| = |x-(x-x')| \ge |x| - |x'-x| \ge |x|-1 \ge |x|/2$. Therefore, when $|x| > 2$ we have

$$|\partial^\alpha \Phi(x)| \le \frac{C(n,a,k)}{|x|^{n+1-a}}.$$

On the other hand, when $|x| \le 2$, then the triangle inequality gives $B(x,1) \subset B(0,3)$, and therefore

$$\begin{aligned} |\partial^\alpha \Phi(x)| &\le C(n,a,k)\int_{|y|\le 1}\int_{|x'|\le 3}\frac{|y|}{(y^2+|x'|^2)^{\frac{n+1-a}{2}}}dx'dy \\ &\le C(n,a,k)\int_{|y|\le 1}|y|^a\int_{|z|\le \frac{3}{y}}\frac{dz}{(1+|z|^2)^{\frac{n+1-a}{2}}}dy \\ &\le C(n,a,k)\int_{|y|\le 1}|y|^a dy\int_{\mathbb{R}^n}\frac{dz}{(1+|z|^2)^{\frac{n+1-a}{2}}} = \overline{C}(n,a,k) < \infty, \end{aligned}$$

since $|a| < 1$. These estimates prove (ii).

□

THEOREM 12.15 ([**ABG15**]). *Let* $u \in \mathscr{L}_s(\mathbb{R}^n)$ *be such that* $(-\Delta)^s u = 0$ *in* $\mathscr{D}'(\Omega)$ *for a given open set* $\Omega \subset \mathbb{R}^n$. *Then, for a.e.* $x \in \Omega$ *and* $0 < r < \operatorname{dist}(x,\partial\Omega)$, *one has*

$$u(x) = \Phi_r \star u(x) = \int_{\mathbb{R}^n} u(x')\Phi_r(x-x')dx', \tag{12.23}$$

where Φ *is the function in* (12.22), *and we have let* $\Phi_r(x) = r^{-n}\Phi(x/r)$.

PROOF. Consider the extension problem (10.1) above with Dirichlet datum u, and denote by

$$U(x,y) = P_s(\cdot,y) \star u(x) = \int_{\mathbb{R}^n} P_s(x-x',y)u(x')dx'$$

its solution in $\mathbb{R}^n \times [0,\infty)$. Consider the symmetric extension of U to the whole $\mathbb{R}^n \times \mathbb{R}$ defined by

$$\tilde{U}(x,y) = U(x,|y|) = \int_{\mathbb{R}^n} P_s(x-x',|y|)u(x')dx'.$$

By the Dirichlet-to-Neumann condition (10.2) we know that at a.e. $x \in \Omega$ we have

$$-\frac{2^{2s-1}\Gamma(s)}{\Gamma(1-s)} \lim_{y\to 0^+} y^{1-2s}\frac{\partial U}{\partial y}(x,y) = (-\Delta)^s u(x) = 0. \tag{12.24}$$

We can thus invoke Lemma 4.1 in [**CS07**] to infer that $\tilde{U}$ is a weak solution to the equation $L_a\tilde{U} = 0$ in $D = \Omega\times(-d,d)$, where $d = \operatorname{diam}(\Omega)$. By the Harnack inequality for weak solutions in [**FKS81**] we can redefine $\tilde{U}$ on a set of $(n+1)$-dimensional measure zero in D so that it be locally Hölder continuous in D. Applying (12.21) in Proposition 12.13 we find for every $x \in \Omega$ and $0 < r < \operatorname{dist}(x,\partial\Omega)$

$$\begin{aligned} u(x) = \tilde{U}(x,0) &= \int_{\mathbb{R}^n\times\mathbb{R}} \tilde{U}(x',y)\varphi_r(x-x',y)|y|^a dx'dy \\ &= \int_{\mathbb{R}^n\times\mathbb{R}} \left(\int_{\mathbb{R}^n} P_s(x'-x'',|y|)u(x'')dx''\right)\varphi_r(x-x',y)|y|^a dx'dy \\ &= \int_{\mathbb{R}^n} u(x'')\left(\int_{\mathbb{R}^n\times\mathbb{R}} \varphi_r(x-x',y)P_s(x'-x'',|y|)|y|^a dx'dy\right)dx'' \\ &= \int_{\mathbb{R}^n} \Psi_r(x,x'')u(x'')dx'', \end{aligned}$$

where we have let

$$\Psi_r(x,x'') = \int_{\mathbb{R}^n\times\mathbb{R}} \varphi_r(x-x',y)P_s(x'-x'',|y|)|y|^a dx'dy.$$

The exchange of order of integration (Fubini's theorem) is justified by the fact that, since $a = 1-2s$, the assumption $u \in \mathscr{L}_s(\mathbb{R}^n)$ can be reformulated as $\int_{\mathbb{R}^n} \frac{|u(x'')|}{(1+|x''|^2)^{\frac{n+1-a}{2}}}dx'' < \infty$. On the other, we claim that for any fixed $x \in \mathbb{R}^n$ we have

$$|\Psi_r(x,x'')| \le \frac{C_r}{(1+|x''|^2)^{\frac{n+1-a}{2}}}. \tag{12.25}$$

Assuming the claim, we thus see that $\int_{\mathbb{R}^n} |\Psi_r(x,x'')||u(x'')|dx'' < \infty$, and thus the application of Fubini's theorem would be justified in view of Tonelli's theorem. We thus need to prove (12.25). For this, recall that from our choice of $\varphi(X)$ we know that supp $\varphi \subset \overline{B}_e(0,3/4) \setminus B_e(0,1/4) \subset \overline{B}_e(0,1) \subset \mathbb{R}^n_x \times \mathbb{R}_y$, and thus we have

$$\begin{aligned}|\Psi_r(x,x'')| &\le \int_{-1}^{1}\left(\int_{\mathbb{R}^n} |\varphi_r(x-x',y)|P_s(x'-x'',|y|)dx'\right)|y|^a dy\\ &\le ||\varphi_r||_{L^\infty(\mathbb{R}^n\times\mathbb{R})}\int_{-1}^{1} ||P_s(\cdot - x'',|y|)||_{L^1(\mathbb{R}^n)}|y|^a dy\\ &= ||\varphi_r||_{L^\infty(\mathbb{R}^n\times\mathbb{R})}\int_{-1}^{1}|y|^a dy = C_r,\end{aligned}$$

where in the last equality we have used (10.17). On the other hand, if $x'' \notin \overline{B}(x,2)$ and $|x'-x|<1$, we have $|x''-x'| \ge |x''-x| - |x-x'| \ge |x''-x|-1 > |x''-x|/2$. We thus obtain from (10.8)

$$\begin{aligned}|\Psi_r(x,x'')| &\le C(n,s)||\varphi_r||_{L^\infty(\mathbb{R}^n\times\mathbb{R})}\int_{B_e((x,0),1)} \frac{|y|^{1-a}}{(y^2+|x''-x'|^2)^{\frac{n+1-a}{2}}}|y|^a dx'dy\\ &\le C(n,s)||\varphi_r||_{L^\infty(\mathbb{R}^n\times\mathbb{R})}\int_{B_e((x,0),1)} \frac{1}{|x''-x'|^{n+1-a}}dx'dy\\ &\le \frac{C(n,s)||\varphi_r||_{L^\infty(\mathbb{R}^n\times\mathbb{R})}}{1+|x''-x|^{n+1-a}}\\ &\lesssim \frac{C_r}{1+|x''|^{n+1-a}}.\end{aligned}$$

This proves (12.25).

In order to complete the proof of (12.23), we are thus left with showing that

$$\Psi_r(x,x'') = \frac{1}{r^n}\Phi^\star\left(\frac{|x-x''|}{r}\right).$$

This easily follows from a change of variable. We have in fact

$$\begin{aligned}&\frac{1}{r^n}\Phi^\star\left(\frac{|x-x''|}{r}\right)\\ &\quad= \frac{1}{r^n}\int_{\mathbb{R}^n\times\mathbb{R}} \varphi(\frac{x-x''-rx'}{r},y)P_s(x',|y|)|y|^a dx'dy \qquad (z = x-rx', t = ry)\\ &\quad= \frac{1}{r^{2n+1+a}}\int_{\mathbb{R}^n\times\mathbb{R}} \varphi(\frac{z-x''}{r},\frac{t}{r})P_s(\frac{z-x}{r},\frac{|t|}{r})|t|^a dzdt\\ &\quad= \int_{\mathbb{R}^n\times\mathbb{R}} \varphi_r(z-x'',t)P_s(z-x,|t|)|t|^a dzdt\\ &\quad= \int_{\mathbb{R}^n\times\mathbb{R}} \varphi_r(x''-z,t)P_s(z-x,|t|)|t|^a dzdt = \Psi_r(x'',x),\end{aligned}$$

where in the second to the last equality we have used the fact that $P_s(\frac{z-x}{r},\frac{|t|}{r}) = r^n P_s(z-x,|t|)$. Since $\Phi^\star\left(\frac{|x-x''|}{r}\right) = \Phi^\star\left(\frac{|x-x''|}{r}\right)$, the proof is completed.

□

REMARK 12.16. Formula (12.23) can be seen as a nonlocal counterpart of (12.1) for classical harmonic functions.

We stress that from the validity of (12.23) we also indirectly infer that

$$\int_{\mathbb{R}^n} \Phi(x)dx = 1.$$

We are now ready to state our first result about the hypoellipticity of $(-\Delta)^s$.

THEOREM 12.17 (Nonlocal Caccioppoli-Cimmino-Weyl lemma). *Suppose that $u \in \mathscr{L}_s(\mathbb{R}^n)$, and that on a given open set $\Omega \subset \mathbb{R}^n$ one has $(-\Delta)^s u = 0$ in $\mathscr{D}'(\Omega)$. Then, u can be modified on a set of measure zero in Ω so to coincide with a function in $C^\infty(\Omega)$.*

PROOF. By Theorem 12.15 we know that for a.e. $x \in \Omega$ and $0 < r < \operatorname{dist}(x, \partial\Omega)$, one has

$$u(x) = \Phi_r \star u(x).$$

To complete the proof it suffices to observe that the right-hand side $\Phi_r \star u$ defines a function in $C^\infty(\Omega)$. But this follows combining part (ii) of Lemma 12.14 with the assumption $u \in \mathscr{L}_s(\mathbb{R}^n)$ which can be reformulated as

$$\int_{\mathbb{R}^n} \frac{|u(x)|}{1 + |x|^{n+1-a}} dx < \infty.$$

These two facts allow us to differentiate under the integral sign in the expression

$$\Phi_r \star u(x) = \int_{\mathbb{R}^n} u(x')\Phi_r(x - x')dx',$$

to conclude that $\Phi_r \star u \in C^\infty(\Omega)$.

□

REMARK 12.18. A similar approach based on the nonlocal Poisson kernel in Section 15 below can be found in Theorem 3.12 in [**BB99'**].

As we have already mentioned, Theorem 12.17 admits a stronger form which is contained in the next result, whose use has been kindly suggested to us by Camelia Pop. We also acknowledge a nice conversation with Fausto Segala, who pointed us to the classical reference [**Ho66**] of the "founding father". In the next result the notation $S^m_{1,0}$ indicates the Kohn-Nirenberg class of symbols introduced in Section 5 above.

THEOREM 12.19. *Let $m \in \mathbb{R}$ and let $p(x,\xi) \in S^m_{1,0}$. If the pseudodifferential operator $Pu(x) = \mathscr{F}^{-1}(p(\cdot,\xi))(x)$ is elliptic, then P is hypoelliptic. In particular, $(-\Delta)^s$ is hypoelliptic. Therefore, if for given $f \in \mathscr{S}'(\mathbb{R}^n)$ the distribution $u \in \mathscr{S}'(\mathbb{R}^n)$ solves the equation*

$$(-\Delta)^s u = f,$$

then, $u \in C^\infty(\Omega)$ on every open set $\Omega \subset \mathbb{R}^n$ where $f \in C^\infty(\Omega)$.

PROOF. By (5.2) in Proposition 5.2 we know that $(-\Delta)^s$ is an elliptic pseudodifferential operator with symbol in the Kohn-Nirenberg class $S^{2s}_{1,0}$. Therefore, the second part of the theorem is an immediate consequence of the former. To prove the first part, we recall that the *singular support* of a distribution is defined as the complement of the open set where the distribution is smooth. The *pseudolocal property* of pseudodifferential operators states that if Q is such an operator, then:

$$\text{singsupp } Q(u) \subset \text{singsupp } u. \tag{12.26}$$

On the other hand, a basic theorem in microlocal analysis states that every elliptic pseudodifferential operator P is invertible, in the sense that there exists a pseudodifferential operator Q, and a C^∞ smoothing operator R such that

$$Q \circ P = I + R,$$

where I is the identity operator, see section 4 in Chapter 7 in [**Ta11**]. From this identity it follows that

$$\text{singsupp } (Q \circ P)(u) = \text{singsupp } u. \tag{12.27}$$

Now, by a repeated application of (12.26) we find

$$\text{singsupp } (Q \circ P)(u) \subset \text{singsupp } P(u) \subset \text{singsupp } u. \tag{12.28}$$

From (12.27) and (12.28) we conclude that for any elliptic pseudodifferential operator P we must have

$$\text{singsupp } P(u) = \text{singsupp } u. \tag{12.29}$$

This equation proves that the set where $P(u) \in C^\infty$ coincides with that where $u \in C^\infty$, or, equivalently, that the operator P is hypoelliptic. For (12.29) one should see Proposition 4.1 of [**Ta11**], or also Hörmander's seminal work [**Ho66**]. We mention that the above proof also works for the more general class of Hörmander symbols $S^m_{\rho,\delta}$, with $0 \leq \delta < \rho \leq 1$.

□

13. Regularity at the boundary

The results in the previous section show that, as far as interior regularity is concerned, the nonlocal operator $(-\Delta)^s$ behaves much as its local predecessor $-\Delta$. But what about global regularity in boundary value problems? We will see in the present section that here the situation departs dramatically from the local case.

A central focus of investigation in the classical theory of elliptic and parabolic equations is the Dirichlet problem. For instance, for the Laplacean, given a bounded open set $\Omega \subset \mathbb{R}^n$ and a function $\varphi \in C(\partial\Omega)$, one seeks a function $u \in C^2(\Omega) \cap C(\overline{\Omega})$ such that

$$\begin{cases} \Delta u = 0 & \text{in } \Omega, \\ u = \varphi & \text{on } \partial\Omega. \end{cases} \tag{13.1}$$

From Corollary 12.4 we know that, when such a u exists, we have in fact $u \in C^\infty(\Omega)$. One of the culminating achievements of classical potential theory is the Wiener-Perron-Brelot method which, for any bounded open set Ω, allows to construct a generalized solution u to (13.1). Watch out! Here generalized means that, although $u \in C^\infty(\Omega)$ and $\Delta u = 0$ in Ω, it is not guaranteed that $u \in C(\overline{\Omega})$. A famous example is provided by the so-called *Lebesgue spine*, see for instance [**He69**].

However, the celebrated Wiener's criterion guarantees that when, for instance, Ω is a Lipschitz domain (much less regularity does in fact suffice), then the generalized Wiener-Perron-Brelot solution to the problem (13.1) is in fact Hölder continuous up to the boundary, if such is φ on $\partial\Omega$. If Ω is better than Lipschitz, then more regularity is expected for u. For instance, a well-known fact is that when Ω

is a $C^{1,1}$ domain and $\varphi = 0$ on a portion of $\partial\Omega$, if $u \geq 0$ in Ω then nearby a point $x_0 \in \partial\Omega$ where φ vanishes one has for some universal constant $C > 0$,

$$C\frac{d(x,\partial\Omega)}{r} \leq \frac{u(x)}{u(A_r(x_0))} \leq C^{-1}\frac{d(x,\partial\Omega)}{r}, \tag{13.2}$$

where $d(x,\partial\Omega) = \operatorname{dist}(x,\partial\Omega)$, and $A_r(x_0) \in \Omega$ is a so-called nontangential point attached to x_0, see for instance [**G84**]. The estimate (13.2) in essence says that u has a linear growth near the boundary, and thus it is Lipschitz continuous up to $\partial\Omega$. We observe the important fact that, since the constant C in (13.2) is independent of the particular harmonic function u, we conclude that for any two nonnegative harmonic functions u, v, which continuously vanish on a portion of the boundary, the so-called *Boundary Harnack Principle* holds:

$$C\frac{u(A_r(x_0))}{v(A_r(x_0))} \leq \frac{u(x)}{v(x)} \leq C^{-1}\frac{u(A_r(x_0))}{v(A_r(x_0))}. \tag{13.3}$$

Thus, *all nonnegative harmonic functions which vanish on a portion of the boundary in a $C^{1,1}$ domain, must do so at the same linear rate.*

We note that for nonnegative solutions of $(-\Delta)^s$ in a Lipschitz domain the estimate (13.3) has been proved in Theorem 1 in [**Bo97**], thus in a Lipschitz domain the Boundary Harnack Principle does hold for $(-\Delta)^s$ (however, for the correct formulation of the Dirichlet problem in the nonlocal case, see (13.4) below).

However, even in the classical local case of the Laplacean the Boundary Harnack Principle (13.3) <u>does not</u> imply the linear decay estimate! The reason for this is that, whereas (13.3) only depends on general *metric properties* of the relevant domain, and therefore does hold for large classes of domains with rough boundaries (for instance, in Lipschitz or even NTA domains, see [**CFMS81**], [**JK82**]), the linear decay estimate (13.2) breaks down if the domain fails to satisfy some bound on its curvatures. For instance, if $0 < \theta_0 < \pi/2$ and we consider in $\mathbb{R}^2$ the convex circular sector $\Omega = \{(r,\theta) \mid 0 < r < 1, |\theta| < \theta_0\}$, where θ indicates the angle formed by the directional vector of the point (x, y) with the positive direction of the y-axis, the function $u(r,\theta) = r^\lambda \cos(\lambda\theta)$ is a nonnegative harmonic function in Ω vanishing on that portion of $\partial\Omega$ corresponding to $|\theta| = \theta_0$ provided that

$$\lambda = \frac{\pi}{2\theta_0}.$$

From our choice, we have $\lambda > 1$ and therefore this example shows that for domains without an interior tangent ball the estimate from below in (13.2) cannot possibly hold in general. Using the same type of domain and function, but this time with $\pi/2 < \theta_0 < \pi$ (a non-convex cone) we see that if the tangent outer ball condition fails, then there exist harmonic functions which vanish at the boundary at best with a Hölder rate < 1. Therefore, the estimate from above in (13.2) cannot possibly hold in general. Now, it is well-known that the uniform tangent ball condition both from inside and outside characterizes $C^{1,1}$ domains, and this degree of smoothness guarantees a linear decay such as that in (13.2). We mention here that (13.2) is also valid in $C^{1,\alpha}$ domains. This is due to the fact that the Hopf's boundary point lemma continues to be valid in such domains, see [**KH73**].

In the nonlocal case the Dirichlet problem is formulated as follows, see 21. on p. 267 in [**La72**]. Let $\Omega \subset \mathbb{R}^n$ be a bounded open set. Given $\varphi \in C(\mathbb{R}^n \setminus \Omega)$, and such that

$$\int_{\mathbb{R}^n\setminus\Omega} \frac{|\varphi(x)|}{1+|x|^{n+2s}} dx < \infty,$$

one seeks $u \in \mathscr{L}_s(\mathbb{R}^n) \cap C(\mathbb{R}^n)$ such that

$$\begin{cases} (-\Delta)^s u = 0 & \text{in } \Omega, \\ u = \varphi & \text{in } \mathbb{R}^n \setminus \Omega. \end{cases} \tag{13.4}$$

As we have seen, there exists a unique solution to (13.4). This follows from the maximum principle in Proposition 3.1 above.

From Theorem 12.17 we know that such u must be in $C^\infty(\Omega)$. But what about its regularity at the boundary of Ω? For instance, if we have a $C^{1,1}$ domain $\Omega \subset \mathbb{R}^n$, then is an estimate such as (13.2) true? The answer to this question is negative! Proposition 13.1 below shows that there exist C^∞ domains (in fact, real analytic), and C^∞ boundary data φ (in fact, real analytic) such that the solution to the Dirichlet problem (13.4) is not any better than $C^{0,s}$ Hölder continuous near $\partial\Omega$. Perhaps this negative phenomenon at the boundary can be understood with the fact that standard smoothness is badly altered if we look at it with the spectacles of the intrinsic scalings (2.10) of $(-\Delta)^s$, which instead preserve only some degree of Hölder regularity.

This is yet another example of the fact, highlighted in Section 11, that the nonlocal operator $(-\Delta)^s$ displays a behavior that is typical of subelliptic operators such as the Baouendi operator $\mathcal{P}_\alpha$, and/or the sub-Laplacean on the Heisenberg group $\mathbb{H}^n$. We recall in this respect that it was shown by D. Jerison in his Ph.D. Dissertation [**Je81**] that there exists real analytic domains $\Omega \subset \mathbb{H}^n$ and real analytic boundary data φ for which the solution to the Dirichlet problem for the sub-Laplacean Δ_H in (11.16) is not any better than Hölder continuous up to the boundary. However, Theorems 13.4 and 13.5 below show that, despite the failure of the estimate (13.2) in $C^{1,1}$ domains, the latter does hold if one replaces $d(x,\partial\Omega)$ with $d(x,\partial\Omega)^s$!

In Euclidean analysis and partial differential equations the function $u(x) = \frac{|x|^2}{2n}$ plays a special role. First of all, it has the remarkable property that $\Delta u \equiv 1$, which means that the global vector field ∇u has constant divergence (this is strongly connected to the *flatness* of Euclidean space). Secondly, and because of this, it serves as a barrier in many boundary value problems. A first beautiful and simple, yet enormously important, instance of this is the weak Maximum Principle for a function v such that $\Delta v \geq 0$. By considering the function $w = v + \varepsilon u$ one reduces matters to the situation when $\Delta w > 0$, in which case the maximum principle trivially follows from calculus.

In connection with the globally defined function $u(x) = \frac{|x|^2}{2n}$, it is important to consider the so-called *torsion function* of a connected, bounded open set $\Omega \subset \mathbb{R}^n$, i.e., the unique solution of the Dirichlet problem

$$\begin{cases} -\Delta u = 1 & \text{in } \Omega, \\ u = 0 & \text{on } \partial\Omega. \end{cases} \tag{13.5}$$

A celebrated theorem of Serrin in [**Se71**] states that Ω is a ball if and only if $\frac{\partial u}{\partial \nu} = -\kappa$ on $\partial\Omega$. In fact, when $\Omega = B(x_0, R)$, then one knows that

$$u(x) = \frac{R^2 - |x - x_0|^2}{2n}, \tag{13.6}$$

for which we easily recognize that $\frac{\partial u}{\partial \nu} = -\frac{R}{n}$. An alternative approach, based on the strong maximum principle and a beautiful integral identity of Rellich, was proposed by Weinberger in [**We71**] simultaneously to Serrin's paper. One should also see the later work [**GLe89**], in which Weinberger's ideas were generalized to nonlinear degenerate elliptic equations such as the p-Laplacean.

It is remarkable that the mere existence of a smooth solution to (13.5), or to the related Neumann problem

$$\begin{cases} \Delta u = 1 & \text{in } \Omega, \\ \frac{\partial u}{\partial \nu} = \frac{|\Omega|}{|\partial\Omega|} & \text{on } \partial\Omega, \end{cases} \tag{13.7}$$

have deep implications in geometry. For instance, in his note [**Re82**] R. Reilly using (13.5) and an adaption of Weinberger's ideas in [**We71**] provided a beautiful proof of A.D. Alexandrov's *soap bubble theorem* stating that the only (sufficiently) smooth compact hypersurface in $\mathbb{R}^n$ having constant mean curvature must be a ball. Another beautiful result is Cabré's proof of the isoperimetric inequality which combines the Alexandrov-Bakelman-Pucci maximum principle with the solution of problem (13.7), see [**Ca08**] and [**Ca17**]. Section 2.1.2 of [**Ca08**] also contains an interesting account of the probabilistic interpretation of the torsion function $u(x)$ in (13.5) above as the expected time for a particle located at $x \in \Omega$ to hit the boundary. For this one should also see Getoor's original paper on the subject [**Ge61**].

In connection with the fractional Laplacean we mention that an interesting nonlocal version of Serrin's theorem was obtained in the work [**FJ15**], based on an adaption of the moving plane method. Whereas such method lends itself well to the fractional setting, we strongly feel that Weinberger's approach should be equally investigated because of its potential geometric implications. In this respect one should see the interesting work [**ROS14'**], in which the authors establish a nonlocal version of the Rellich-Pohozaev identity as a consequence of their regularity results in [**ROS14**]. It is an intriguing and challenging question whether such result can be employed to provide a proof alternative to that in [**FJ15**]. In fact, this aspect is deeply connected to the questions we raise at the end of Section 20 below.

In light of the above considerations it is natural to seek a function analogous to (13.6) for the non-local operator $(-\Delta)^s$, i.e., a solution to the equation $(-\Delta)^s u = 1$ in ball, with zero boundary data. In what follows we construct such a function. Our presentation is based on Lemma 7.1 above, and it differs in part from the original one in the above cited paper [**Ge61**]. The reader should also see [**D12**] and [**DKK17**] for more general results and the discussion in [**BuV16**] of the one-dimensional case.

PROPOSITION 13.1 (Torsion function for the ball). *For $0 < s < 1$ consider the non-homogeneous Dirichlet problem*

$$\begin{cases} (-\Delta)^s u = 1 & \text{in } B(x_0, R), \\ u = 0 & \text{in } \mathbb{R}^n \setminus B(x_0, R). \end{cases} \tag{13.8}$$

Then, the unique solution to (13.8) *is provided by*

$$u(x) = \frac{\Gamma(\frac{n}{2})}{4^s\Gamma(\frac{n+2s}{2})\Gamma(s+1)} \left(R^2 - |x-x_0|^2\right)_+^s. \tag{13.9}$$

PROOF. By dilation and translation it suffices to consider the case $x_0 = 0$ and $R = 1$. For $0 < s < 1$ consider the equation

$$(-\Delta)^s B_s = f, \tag{13.10}$$

where

$$B_s(x) = \frac{\left(1-|x|^2\right)_+^s}{\Gamma(s+1)}$$

is the Bochner-Riesz kernel introduced in (4.34) above. Applying the Fourier transform to both sides of (13.10), and using (5.2) in Proposition 5.1, we find

$$(4\pi^2|\xi|^2)^s \hat{B}_s(\xi) = \hat{f}(\xi).$$

If in this formula we use (4.35) in Lemma 4.5 with $z = s$ to compute $\hat{B}_s(\xi)$, we obtain

$$\begin{aligned}\hat{f}(\xi) &= (4\pi^2|\xi|^2)^s \pi^{-s}|\xi|^{-(\frac{n}{2}+s)} J_{\frac{n}{2}+s}(2\pi|\xi|)\\ &= 4^s\pi^s|\xi|^{-\frac{n}{2}+s} J_{\frac{n}{2}+s}(2\pi|\xi|).\end{aligned}$$

We next use Theorem 4.4 again to recover $f(x)$ from the latter formula

$$\begin{aligned}f(x) &= 2\pi 4^s\pi^s|x|^{-\frac{n-2}{2}} \int_0^\infty t^{\frac{n}{2}} t^{-\frac{n}{2}+s} J_{\frac{n}{2}+s}(2\pi t) J_{\frac{n-2}{2}}(2\pi|\xi|t)dt.\\ &= 2\pi 4^s\pi^s|x|^{-\frac{n-2}{2}} \int_0^\infty t^s J_{\frac{n}{2}+s}(2\pi t) J_{\frac{n-2}{2}}(2\pi|\xi|t)dt.\end{aligned}$$

Comparing this formula with Lemma 4.8 we see that we presently need to have

$$\lambda = -s, \quad \nu = \frac{n}{2}+s, \quad \mu = \frac{n-2}{2}, \qquad a = 2\pi, \quad b = 2\pi|x|.$$

Since $0 < s < 1$, we clearly have $\lambda > -1$. Furthermore, since we are interested in values of x such that $0 < |x| < 1$, we have $0 < b < a$. With the above choices of λ, ν, μ and a we obtain

$$\nu + \mu - \lambda + 1 = \frac{n}{2} + s + \frac{n}{2} - 1 + s + 1 = n + 2s,$$

and

$$\begin{aligned}-\nu + \mu - \lambda + 1 &= -\frac{n}{2} - s + \frac{n}{2} - 1 + s + 1 = 0,\\ \mu - \lambda + 1 &= \frac{n}{2} - 1 + s + 1 = \frac{n+2s}{2},\\ \nu - \mu + \lambda + 1 &= \frac{n}{2} + s - \frac{n}{2} + 1 - s + 1 = 2.\end{aligned}$$

Applying Lemma 4.8 above we thus find for $0 < |x| < 1$

$$f(x) = 2\pi 4^s\pi^s|x|^{-\frac{n-2}{2}}(2\pi|x|)^{\frac{n-2}{2}} \frac{\Gamma(\frac{n+2s}{2})}{2^{-s}(2\pi)^{\frac{n}{2}+s}\Gamma(1)\Gamma(\frac{n}{2})} F\left(\frac{\nu+\mu-\lambda+1}{2}, 0; \frac{n}{2}; |x|^2\right),$$

where $F(a, b; c; z)$ is Gauss' hypergeometric function in Definition 4.7 above. Since thanks to (4.41) we have

$$F\left(\frac{\nu+\mu-\lambda+1}{2}, 0; \frac{n}{2}; |x|^2\right) = 1,$$

we finally conclude that for any $|x|<1$

$$f(x) \equiv \frac{4^s\Gamma(\frac{n+2s}{2})}{\Gamma(\frac{n}{2})}.$$

If we substitute such f in (13.10) above, we see that we have proved that for $|x|<1$

$$(-\Delta)^s B_s(x) \equiv \frac{4^s\Gamma(\frac{n+2s}{2})}{\Gamma(\frac{n}{2})}.$$

Keeping in mind the above definition of B_s it is thus clear that the function

$$u(x) = \frac{\Gamma(\frac{n}{2})}{4^s\Gamma(\frac{n+2s}{2})\Gamma(s+1)} \left(1-|x|^2\right)_+^s.$$

provides the desired solution (13.9) to the problem (13.8).

□

REMARK 13.2. Let us note explicitly that as $s \to 1$ the constant

$$\frac{\Gamma(\frac{n}{2})}{4^s\Gamma(\frac{n+2s}{2})\Gamma(s+1)} \longrightarrow \frac{1}{2n}.$$

Thus, in the local case when $s=1$ we recover the standard torsion function in (13.6) above.

REMARK 13.3. We want to bring to the reader's attention an interesting connection between the nonlocal torsion function u in Proposition 13.1, and the recent paper [**BIK15**] in which the authors construct an analogue of the self-similar Barenblatt-Kompaneets-Pattle-Zeldovich solution for the following nonlocal porous medium equation

$$\partial_t U = \operatorname{div}(|U|\nabla^{2s-1}(|U|^{m-2}U)),$$

where $\nabla^{2s-1} \overset{def}{=} \nabla(-\Delta)^{s-1}$. In their Theorem 2.2 they prove that such self-similar solution is obtained in the form

$$U(x,t) = \frac{1}{t^{n\lambda}}\Phi\left(\frac{x}{t^\lambda}\right), \qquad \lambda = \frac{1}{n(m-1)+2s},$$

where $\Phi(x) = C(n,m,s)u(x)^{\frac{1}{m-1}}$, $u(x)$ is the function in Proposition 13.1, and $C(n,m,s)$ is an explicit constant.

Concerning Proposition 13.1 the reader should notice that, since for $B(0,1)$ we have $d(x) = \operatorname{dist}(x, \partial B(0,1)) = 1-|x|$, the torsion function u in (13.9) satisfies in an obvious way the property that

$$\frac{u}{d^s} \in C^\infty(\overline{B}(0,1) \setminus \{0\}).$$

In the framework of $C^{1,1}$ domains, in Theorem 1.2 of their cited paper [**ROS14**] the authors prove the following general result.

THEOREM 13.4. *Let Ω be a bounded $C^{1,1}$ domain, $f \in L^\infty(\Omega)$, and u be a weak solution to the non-homogeneous Dirichlet problem*

$$\begin{cases} (-\Delta)^s u = f & \text{in } \Omega, \\ u = 0 & \text{in } \mathbb{R}^n \setminus \Omega. \end{cases} \tag{13.11}$$

Then, $u/d^s \in C^{0,\alpha}(\overline{\Omega})$ *for some* $0 < \alpha < \min\{s, 1-s\}$*, and one has*

$$||u/d^s||_{C^{0,\alpha}(\overline{\Omega})} \leq C||f||_{L^\infty(\Omega)},$$

with α, C *depending only on* Ω, n, s.

The reader is also encouraged to see the beautiful survey paper [**RO17'**] on boundary regularity and the nonlocal Pohozaev identity. We also thank X. Ros-Oton for pointing to our attention the following interesting result of Grubb, see [**Gr14**], [**Gr15**].

THEOREM 13.5. *Let* $\Omega \subset \mathbb{R}^n$ *be a* C^∞ *domain, and let* $f \in C^\infty(\overline{\Omega})$. *If* u *solves* (13.11), *then* $u/d^s \in C^\infty(\overline{\Omega})$.

In Theorem 13.5 the notation $d = d(x)$ represents a positive function which coincides with the distance to the boundary of Ω in a neighborhood of $\partial\Omega$.

14. Monotonicity formulas and unique continuation for $(-\Delta)^s$

We open this section with a quote from the paper [**CS07**]: "Monotonicity formulas are a very powerful tool in the study of the regularity properties of elliptic PDEs. They have been used in a number of problems to exploit the local properties of the equations by giving information about the blowup configurations". In what follows we discuss some monotonicity formulas related to nonlocal operators.

In order to provide some motivation and historical perspective we recall that the most fundamental property of harmonic functions is the so-called *strong unique continuation property*, which we will abbreviate in (sucp) henceforth. It states that a solution of $\Delta u = 0$ in a connected open set cannot vanish to infinite order at one point, unless it vanishes identically. Here, vanishing to infinite order means, for instance, that for any $N \in \mathbb{N}$ one has as $r \to 0^+$

$$\sup_{B(x_0,r)} |u| = O(r^N).$$

The weaker *unique continuation property* (ucp) states that if u vanishes in an open subset, then u must vanish everywhere. These properties are shared by large classes of differential operators, such as for instance the stationary Schrödinger operator $H = -\Delta + V$ which plays a central role in quantum mechanics. At the basis of J. Von Neumann's axiomatic formulation there is the assumption of the absence of eigenvalues of H embedded in the continuous spectrum. However, a famous example of Wigner and Von Neumann (1954) showed that in the presence of tunneling effects such eigenvalues can in fact arise, see [**RS75**]. It thus became important to understand assumptions on V that would rule out such possibility.

On the other hand, a famous theorem of F. Rellich [**Rel40**] states that for every $R > 0$ there cannot exist eigenfunctions of the Laplacean in $L^2(\mathbb{R}^n \setminus B(0, R))$ corresponding to $\lambda > 0$. This is where the (ucp) for the operator $H = -\Delta + V$, with $V \in L^\infty_{loc}(\mathbb{R}^n)$ comes into the play. Assuming for simplicity that V be compactly supported, it is easy to see that

$$\text{(ucp)} + \text{Rellich} \implies \text{Absence of embedded eigenvalues.}$$

In 1939 T. Carleman introduced a powerful method to prove the (ucp) or even the (sucp) for $H = -\Delta + V$ based on weighted a priori estimates of the type

$$||e^{t\Phi(x)}u||_{L^q(\mathbb{R}^n)} \leq C(n)||e^{t\Phi(x)}\Delta u||_{L^p(\mathbb{R}^n)},$$

where $1 \leq p \leq q < \infty$, $t \in \mathbb{R}$ is a parameter which is allowed to go to $+\infty$ avoiding a discrete set, and $u \in C_0^\infty(\mathbb{R}^n \setminus \{0\})$. Notice that when $\Phi \equiv 0$ and $\frac{1}{p} - \frac{1}{q} = \frac{2}{n}$, the above estimate is just the Sobolev embedding theorem (for this theorem see e.g. [**St70**]). When $\Phi \not\equiv 0$ instead, this gap allows a potential $V \in L^{n/2}_{loc}(\mathbb{R}^n)$, which is sharp for (sucp), see the celebrated work of D. Jerison and C. Kenig [**JK85**]. However, when $V \in L^\infty_{loc}(\mathbb{R}^n)$, then no gap is necessary. As a consequence, one can take $p = q = 2$ and use Hilbert spaces framework (integration by parts).

A different method to establish the (sucp) for more general elliptic operators of the type

$$Lu = \operatorname{div}(A(x)\nabla u) + < b, \nabla u > + Vu, \tag{14.1}$$

was developed by F.H. Lin and the author in [**GL86**], [**GL87**]. Their approach is energy based and it establishes directly the (sucp), and in fact more, without considerable technical complexities. The central idea is a remarkable monotonicity property that was discovered by F. Almgren in [**A79**] in his approach to the regularity of mass minimizing currents.

THEOREM 14.1 (Almgren's monotonicity formula). *Let $\Delta u = 0$ in $B(0,1)$. Then, the so-called* frequency *of u*

$$r \to N(u,r) = \frac{rD(u,r)}{H(u,r)} = \frac{r\int_{B_r} |\nabla u|^2}{\int_{S_r} u^2}$$

is monotonically increasing. Moreover, $N(u,r) \equiv \kappa$ if and only if u is homogeneous of degree κ.

The name *frequency* comes from the fact that when $u = P_\kappa$, a harmonic polynomial having frequency κ, then $N(u,r) \equiv \kappa$. One important consequence of the monotonicity of the frequency (in fact, only its boundedness suffices!) is the following *doubling condition*: for every $0 < r < 1$ one has

$$\int_{B_{2r}} u^2 dx \leq C(n, ||u||_{W^{1,2}(B_1)}) \int_{B_r} u^2 dx.$$

It is well know, see [**GL86**], that:

$$\text{Doubling condition} \implies \text{(sucp)}\,.$$

In the cited papers [**GL86**] and [**GL87**] Theorem 14.1 was generalized to solutions of elliptic equations $Lu = 0$, with L as in (14.1) and the coefficients of the leading part are Lipschitz continuous. This is known to be the minimal smoothness for the unique continuation to be true. For counterexamples see the groundbreaking paper by Plis [**Pl63**] for nondivergence form equations, and the subsequent work of Miller [**Mi74**] for divergence form operators.

In view of what has been said up to this point, it is natural to ask whether the (sucp) holds for nonlocal operators. Concerning this question in this section we will prove the following result.

THEOREM 14.2 (Strong unique continuation for $(-\Delta)^s$). *Let $u \in \mathscr{L}_s(\mathbb{R}^n) \cap C(\mathbb{R}^n)$ be a weak solution of $(-\Delta)^s u = 0$ in a connected open set $\Omega \subset \mathbb{R}^n$. If u vanishes to infinite order at a point $x_0 \in \Omega$, then $u \equiv 0$ in Ω.*

REMARK 14.3. The reader should bear in mind that there is a way of dealing with Theorem 14.2 different from the one that we are going to present below. In fact, similarly to the local case, solutions of $(-\Delta)^s u = 0$ in an open set $\Omega \subset \mathbb{R}^n$ are not only $C^\infty(\Omega)$, see Theorems 12.17 and 12.19, but in fact real analytic! We will however not pursue this interesting aspect in these notes since, besides not being the real-analyticity an easy fact to establish, it would also confine our discussion to a special situation.

Since there presently exists no direct Almgren type monotonicity formula, nor to the best of our knowledge there exists a direct Carleman estimate for solutions of $(-\Delta)^s u = 0$, in order to prove Theorem 14.2 one has to proceed indirectly and resort to the extension problem (10.1) above. As we will see, this imposes a delicate detour since it is not a priori true that the information of the vanishing to infinite order transfers from the solution of $(-\Delta)^s u = 0$, to that of the extension problem. In light of this it would be desirable (and also quite interesting) to know whether a direct nonlocal analogue of Theorem 14.1 be possible. Perhaps the Rellich-Pohozaev identity in the cited work [**ROS14'**] might prove useful in this direction. Concerning nonlocal integral identities we also mention the work [**Gr16**] which contains a generalization to pseudodifferential operators of the results in [**ROS14'**].

In what follows we sketch a proof of Theorem 14.2 based on monotonicity formulas. This is a special case of the work [**FF14**], in which the authors prove a general result for nonlocal semilinear equations combining the extension procedure with the approach in [**GL86**], [**GL87**]. We mention that a different approach that combines the extension procedure with Carleman estimates was developed in [**Ru15**].

In the sequel we continue to use the notation of Section 11, and will assume that $-1 < a < 1$ is given. We will need the following simple lemma which provides a trace inequality (and a reverse form of it) for functions in the Sobolev space $W^{1,2}(B_e(r), |y|^a dX)$.

LEMMA 14.4. *For $r > 0$ let $\tilde{U} \in W^{1,2}(B_e(r), |y|^a dX)$. There exists a constant $C(n, a) > 0$ such that*

$$\int_{S_e(r)} \tilde{U}^2 |y|^a d\sigma \leq C(n,a) \left(\frac{1}{r} \int_{B_e(r)} \tilde{U}^2 |y|^a dX + r \int_{B_e(r)} |\nabla \tilde{U}|^2 |y|^a dX \right). \tag{14.2}$$

and

$$\frac{1}{r} \int_{B_e(r)} \tilde{U}^2 |y|^a dX \leq C(n,a) \left(\int_{S_e(r)} \tilde{U}^2 |y|^a d\sigma + r \int_{B_e(r)} |\nabla \tilde{U}|^2 |y|^a dX \right). \tag{14.3}$$

PROOF. Since $\omega(X) = |y|^a$ is an A_2 weight of Muckenhoupt, from the general result in Theorem 6.1 in [**Chu92**] we know that $C^\infty(\overline{B}_e(r))$ is dense in $W^{1,2}(B_e(r), |y|^a dX)$. Therefore, in order to prove (14.2) or (14.3) it suffices to assume that $\tilde{U} \in C^\infty(\overline{B}_e(r))$. For any such function, if we indicate with ν the outer unit normal to $S_e(r)$, applying the divergence theorem we obtain

$$\begin{aligned} \int_{S_e(r)} \tilde{U}^2 |y|^a d\sigma &= \frac{1}{r} \int_{S_e(r)} < \tilde{U}^2 |y|^a X, \nu > d\sigma = \int_{B_e(r)} \operatorname{div}(\tilde{U}^2 |y|^a X) dX \\ &= \frac{n+a+1}{r} \int_{B_e(r)} \tilde{U}^2 |y|^a dX + \frac{2}{r} \int_{B_e(r)} \tilde{U} < \nabla \tilde{U}, X > |y|^a dX \end{aligned}$$

From this identity the desired conclusions (14.2) and (14.3) easily follow. □

We now assume that $\tilde{U}(x,y) \in W^{1,2}(B_e(R), |y|^a dX)$ is a weak solution of the equation $L_a\tilde{U} = 0$ in a ball $B_e(R)$, and we assume that $\tilde{U}$ is even in the variable y. For any $r \in (0,R)$ we consider the quantities

$$\tilde{H}(\tilde{U},r) = \int_{S_e(r)} \tilde{U}^2 |y|^a dH_n, \tag{14.4}$$

$$\tilde{D}(\tilde{U},r) = (1-a)^{2a} \int_{B_e(r)} |\nabla \tilde{U}|^2 |y|^a dX, \quad 0 < r < R, \tag{14.5}$$

$$\tilde{N}(\tilde{U},r) = \frac{r\tilde{D}(\tilde{U},r)}{\tilde{H}(\tilde{U},r)}. \tag{14.6}$$

In Theorem 6.1 in [**CS07**] Caffarelli and Silvestre proved the following important result.

THEOREM 14.5 (Monotonicity formula of Almgren type for L_a). *Suppose that for no $r \in (0,R_0)$ we have $\tilde{H}(\tilde{U},r) = 0$. Then, the function $r \to \tilde{N}(\tilde{U},r)$ is nondecreasing on $(0,R_0)$. Furthermore, $\tilde{N}(\tilde{U},\cdot) \equiv \tilde{\kappa}$ if and only if $\tilde{U}$ is homogeneous of degree $\tilde{\kappa}$ with respect to the standard Euclidean dilations $\delta_\lambda(x,y) = (\lambda x, \lambda y)$.*

It is worth noting that Theorem 14.5 was already contained in the results in the paper [**GL87**]. In fact, it was mentioned in the Remark after Theorem 1.4 on p. 351 in [**GL87**] that an Almgren type monotonicity formula continues to be valid for weak solutions of divergence equations in $\mathbb{R}^n$ of the type

$$\operatorname{div}(\omega(x)\nabla u) = 0, \tag{14.7}$$

provided that $\omega \geq 0$ is a (possibly) degenerate weight satisfying the condition

$$| < \nabla\omega(x), x > | \leq C\, \omega(x), \tag{14.8}$$

for some constant $C > 0$. Under (14.8) it was noted in [**GL87**] that, with

$$H(u,r) = \int_{\partial B(r)} u^2\omega, \qquad D(u,r) = \int_{B(r)} |\nabla u|^2 \omega,$$

then the frequency $N(u,r) = \frac{rD(u,r)}{H(u,r)}$ is *almost* monotone, in the sense that there exists a constant $C' \geq 0$, depending on n and C in (14.8), such that $r \to e^{C'r}N(u,r)$ is monotone nondecreasing. As it is well-known by now such adjusted monotonicity is all that is needed to carry the blow-up analysis. In the special case in which instead of (14.8) one has pure homogeneity of the weight, i.e.,

$$< \nabla\omega(x), x > = a\, \omega(x), \tag{14.9}$$

for some $a \in (-n,n)$ (this restriction on a guarantees local integrability of ω and ω^{-1}), then the calculations in [**GL86**], [**GL87**] give the exact formulas

$$H'(u,r) = \frac{n+a-1}{r} H(u,r) + 2D(u,r),$$

$$D'(u,r) = \frac{n+a-2}{r} D(u,r) + 2\int_{\partial B(r)} u_\nu^2\, \omega,$$

and therefore one obtains

$$\frac{N'(u,r)}{N(u,r)} = \frac{1}{r} + \frac{D'(u,r)}{D(u,r)} - \frac{H'(u,r)}{H(u,r)} = 2\frac{\int_{\partial B(r)} u_\nu^2 \omega}{D(u,r)} - 2\frac{D(u,r)}{H(u,r)}.$$

At this point the pure monotonicity of $r \to N(u,r)$ follows from the Cauchy-Schwarz inequality after one notes that from the equation (14.7) one has

$$D(u,r) = \int_{\partial B(r)} u u_\nu \omega.$$

This discussion of course includes the A_2 weight in $\mathbb{R}^n$ given by $\omega(x) = |x|^a$, with $-n < \alpha < n$, but also the A_2 weight in $\mathbb{R}^{n+1}$ given by $\omega(X) = \omega(x,y) = |y|^a$, with $-1 < a < 1$, in the extension operator

$$L_a = \operatorname{div}_X(|y|^a \nabla_X).$$

In was also shown in [**GRO17**] that, at least in the range $0 \le a < 1$, Theorem 14.5 can be directly derived by the Almgren type monotonicity formula for solutions of $\mathcal{P}_\alpha U = 0$ established in [**G93**]. In fact, with a in such range consider the number $\alpha \ge 0$ defined by $\alpha = \frac{a}{1-a}$, and for $U(x,z)$ defined by the equation (11.3), consider the height function, the Dirichlet integral and the frequency of U respectively defined by:

$$H(U,r) = \int_{S_{\rho_\alpha}(r)} U^2 \frac{\psi_\alpha}{|\nabla \rho_\alpha|}\, dH_n, \quad 0 < r < R, \tag{14.10}$$

$$D(U,r) = \int_{B_{\rho_\alpha}(r)} |\nabla_\alpha U|^2 dx dz \tag{14.11}$$

$$N(U,r) = \frac{r D(U,r)}{H(U,r)}, \tag{14.12}$$

where $|\nabla_\alpha U|^2$ is the degenerate *carré du champ* introduced in (12.3) above. One has the following transformation formulas contained in Lemmas 2.7 and 2.8 in [**GRO17**].

PROPOSITION 14.6. *For every $r > 0$ one has*

$$\tilde{H}(\tilde{U},r) = r^a H(U,h(r)), \qquad \tilde{D}(\tilde{U},r) = (1-a)^a D(U,h(r)). \tag{14.13}$$

As a consequence of (14.13) we obtain the remarkable formula

$$\tilde{N}(\tilde{U},r) = (1-a) N(U,h(r)). \tag{14.14}$$

We next state a result which is a special case of Theorem 4.2 in [**G93**].

THEOREM 14.7 (Almgren type monotonicity formula for $\mathcal{P}_\alpha$). *Suppose that for no $r \in (0,R_0)$ we have $H(U,r) = 0$. Then, the function $r \to N(U,r)$ is nondecreasing in $(0,R_0)$. Furthermore, $N(U,\cdot) \equiv \kappa$ if and only if U is homogeneous of degree κ with respect to the nonisotropic dilations* (11.20).

Having stated Theorem 14.7, we now make an interesting observation.

THEOREM 14.8. *The following is true when $0 \le a < 1$:*

$$\textit{Theorem } 14.7 \iff \textit{Theorem } 14.5.$$

PROOF. It is enough to observe that the function $r \to h(r)$ in (11.6) is monotonically increasing. The desired conclusion thus follows immediately by combining (14.14) with Theorem 14.7, keeping in mind that $\alpha = \frac{a}{1-a}$, and therefore $a = \frac{\alpha}{1+\alpha}$. □

Theorem 14.8 was not noted in [**CS07**], but it was subsequently observed in Remark 3.2 in [**CSS08**]. We also observe that the same conclusion continues to work also in the range $-1 < a < 0$, since using the regularization procedures employed in [**G93**] one can extend the validity of the results there to the range $-\frac{1}{2} < \alpha < 0$.

Let us return to the central objective of this section, namely Theorem 14.2. We note that one important corollary of Theorem 14.5 is the following.

COROLLARY 14.9. *Let $\tilde{U}$ be as in the hypothesis of Theorem 14.5. Then, $\lim_{r\to 0^+} N(\tilde{U}, r) = \tilde{\kappa}$ exists finite.*

Another crucial consequence is the following result.

THEOREM 14.10 (Non-degeneracy). *Under the assumptions in Theorem 14.5, given $R \in (0, R_0)$, one has for every $0 < r < R$*

$$\tilde{H}(\tilde{U}, r) \geq \tilde{H}(\tilde{U}, R)\left(\frac{r}{R}\right)^{n+a+2||\tilde{N}(\tilde{U},\cdot)||_{L^\infty(0,R)}}. \tag{14.15}$$

PROOF. By Theorem 14.5 we know that the function $t \to \tilde{N}(\tilde{U}, t)$ is monotonically increasing on $(0, R)$. On the other hand, using the equation $L_a\tilde{U} = 0$ and integration by parts, it is not difficult to prove that

$$\tilde{H}'(\tilde{U}, t) = \frac{n+a}{t}\tilde{H}(\tilde{U}, t) + 2\tilde{D}(\tilde{U}, t).$$

Keeping the definition (14.6) in mind, we can rewrite this equation in the following way

$$\frac{d}{dt}\log\frac{\tilde{H}(\tilde{U}, t)}{t^{n+a}} = 2\frac{\tilde{N}(\tilde{U}, t)}{t}.$$

Integrating this formula on the interval (r, R), and using the monotonicity of $r \to \tilde{N}(\tilde{U}, r)$ (in fact, just the boundedness of this function suffices in this argument), we obtain the desired conclusion (14.15).

□

COROLLARY 14.11. *If we have $\tilde{H}(\tilde{U}, R_0) \neq 0$, then we must have $\tilde{H}(\tilde{U}, r) \neq 0$ for all $0 < r < R_0$.*

PROOF. We argue by contradiction and assume that there exist $0 < \overline{r} < R_0$ such that $\tilde{H}(\tilde{U}, \overline{r}) = 0$. Define

$$\rho = \sup\{r \leq R_0 \mid \tilde{H}(\tilde{U}, r) = 0\}.$$

Since by the hypothesis $\tilde{H}(\tilde{U}, R_0) \neq 0$, we must have $0 < \rho < R_0$. But then, we have $\tilde{H}(\tilde{U}, r) \neq 0$ for $r \in (\rho, R_0]$. Applying Theorem 14.10 we obtain for $r \in (\rho, R_0]$

$$\tilde{H}(\tilde{U}, r) \geq \tilde{H}(\tilde{U}, R_0)\left(\frac{r}{R_0}\right)^{n+a+2||\tilde{N}(\tilde{U},\cdot)||_{L^\infty(0,R_0)}} > 0.$$

Letting $r \to \rho^+$ this leads to a contradiction since $\tilde{H}(\tilde{U}, \rho) = 0$.

□

We next define an important family of non-homogeneous rescalings. In a different context, they were introduced the first time in [**ACS08**] in the blowup analysis of the Signorini problem. In what follows, we indicate with $X = (x, y)$ the generic point in $\mathbb{R}^n \times \mathbb{R}$.

DEFINITION 14.12. Let $R_0 > 0$ be such that $\tilde{H}(\tilde{U}, r) > 0$ for every $0 < r < R_0$. We define the *Almgren rescalings* of a function $\tilde{U}$ as

$$\tilde{U}_r(X) = \frac{r^{\frac{n+a}{2}}\tilde{U}(\tilde{\delta}_r(X))}{\sqrt{\tilde{H}(\tilde{U}, r)}} = \frac{r^{\frac{n+a}{2}}\tilde{U}(rX)}{\sqrt{\tilde{H}(\tilde{U}, r)}}. \tag{14.16}$$

Obviously, if $\tilde{U}$ is a solution of $L_a\tilde{U} = 0$ in $B_e(R_0)$, then $L_a\tilde{U}_r = 0$ in $B_e(R_0/r)$. An elementary, yet crucial property, of the Almgren rescalings which follows from (14.4) and a change of variable is that

$$\tilde{H}(\tilde{U}_r, 1) = \int_{S_e(1)} \tilde{U}_r^2(X)|y|^a dH_n(X) = 1. \tag{14.17}$$

Another basic property is the following identity

$$\tilde{N}(\tilde{U}_r, \rho) = \tilde{N}(\tilde{U}, r\rho), \qquad r, \rho > 0. \tag{14.18}$$

The next lemma plays a key role in the proof of Theorem 14.2. Its proof hinges crucially on the monotonicity formula in Theorem 14.5.

LEMMA 14.13. *Suppose $\tilde{H}(\tilde{U}, R_0) \neq 0$. With $\tilde{U}_r$ as in (14.16), there exists a subsequence $r_j \to 0$, and a function $\tilde{U}_0 : \mathbb{R}^n \times \mathbb{R} \to \mathbb{R}$, such that $\tilde{U}_{r_j} = \tilde{U}_j$ converges uniformly to $\tilde{U}_0$ and $\nabla\tilde{U}_j \to \nabla\tilde{U}_0$ weakly in $L^2(\mathbb{R}^n \times \mathbb{R}, |y|^a dX)$ on compact subsets of $\mathbb{R}^n \times \mathbb{R}$. Moreover, $\tilde{U}_0$ is a weak solution (even in the variable y) to*

$$\begin{cases} \operatorname{div}(|y|^a \nabla\tilde{U}_0) = 0, \\ \lim\limits_{y\to 0} y^a \partial_y \tilde{U}_0 = 0, \end{cases} \tag{14.19}$$

on every compact subset of $\mathbb{R}^n \times \mathbb{R}$. Finally, $\tilde{U}_0$ is homogeneous of degree $\tilde{\kappa} = \lim\limits_{r\to 0^+} N(\tilde{U}, r)$.

PROOF. Using (14.17) and (14.18) we obtain for $0 < r < R_0$

$$\int_{B_e(1)} |\nabla\tilde{U}_r(X)|^2 |y|^a dX = \tilde{N}(\tilde{U}_r, 1) = \tilde{N}(\tilde{U}, r) \leq \tilde{N}(\tilde{U}, R_0) < \infty, \tag{14.20}$$

where in the second to the last inequality we have used the monotonicity of $r \to \tilde{N}(\tilde{U}, r)$ in Theorem 14.5. We note that since we are assuming $\tilde{H}(\tilde{U}, R_0) \neq 0$, Corollary 14.11 allows to infer that $\tilde{H}(\tilde{U}, r) \neq 0$ for every $0 < r < R_0$, and thus we are in the hypothesis of Theorem 14.5.

At this point we combine the trace inequality (14.3) with (14.17) and (14.20) to infer that for every $0 < r < R_0$

$$\int_{B_e(1)} \tilde{U}_r^2 |y|^a dX < \infty. \tag{14.21}$$

Combining (14.21) with (14.20), we conclude that for every $0 < r < R_0$

$$||\tilde{U}_r||_{W^{1,2}(B_e(1),|y|^a dX)} \leq C < \infty,$$

for some constant independent of r. This implies the existence of a sequence $r_j \searrow 0$ and of a function $\tilde{U}_0 \in W^{1,2}(B_e(1), |y|^a dX)$ such that

$$\tilde{U}_{r_j} \longrightarrow \tilde{U}_0$$

weakly in $W^{1,2}(B_e(1), |y|^a dX)$. By the local Hölder continuity of solutions of $L_a\tilde{U} = 0$, we conclude that, possibly on a subsequence, which we still denote $\tilde{U}_{r_j}$, we have convergence of $\tilde{U}_{r_j} \longrightarrow \tilde{U}_0$ in $C^{0,\alpha}$ norm on compact subset of $\mathbb{R}^n \times \mathbb{R}$. The Caccioppoli inequality (12.19) then implies that for any $0 < s < t$

$$\int_{B_e(s)} |\nabla \tilde{U}_{r_j} - \nabla \tilde{U}_0|^2 |y|^a dX \leq \frac{C(n,a)}{t-s} \int_{S_e(t)} |\tilde{U}_{r_j} - \tilde{U}_0|^2 |y|^a dH_n \longrightarrow 0, \tag{14.22}$$

as $j \to \infty$. The inequality (14.22) and the $C^{0,\alpha}_{loc}$ estimates for $\tilde{U}$ imply that $\tilde{U}_0 \in W^{1,2}_{loc}(\mathbb{R}^n \times \mathbb{R}, |y|^a dX)$, and that $\tilde{U}_0$ is a weak solution of (14.19) on every compact subset of $\mathbb{R}^n \times \mathbb{R}$. Moreover, (14.22) also implies that for any $\rho > 0$

$$\tilde{D}(\tilde{U}_{r_j}, \rho) \longrightarrow \tilde{D}(\tilde{U}_0, \rho). \tag{14.23}$$

Since from the uniform convergence of $\tilde{U}_{r_j} \longrightarrow \tilde{U}_0$ in $C^{0,\alpha}$ we have

$$\tilde{H}(\tilde{U}_{r_j}, \rho) \longrightarrow \tilde{H}(\tilde{U}_0, \rho). \tag{14.24}$$

Next, we claim that for every $0 < \rho < 1$ we must have

$$\tilde{H}(\tilde{U}_0, \rho) > 0. \tag{14.25}$$

This follows from the estimate (14.15) in Theorem 14.10 that gives for every $0 < r < R_0$ and $0 < \rho < 1$

$$\tilde{H}(\tilde{U}, r\rho) \geq \tilde{H}(\tilde{U}, r) \rho^{n+a+2||\tilde{N}(\tilde{U},\cdot)||_{L^\infty(0,R_0)}}. \tag{14.26}$$

On the other hand, an easy change of variable gives

$$\tilde{H}(\tilde{U}_r, \rho) = \frac{\tilde{H}(\tilde{U}, r\rho)}{\tilde{H}(\tilde{U}, r)},$$

and from this observation and (14.26) we find

$$\tilde{H}(\tilde{U}_{r_j}, \rho) \geq \rho^{n+a+2||\tilde{N}(\tilde{U},\cdot)||_{L^\infty(0,R_0)}}.$$

Letting $j \to \infty$ and using (14.24) we conclude that (14.25) holds. Now that we know (14.25), from (14.23) and (14.24) we conclude that as $j \to \infty$

$$\tilde{N}(\tilde{U}_{r_j}, \rho) \longrightarrow \tilde{N}(\tilde{U}_0, \rho).$$

On the other hand, (14.18) and Corollary 14.9 give

$$\tilde{N}(\tilde{U}_{r_j}, \rho) = \tilde{N}(\tilde{U}, r_j \rho) \to \tilde{\kappa},$$

as $j \to \infty$. Therefore, for every $0 < \rho < 1$ we must have

$$\tilde{N}(\tilde{U}_0, \rho) \equiv \tilde{\kappa}. \tag{14.27}$$

Once we know this, we finish by invoking the second part of Theorem 14.5 that allows to infer that $\tilde{U}_0$ must be homogeneous of degree $\tilde{\kappa}$ in $B_e(1)$.

□

DEFINITION 14.14. We call any function $\tilde{U}_0$ as in Lemma 14.13 an *Almgren blowup* of U at $x_0 = 0$.

We are now ready to give the

PROOF OF THEOREM 14.2. Let u be as in the statement of the theorem. Without loss of generality we can assume that the point x_0 at which u vanishes to infinite order be $x_0 = 0$. We want to show that $u \equiv 0$ in Ω. We would like to show that there exists $R_0 > 0$ such that $u \equiv 0$ in $B_e(R_0)$. A standard connectedness argument would then imply that $u \equiv 0$ in Ω. Consider the extension problem (10.1) above with Dirichlet datum u, and denote by $U(x,y) = P_s(\cdot, y) \star u(x)$ its solution in $\mathbb{R}^n \times \mathbb{R}^+$. Arguing as in the proof of Theorem 12.15 we know that $\tilde{U}(x,y) = U(x,|y|)$, is a weak solution of $L_a\tilde{U} = 0$ in $\Omega \times (-d,d)$, where $a = 1-2s$.

Let $R_0 = \operatorname{dist}(x_0, \partial\Omega)$. If we show that $\tilde{H}(\tilde{U}, R_0) = 0$, then by the Caccioppoli inequality (12.19) we would infer that $\int_{B_e(s)} |\nabla \tilde{U}(x',y)|^2 |y|^a dx'dy = 0$ for every $0 < s < R_0$ and thus $\tilde{U}$ would have to be constant in $B_e(R_0)$. But $\tilde{H}(\tilde{U}, R_0) = 0$ would force the constant to be zero. This would imply in particular that $u(x) = \tilde{U}(x,0) = 0$ for every $|x| < R_0$, and we would be done.

We thus assume that $\tilde{H}(\tilde{U}, R_0) > 0$ and show that this leads to a contradiction. Now, this assumption and Corollary 14.11 imply that $\tilde{H}(\tilde{U}, r) > 0$ for every $0 < r < R_0$. Therefore, by Theorem 14.5 we infer that the function $r \to \tilde{N}(\tilde{U}, r)$ is nondecreasing on $(0, R_0)$. Let $\tilde{U}_0$ be an Almgren blowup of $\tilde{U}$ at $x_0 = 0$. If $\tilde{\kappa} = \lim_{r\to 0^+} N(\tilde{U}, r)$, we know from Lemma 14.13 that $\tilde{U}_0$ is a global solution of $L_a\tilde{U}_0 = 0$, even in y, and which is homogeneous of degree $\tilde{\kappa}$. At this point we make the important observation that $\tilde{U}_0 \not\equiv 0$. This follows from its homogeneity and the non-degeneracy estimate (14.25) above. We now make the following

Claim: $\tilde{U}_0$ restricted to $\{y = 0\}$ is not identically zero.

The proof of this claim proceeds by contradiction and it is based on an argument in Step 1 and 2 in the proof of Proposition 2.2 in [**Ru15**], and we refer the reader to that source. What Rüland shows is that, if the claim is not true, then $\tilde{U}_0$ must vanish to infinite order at $(0,0)$ in the thick space $\mathbb{R}^n \times \mathbb{R}$. However, such vanishing to infinite order at $(0,0)$ in the thick space $\mathbb{R}^n \times \mathbb{R}$ would contradict the homogeneity of $\tilde{U}_0$. Therefore, the claim must be true.

From the claim we know that $\tilde{U}_0(\cdot, 0) \not\equiv 0$. The final step is now proving that this fact contradicts the assumption that $u(x) = \tilde{U}(x,0)$ vanishes to infinite order at $x_0 = 0$.

Since $\tilde{U}_0(\cdot, 0) \not\equiv 0$, there exist $C, \bar{r} > 0$ such that

$$||\tilde{U}_0||_{L^\infty(B_e(\bar{r}) \cap \{y=0\})} = C.$$

From the uniform convergence of $\tilde{U}_{r_j} \longrightarrow \tilde{U}_0$ on $B_e(\bar{r}) \cap \{y = 0\}$, we know that for sufficiently large $j \in \mathbb{N}$ we must have

$$||\tilde{U}_{r_j}||_{L^\infty(B_e(\bar{r}) \cap \{y=0\})} \geq \frac{C}{2}.$$

Using the definition (14.16) of $\tilde{U}_{r_j}$, we obtain from the latter inequality and the fact that $\tilde{U}(x,0) = u(x)$,

$$||u||_{L^\infty(B(r_j\bar{r}) \cap \{y=0\})} \geq \frac{C}{2} r_j^{-\frac{n+a}{2}} \sqrt{\tilde{H}(\tilde{U}, r_j)}.$$

Finally, applying Theorem 14.10 we find

$$\tilde{H}(\tilde{U}, r_j) \geq \tilde{H}(\tilde{U}, R_0) \left(\frac{r_j}{R_0}\right)^{n+a+2||\tilde{N}(\tilde{U},\cdot)||_{L^\infty(0,R_0)}}.$$

We thus conclude

$$||u||_{L^\infty(B(r_j\bar{r}) \cap \{y=0\})} \geq \frac{C}{2} r_j^{-\frac{n+a}{2}} \sqrt{\tilde{H}(\tilde{U}, R_0)} \left(\frac{r_j}{R_0}\right)^{\frac{n+a+2||\tilde{N}(\tilde{U},\cdot)||_{L^\infty(0,R_0)}}{2}}.$$

This contradicts the assumption that u vanishes to infinite order at $x_0 = 0$, unless of course $\tilde{H}(\tilde{U}, R_0) = 0$. By the monotonicity of $r \to \tilde{H}(\tilde{U}, r)$ we conclude that it must be $\tilde{U} \equiv 0$ in $B_e(R_0)$, and therefore $u \equiv 0$ in $B(R_0)$. By connectedness we infer that it must be $u \equiv 0$ in Ω.

□

In connection with Theorem 14.2 above we mention the recent work [**BG17**] in which the authors establish a delicate theorem of strong unique continuation backward in time for solutions of the nonlocal equation

$$(\partial_t - \Delta)^s u = V(x,t)u, \qquad 0 < s < 1. \tag{14.28}$$

They assume that the potential $V : \mathbb{R}^{n+1} \to \mathbb{R}$ is such that for some $K > 0$

$$\begin{cases} ||V||_{C^1(\mathbb{R}^{n+1})} \le K, & \text{if } 1/2 \le s < 1, \\ \\ ||V||_{C^2(\mathbb{R}^{n+1})}, \quad || < \nabla_x V, x > ||_{L^\infty(\mathbb{R}^{n+1})} \le K, & \text{if } 0 < s < 1/2. \end{cases} \tag{14.29}$$

The main result in [**BG17**] is as follows.

THEOREM 14.15 (Space-time strong unique continuation property). *Let $u \in$ Dom(H^s) be a solution to* (14.28) *with V satisfying* (14.29). *If u vanishes to infinite order backward in time at some point (x_0, t_0) in $\mathbb{R}^{n+1}$, then $u(\cdot, t) \equiv 0$ for all $t \le t_0$.*

One of the central tools in the proof of Theorem 14.15 is Theorem 14.16 below. In it the function U represents the solution to the following problem:

$$\begin{cases} y^a \partial_t U(X,t) = \operatorname{div}_X(y^a \nabla_X U)(X,t), \\ U(x,0,t) = u(x,t), \\ \lim_{y \to 0^+} y^a \frac{\partial U}{\partial y}(x,y,t) = -V(x,t)u(x,t), \end{cases} \tag{14.30}$$

where $y^a \partial_t U(X,t) = \operatorname{div}_X(y^a \nabla_X U)(X,t)$ is the parabolic extension operator, introduced by Nyström and Sande in [**NS16**] and independently by Stinga and Torrea in [**ST17**], whereas $N(U,r) = I(U,r)/H(U,r)$ represents a suitable frequency function in Gaussian space, see Definition 6.2 in [**BG17**].

THEOREM 14.16 (Monotonicity of the adjusted frequency). *Let $u \in$ Dom(H^s) be a solution to* (14.28) *with V satisfying* (14.29). *There exist universal constants $C, t_0 > 0$, depending only on n, s and the number K in* (14.29), *such that with $r_0 = \sqrt{t_0}$ and $a = 1 - 2s$, under the assumption that $H(U,r) \neq 0$ for all $0 < r \le r_0$, then*

$$r \to \exp\left\{C \int_0^r t^{-a} dt\right\} \left(N(U,r) + C \int_0^r t^{-a} dt\right) \tag{14.31}$$

is monotone increasing on $(0, r_0)$. Furthermore, when the potential $V \equiv 0$, then the constant C in (14.31) *can be taken equal to zero and we have pure monotonicity of the function $r \to N(U,r)$. In such case, $N(U,r) \equiv \kappa$ for $0 < r < R$ if and only if U is homogeneous of degree 2κ in $\mathbb{S}_R^+$ with respect to the parabolic dilations $\delta_\lambda(X,t) = (\lambda X, \lambda^2 t)$.*

Theorem 14.16 generalizes to the case $s \neq 1/2$ a related monotonicity result that was obtained in [**DGPT17**] for the case $s = 1/2$ to establish the optimal regularity of the solution of the Signorini problem for the heat equation. We also mention the recent paper [**Yu17**] in which the author uses a generalization of the monotonicity formula in Theorem 14.5 to establish the (sucp) for solutions of the nonlocal equation $(-L)^s u = 0$, where L is a divergence form elliptic operator with Lipschitz continuous coefficients.

Concerning monotonicity formulas we cannot fail to mention the considerable developments that have taken place over the last decade in connection with free boundary problems for nonlocal operators, starting with the pioneering paper [**ACS08**]. In that paper, for the first time, the role of the Almgren monotonicity formula in Theorem 14.1 was recognized as central to the analysis of both the optimal regularity of the solution and of the regular part of the free boundary in the nonlocal obstacle problem: *given a smooth function* $\psi : \mathbb{R}^n \to \mathbb{R}$, *find a function* $u : \mathbb{R}^n \to \mathbb{R}$ *such that*

$$\begin{cases} \min\{u - \psi, (-\Delta)^s u\} = 0 & \text{in } \mathbb{R}^n, \\ \lim_{|x|\to\infty} u(x) = 0. \end{cases} \tag{14.32}$$

Via the extension procedure described above, this problem becomes equivalent to the following *thin obstacle problem* (the name "thin" comes from the fact that now the obstacle lives on the thin manifold $M = \mathbb{R}^n_x \times \{0\}$ in the thick space $\mathbb{R}^n_x \times \mathbb{R}_y$) for the extension operator L_a, with $a = 1 - 2s$ and for the function $\tilde{U}(x, y)$:

(14.33)

$$\begin{cases} L_a \tilde{U} = \operatorname{div}_{x,y}(|y|^a \nabla_{x,y} \tilde{U}) = 0 & \text{in } \mathbb{R}^{n+1}_+ \cup \mathbb{R}^{n+1}_-, \\ \tilde{U}(x, -y) = \tilde{U}(x, y), & \text{for } x \in \mathbb{R}^n, y \in \mathbb{R}, \\ \tilde{U}(x, 0) \geq \psi(x), & \text{for } x \in \mathbb{R}^n, \\ -\lim\limits_{y\to 0^+} y^a D_y \tilde{U}(x, y) \geq 0, & \text{for } x \in \mathbb{R}^n, \\ \lim\limits_{y\to 0^+} y^a D_y \tilde{U}(x, y) = 0, & \text{on the set where } \tilde{u}(x, 0) > \psi(x). \end{cases}$$

We emphasize that when $s = 1/2$, we have $a = 1 - 2s = 0$, and thus the extension operator L_a is simply the Laplacean in the variables (x, y). In such case the problem (14.33) is known as the famous problem in elasticity posed in the 50's by Antonio Signorini, an engineer and mathematician: *what is the equilibrium configuration of a spherically shaped elastic body resting on a rigid frictionless plane.* Thus, the nonlocal obstacle problem (14.32) for $(-\Delta)^{1/2}$ is equivalent to the Signorini problem for the operator Δ in $\mathbb{R}^{n+1}$. For a discussion of the latter one should see the beautiful book [**PSU12**]. A remarkable up-to-date account on obstacle problems for fractional operators is the recent survey paper [**DS17**].

Returning to (14.32), in [**ACS08**] the authors considered the case of zero obstacle ψ and critical exponent $s = 1/2$, and using Theorem 14.1 they proved:

1) the optimal regularity $C^{1,1/2}_{loc}$ of the solution from either side of the thin manifold (a different proof of such result had first been found in [**AC04**]);
2) that the *regular part* of the free boundary is locally a $C^{1,\alpha}$ hypersurface.

In [**CSS08**] the authors using an almost monotonicity formula that generalizes Theorem 14.5 above were able to extend the results in [**ACS08**] to the full range $0 < s < 1$, and to the case of general obstacle. In [**GP09**] some new one-parameter families of monotonicity formulas of Weiss and Monneau type were discovered. With such formulas the authors were able to analyze, for the fractional exponent $s = 1/2$ in (14.32) and for a general obstacle, the so-called *singular part* of the free boundary. They classified singular points and proved the rectifiability of the singular part of the free boundary. Using some new monotonicity formulas their results have been recently generalized to the full range $0 < s < 1$ in [**GRO17**], see also the interesting paper [**FoS17**] where methods from geometric measure theory are used to achieve a fine analysis of the free boundary. In the above cited work [**DGPT17**] the authors developed an extensive analysis of the obstacle problem

$$\begin{cases} \min\{u-\psi,\, (\partial_t-\Delta)^{1/2}u\}=0 & \text{in } \mathbb{R}^n, \\ \lim_{|(x,t)|\to\infty} u(x,t)=0, \end{cases} \tag{14.34}$$

for the nonlocal heat equation corresponding to the value $s = 1/2$ of the fractional exponent and for a general obstacle ψ. They establish the optimal regularity of the solution, the regularity of the regular part of the free boundary, they classify the singular set and prove its regularity. Some of the central tools in their analysis are some new Almgren type monotonicity formulas inspired to those found by Poon in [**Po96**] for the standard heat equation, as well as one-parameter parabolic monotonicity formulas of Weiss and Monneau type.

Finally, the papers [**GS14**], [**PP15**], [**GPS16**], and [**GPPS17**] contain various new monotonicity formulas of Almgren, Weiss and Monneau type which are applied to nonlocal obstacle problems such as (14.32) above, but in which either $(-\Delta)^{1/2}$ is replaced by $(-L)^{1/2}$, where L is a divergence form elliptic operator with Lipschitz continuous coefficients (for a previous result for $C^{1,\gamma}$ coefficients, see [**Gui09**]), or $(-\Delta)^s$ is replaced by $(-\Delta)^s + < b(x), \nabla >$. The authors establish for the relevant problems the optimal regularity of the solution, the $C^{1,\alpha}$ smoothness of the regular free boundary and the regularity of the singular part of the latter. We note that being able to treat variable coefficient operators in the obstacle problem is crucial to understanding (14.33) above when the separating manifold is non-flat. In such situation, the natural approach is to flatten the manifold. If the latter is, for instance, $C^{1,1}$, one is thus lead to the study of a Signorini problem for a variable coefficient operator with Lipschitz coefficients and flat thin manifold. A different approach to the Signorini problem for variable coefficient operators, based on Carleman estimates, was found in [**KRS16**], [**KRS17**].

15. Nonlocal Poisson kernel and mean-value formulas

The most fundamental property of classical harmonic functions is Gauss' mean-value property: if $\Delta u = 0$ in an open set $\Omega \subset \mathbb{R}^n$, then for every $x \in \Omega$ and every $0 < r < \operatorname{dist}(x, \partial\Omega)$, one has (12.1) above. Classical potential theory, i.e., the study of subharmonic functions, can be entirely developed starting from the corresponding sub-mean value formula for subharmonic functions, see for instance [**He69**] and [**DP70**]. It is thus not surprising that a suitable analogue of (12.1) should play an equally important role in the potential theory of the fractional Laplacean. In this respect one should keep in mind that one way (admittedly, not the simplest one!) of obtaining the spherical mean-value formula in (12.1) is by choosing $x = 0$ in the Poisson representation formula

$$u(x) = \int_{S_r} P_r(x,y)\varphi(y)d\sigma(y), \qquad x \in B_r, \tag{15.1}$$

where we have denoted with

$$P_r(x,y) = \frac{1}{\sigma_{n-1} r}\frac{r^2-|x|^2}{|y-x|^n}, \qquad x \in B_r,\ y \in S_r,$$

the Poisson kernel for the ball B_r. We recall that the function u in (15.1) provides the unique solution to the Dirichlet problem for the ball B_r. In his seminal paper

[**R38**] using the fundamental solution $E_s(x)$ in (8.3) of Theorem 8.4 above, and the *nonlocal Kelvin transform* of a function u, defined by

$$\tilde{u}(x) = E_s(x) u\left(\frac{x}{|x|^2}\right), \tag{15.2}$$

M. Riesz constructed the nonlocal Poisson kernel (15.4) below, and with it he was able to solve the Dirichlet problem

$$\begin{cases} (-\Delta)^s u = 0 & \text{in } B_r, \\ u = \varphi & \text{in } \mathbb{R}^n \setminus B_r. \end{cases} \tag{15.3}$$

In the following definition we recall the nonlocal counterpart of (15.1) discovered by M. Riesz, see formula (3) on p. 17 in [**R38**], but also (1.6.11') and (1.6.2) on pages 122 and 112 in [**La72**].

DEFINITION 15.1. For every $0 < s < 1$ and $r > 0$ we define the *nonlocal interior Poisson kernel* for B_r as

$$P_r^{(s)}(x,y) = c(n,s)\left(\frac{r^2 - |x|^2}{|y|^2 - r^2}\right)^s \frac{1}{|y-x|^n}, \qquad |x| < r, |y| > r, \tag{15.4}$$

where

$$c(n,s) = \frac{\sin(\pi s)\Gamma(\frac{n}{2})}{\pi^{\frac{n}{2}+1}} = \frac{2\sin(\pi s)}{\pi \sigma_{n-1}}. \tag{15.5}$$

When $x = 0$ is the center of the ball B_r, then we use the notation

$$A_r^{(s)}(y) = P_r^{(s)}(0,y) = \begin{cases} c(n,s)\frac{r^{2s}}{(|y|^2-r^2)^s |y|^n}, & |y| > r, \\ 0 & |y| \le r. \end{cases} \tag{15.6}$$

We call $A_r^{(s)}(y)$ the kernel of the *nonlocal mean-value operator*

$$\mathscr{A}_r^{(s)} u(x) = A_r^{(s)} \star u(x). \tag{15.7}$$

The mean-value operator defined by (15.7) is the nonlocal counterpart of the spherical mean (4.38). In Proposition 15.4 below we show that as $s \nearrow 1$ then $\mathscr{A}_r^{(s)} u(x) \to \mathscr{M}_r u(x)$.

Returning to (15.4), the following theorem of M. Riesz's provides the unique solution to the nonlocal Dirichlet problem (15.3).

THEOREM 15.2. *Let $\varphi \in \mathscr{L}_s(\mathbb{R}^n) \cap C(\mathbb{R}^n)$. Consider the function u in $\mathbb{R}^n$ defined by*

$$u(x) = \begin{cases} \int_{\mathbb{R}^n \setminus B_r} P_r^{(s)}(x,y)\varphi(y)dy, & x \in B_r, \\ \varphi(x), & x \in \mathbb{R}^n \setminus B_r. \end{cases} \tag{15.8}$$

Then, u is the unique solution to the Dirichlet problem for the ball B_r.

Let us observe right-away that from (15.6) we have $A_r^{(s)} \in L^1(\mathbb{R}^n)$. Therefore, Young's convolution theorem (or Minkowski integral inequality) implies that

$$\mathscr{A}_r^{(s)} : L^p(\mathbb{R}^n) \longrightarrow L^p(\mathbb{R}^n), \qquad 1 \le p \le \infty,$$

and that

$$||\mathscr{A}_r^{(s)} u||_{L^p(\mathbb{R}^n)} \le ||A_r^{(s)}||_{L^1(\mathbb{R}^n)} ||u||_{L^p(\mathbb{R}^n)}, \qquad u \in L^p(\mathbb{R}^n).$$

The result that follows shows that $\mathscr{A}_r^{(s)}$ is a contraction in $L^p(\mathbb{R}^n)$.

LEMMA 15.3. *For every $0 < s < 1$ and $r > 0$ one has*

$$||A_r^{(s)}||_{L^1(\mathbb{R}^n)} = \int_{\mathbb{R}^n} A_r^{(s)}(y)dy = 1.$$

PROOF. Using (15.6) we find

$$\begin{aligned}\int_{\mathbb{R}^n} A_r^{(s)}(y)dy &= c(n,s)\int_{|y|>r} \frac{r^{2s}}{(|y|^2-r^2)^s|y|^n}dy \\ &= r^n c(n,s)\int_{|z|>1}\frac{dz}{(|z|^2-1)^s r^n|z|^n} = c(n,s)\int_{|z|>1}\frac{dz}{(|z|^2-1)^s|z|^n} \\ &= c(n,s)\sigma_{n-1}\int_1^\infty \frac{1}{(\rho^2-1)^s}\frac{d\rho}{\rho}\end{aligned}$$

Now we make the substitution $\rho = \sec\vartheta$, for which $d\rho = \sec\vartheta\tan\vartheta d\vartheta$. With such substitution we find

$$\begin{aligned}\int_{\mathbb{R}^n} A_r^{(s)}(y)dy =& c(n,s)\sigma_{n-1}\int_0^{\frac{\pi}{2}} \frac{\sec\vartheta\tan\vartheta}{(\tan^2\vartheta)^s}\frac{d\vartheta}{\sec\vartheta} \\ &= c(n,s)\sigma_{n-1}\int_0^{\frac{\pi}{2}} (\sin\vartheta)^{1-2s}(\cos\vartheta)^{2s-1}d\vartheta.\end{aligned}$$

Applying (4.9) above with $y = 1-s$ and $x = s$, we conclude

$$\begin{aligned}\int_{\mathbb{R}^n} A_r^{(s)}(y)dy =& c(n,s)\sigma_{n-1}\int_0^{\frac{\pi}{2}} (\sin\vartheta)^{1-2s}(\cos\vartheta)^{2s-1}d\vartheta \\ &= \frac{c(n,s)\sigma_{n-1}}{2}B(s,1-s) = \frac{c(n,s)\sigma_{n-1}}{2}\Gamma(s)\Gamma(1-s)\end{aligned}$$

Finally, we use the formula (4.3) to obtain

$$\int_{\mathbb{R}^n} A_r^{(s)}(y)dy = \frac{c(n,s)\sigma_{n-1}}{2}\frac{\pi}{\sin\pi s}.$$

If we now recall the formula (4.6) it is easy to see that, by choosing $c(n,s) > 0$ as in (15.5), we reach the desired conclusion that

$$\int_{\mathbb{R}^n} A_r^{(s)}(y)dy = 1.$$

□

The next result shows that, as $s \to 1$, the nonlocal mean-value operator in (15.7) converges to the spherical mean-value operator (2.3) for the sphere.

PROPOSITION 15.4 (Asymptotic behavior of $\mathscr{A}_r^{(s)}u(x)$ as $s \nearrow 1$). *For every function $u \in \mathscr{S}(\mathbb{R}^n)$ one has*

$$\lim_{s\to 1}\mathscr{A}_r^{(s)}u(x) = \mathscr{M}_r u(x),$$

for every $x \in \mathbb{R}^n$, where $\mathscr{M}_r u(x)$ is defined as in (2.3) above.

PROOF. It suffices to prove the result when $x = 0$. By Lemma 15.3 we have

$$\begin{aligned}\mathscr{A}_r^{(s)}u(0) &= c(n,s)\int_{|y|>r}\frac{r^{2s}}{(|y|^2-r^2)^s|y|^n}\left[u(y)-\mathscr{M}_r u(0)\right]dy + \mathscr{M}_r u(0) \\ &= \mathscr{M}_r u(0) + \frac{2\sin(\pi s)}{\pi}r^{2s}\int_r^\infty \frac{1}{(\rho^2-r^2)^s\rho}\left[\mathscr{M}_\rho u(0) - \mathscr{M}_r u(0)\right]d\rho.\end{aligned}$$

Since $\sin(\pi s) \to 0$ as $s \to 1$, to finish the proof it suffices to show that there exists a number $C > 0$ independent of $s \in (0,1)$ such that

$$\left| \int_r^\infty \frac{1}{(\rho^2 - r^2)^s \rho} \left[\mathscr{M}_\rho u(0) - \mathscr{M}_r u(0)\right] d\rho \right| \le C.$$

From (i) of Proposition 12.2 we find

$$\left| \frac{\partial \mathscr{M}_r u}{\partial r}(0) \right| = \left| \frac{1}{\sigma_{n-1} r^{n-1}} \int_{S(0,r)} \frac{\partial u}{\partial \nu}(y) d\sigma(y) \right| \le ||\nabla u||_{L^\infty(\mathbb{R}^n)}.$$

We thus have for a fixed $R > r$

$$\int_r^\infty \frac{1}{(\rho^2 - r^2)^s \rho} \left[\mathscr{M}_\rho u(0) - \mathscr{M}_r u(0)\right] d\rho = \int_r^R \frac{1}{(\rho^2 - r^2)^s \rho} \left[\mathscr{M}_\rho u(0) - \mathscr{M}_r u(0)\right] d\rho$$
$$+ \int_R^\infty \frac{1}{(\rho^2 - r^2)^s \rho} \left[\mathscr{M}_\rho u(0) - \mathscr{M}_r u(0)\right] d\rho = I(s) + II(s).$$

Now, for any $\rho \in (r, R)$ we have from the above estimate

$$|\mathscr{M}_\rho u(0) - \mathscr{M}_r u(0)| = \left| \int_r^\rho \frac{d}{dt} \mathscr{M}_t u(0) dt \right| \le ||\nabla u||_{L^\infty(\mathbb{R}^n)} (\rho - r).$$

This gives

$$|I(s)| \le ||\nabla u||_{L^\infty(\mathbb{R}^n)} \int_r^R \frac{(\rho - r)^{1-s}}{(\rho + r)^s \rho} d\rho \le ||\nabla u||_{L^\infty(\mathbb{R}^n)} 2^{-s} r^{-1-s} \int_r^R (\rho - r)^{1-s} d\rho$$
$$\le ||\nabla u||_{L^\infty(\mathbb{R}^n)} 2^{-s} r^{-1-s} \frac{(R - r)^{2-s}}{2 - s} \le C,$$

with a $C > 0$ independent of $s \to 1$. On the other hand, we have

$$|II(s)| \le 2||u||_{L^\infty(\mathbb{R}^n)} \int_R^\infty \frac{d\rho}{(\rho^2 - r^2)^s \rho} \le C,$$

where again $C > 0$ can be taken independent of $s \to 1$.

□

Proposition 15.4 has been very recently generalized in [**BuSq18**] by allowing in the definition of $\mathscr{A}_r^{(s)} u(x)$ a measure on the unit sphere $\mathbb{S}^{n-1}$ which is bounded both from above and from below (away from zero).

In the next result, we use the space $\mathscr{L}_s(\mathbb{R}^n)$ introduced in Definition 2.12, whereas the notation $C^{2s+\varepsilon}_{loc}$ is that introduced in Definition 2.14.

LEMMA 15.5. *Let $0 < s < 1$ and suppose that $u \in \mathscr{L}_s(\mathbb{R}^n)$. Assume furthermore that for some $0 < \varepsilon < 1$ we have $u \in C^{2s+\varepsilon}_{loc}$ in a neighborhood of $x \in \mathbb{R}^n$. Then,*

(15.9)
$$\lim_{r \to 0^+} \int_{|y|>r} \frac{2u(x) - u(x+y) - u(x-y)}{(|y|^2 - r^2)^s |y|^n} dy = \int_{\mathbb{R}^n} \frac{2u(x) - u(x+y) - u(x-y)}{|y|^{n+2s}} dy.$$

PROOF. For $0 < r < \frac{1}{\sqrt{2}}$ we can write

$$\int_{|y|>r} \frac{2u(x) - u(x+y) - u(x-y)}{(|y|^2 - r^2)^s |y|^n} dy = \int_{r<|y|\le 1} \frac{2u(x) - u(x+y) - u(x-y)}{(|y|^2 - r^2)^s |y|^n} dy$$
$$+ \int_{1<|y|} \frac{2u(x) - u(x+y) - u(x-y)}{(|y|^2 - r^2)^s |y|^n} dy = I_1(r) + I_2(r).$$

Now on the set where $|y| > 1$ we have

$$|y|^2 > 1 > 2r^2,$$

and so

$$\frac{1}{|y|^2 - r^2} < \frac{2}{|y|^2}.$$

This gives

$$I_2(r) < 2^s \int_{1<|y|} \frac{2u(x) - u(x+y) - u(x-y)}{|y|^{n+2s}} dy < \infty,$$

since by assumption $u \in \mathscr{L}_s(\mathbb{R}^n)$. By Lebesgue dominated convergence theorem, we conclude that

$$\lim_{r\to 0^+} I_2(r) = \int_{1<|y|} \frac{2u(x) - u(x+y) - u(x-y)}{|y|^{n+2s}} dy.$$

Next, we evaluate

$$I_1(r) = \int_{r<|y|\le 1} \frac{2u(x) - u(x+y) - u(x-y)}{(|y|^2 - r^2)^s |y|^n} dy.$$

We distinguish two cases:

(i) $0 < s < 1/2$;

(ii) $1/2 \le s < 1$.

In case (i) we know by Definition 2.14 that $u \in C^{0,2s+\varepsilon}$ at x. Thus, for $|y| \le 1$ we have

$$|2u(x) - u(x+y) - u(x-y)| \le C|y|^{2s+\varepsilon}. \tag{15.10}$$

We then find

$$\begin{aligned} I_1(r) &= \int_{r<|y|\le 1} \frac{2u(x) - u(x+y) - u(x-y)}{|y|^{n+2s}} dy \\ &\quad + \left(I_1(r) - \int_{r<|y|\le 1} \frac{2u(x) - u(x+y) - u(x-y)}{|y|^{n+2s}} dy \right) \\ &= \int_{r<|y|\le 1} \frac{2u(x) - u(x+y) - u(x-y)}{|y|^{n+2s}} dy + \tilde{I}_1(r). \end{aligned}$$

Now, (15.10) and dominated convergence give

$$\lim_{r\to 0^+} \int_{r<|y|\le 1} \frac{2u(x) - u(x+y) - u(x-y)}{|y|^{n+2s}} dy = \int_{|y|\le 1} \frac{2u(x) - u(x+y) - u(x-y)}{|y|^{n+2s}} dy.$$

On the other hand, again by (15.10) we have

$$\begin{aligned} |\tilde{I}_1(r)| &\le \int_{r<|y|\le 1} |2u(x) - u(x+y) - u(x-y)| \left| \frac{1}{(|y|^2 - r^2)^s |y|^n} - \frac{1}{|y|^{n+2s}} \right| dy \\ &\le C \int_{r<|y|\le 1} |y|^{2s+\varepsilon} \left| \frac{1}{(|y|^2 - r^2)^s |y|^n} - \frac{1}{|y|^{n+2s}} \right| dy \\ &= C'(n) \int_r^1 t^{2s+\varepsilon} \left| \frac{1}{(t^2 - r^2)^s} - \frac{1}{t^{2s}} \right| \frac{dt}{t} \\ &= C'(n) r^\varepsilon \int_1^{\frac{1}{r}} \tau^{2s+\varepsilon} \left| \frac{1}{(\tau^2 - 1)^s} - \frac{1}{\tau^{2s}} \right| \frac{d\tau}{\tau} \end{aligned}$$

Clearly, the integrand is summable near $\tau = 1$ since $0 < s < 1$. Near $\tau = \infty$ the integrand behaves like

$$\frac{1}{\tau^{1-\varepsilon}}\left|\frac{1}{(1-\tau^{-2})^s} - 1\right| \cong \frac{1}{\tau^{3-\varepsilon}}$$

which is in L^1 provided that $3 - \varepsilon > 1$, which is of course true. Therefore,

$$|\tilde{I}_1(r)| \le C'(n) r^{\varepsilon} \int_1^{\infty} \tau^{2s+\varepsilon}\left|\frac{1}{(\tau^2-1)^s} - \frac{1}{\tau^{2s}}\right| \frac{d\tau}{\tau} = C''(n,\varepsilon) r^{\varepsilon} \longrightarrow 0,$$

as $r \to 0^+$. In conclusion, when $0 < s < \frac{1}{2}$ we obtain

$$\lim_{r\to 0^+} I_1(r) = \int_{|y|\le 1} \frac{2u(x) - u(x+y) - u(x-y)}{|y|^{n+2s}} dy.$$

It follows that

$$\lim_{r\to 0^+} \int_{|y|>r} \frac{2u(x) - u(x+y) - u(x-y)}{(|y|^2 - r^2)^s |y|^n} dy = \lim_{r\to 0^+} I_1(r) + \lim_{r\to 0^+} I_2(r)$$

$$= \int_{|y|\le 1} \frac{2u(x) - u(x+y) - u(x-y)}{|y|^{n+2s}} dy + \int_{1<|y|} \frac{2u(x) - u(x+y) - u(x-y)}{|y|^{n+2s}} dy.$$

This proves (15.9) in the case (i).

Suppose now that case (ii) occurs. By Definition 2.14 the assumption $u \in C^{2s+\varepsilon}_{loc}$ in a neighborhood of x now reads $u \in C^{1,2s+\varepsilon-1}_{loc}$. Using Taylor's formula we now find

$$2u(x) - u(x+y) - u(x-y) = O(|y|^{2s+\varepsilon}),$$

and thus (15.10) is valid again. We can thus argue as we did in the case (i) to reach the conclusion that (15.9) hold. This completes the proof.

□

We next draw an interesting consequence of Lemma 15.5. Recall Proposition 2.2 above. As it is well-known the operators defined by the right-hand sides of (2.2) are at the basis of the development of potential theory of non-smooth subharmonic functions. We next prove a result that generalizes Proposition 2.2 to the nonlocal setting.

PROPOSITION 15.6 (The Blaschke-Privalov fractional Laplacean). *Let $0 < s < 1$ and suppose that $u \in \mathscr{L}_s(\mathbb{R}^n)$ be in $C^{2s+\varepsilon}$ in a neighborhood of $x \in \mathbb{R}^n$, for some $0 < \varepsilon < 1$. One has*

$$(15.11)\qquad (-\Delta)^s u(x) = \gamma(n,s) \lim_{r\to 0^+} \frac{u(x) - A_r^{(s)} \star u(x)}{r^{2s}},$$

where $\gamma(n,s) = \frac{s 2^{2s}\Gamma(\frac{n}{2}+s)}{\pi^{\frac{n}{2}}\Gamma(1-s)}$ is the constant in (5.10) in Proposition 5.6.

PROOF. By the definition (15.7) we have

$$A_r^{(s)} \star u(x) = \int_{\mathbb{R}^n} A_r^{(s)}(y) u(x-y) dy.$$

Changing y in $-y$, and using the fact that $A_r^{(s)}(-y) = A_r^{(s)}(y)$, we also have

$$A_r^{(s)} \star u(x) = \int_{\mathbb{R}^n} A_r^{(s)}(y) u(x+y) dy.$$

On the other hand, Lemma 15.3 gives for $r > 0$

$$2u(x) = \int_{\mathbb{R}^n} A_r^{(s)}(y) 2u(x) dy.$$

Therefore, we obtain

$$u(x) - A_r^{(s)} \star u(x) = \frac{1}{2} \int_{\mathbb{R}^n} A_r^{(s)}(y) \left[2u(x) - u(x+y) - u(x-y)\right] dy$$
$$= \frac{r^{2s}}{2} \int_{|y|>r} \frac{2u(x) - u(x+y) - u(x-y)}{(|y|^2 - r^2)^s |y|^n} dy.$$

This gives

$$\frac{u(x) - A_r^{(s)} \star u(x)}{r^{2s}} = \frac{1}{2} \int_{|y|>r} \frac{2u(x) - u(x+y) - u(x-y)}{(|y|^2 - r^2)^s |y|^n} dy$$

From this identity and from (15.9) in Lemma 15.5, under the given hypothesis on u, we obtain

$$\lim_{r \to 0^+} \frac{u(x) - A_r^{(s)} \star u(x)}{r^{2s}} = \frac{1}{2} \lim_{r \to 0^+} \int_{|y|>r} \frac{2u(x) - u(x+y) - u(x-y)}{(|y|^2 - r^2)^s |y|^n} dy$$
$$= \frac{1}{2} \int_{\mathbb{R}^n} \frac{2u(x) - u(x+y) - u(x-y)}{|y|^{n+2s}} dy = \frac{1}{2} \frac{2}{\gamma(n,s)} (-\Delta)^s u(x)$$
$$= \frac{1}{\gamma(n,s)} (-\Delta)^s u(x),$$

where in the last equality we have used (2.8). This establishes (15.11).

□

We mention that the operator defined by the limit relation (15.11) is known as the *Dynkin operator* to people in probability, see [**Dy65**] and also the interesting paper [**Kwa17**].

The classical Theorem of Köebe 12.3 states that if a continuous function in an open set $\Omega \subset \mathbb{R}^n$ satisfies either one of the mean-value properties (12.1) for every $x \in \Omega$ and $0 < r < \operatorname{dist}(x, \partial\Omega)$, then in fact $u \in C^\infty(\Omega)$ and $\Delta u = 0$. We next establish a Köebe type theorem for the fractional Laplacean.

PROPOSITION 15.7. *Let $0 < s < 1$ and suppose that $u \in \mathscr{L}_s(\mathbb{R}^n)$ and that, for some $0 < \varepsilon < 1$, we have $u \in C^{2s+\varepsilon}_{loc}$ in a neighborhood of $x \in \mathbb{R}^n$. If for every $r > 0$ sufficiently small we have $u(x) = A_r^{(s)} \star u(x)$, then we must have $(-\Delta)^s u(x) = 0$.*

PROOF. By hypothesis we know that for all $r > 0$ arbitrarily small we have

$$u(x) = A_r^{(s)} \star u(x) = \int_{\mathbb{R}^n} A_r^{(s)}(y) u(x-y) dy.$$

Changing y in $-y$, and using the fact that $A_r^{(s)}(-y) = A_r^{(s)}(y)$, we also have

$$u(x) = \int_{\mathbb{R}^n} A_r^{(s)}(y) u(x+y) dy.$$

On the other hand, thanks to Lemma 15.3 we can write for every such $r > 0$

$$u(x) = \int_{\mathbb{R}^n} A_r^{(s)}(y) u(x) dy.$$

We thus have for every $r > 0$ sufficiently small

$$0 = \int_{\mathbb{R}^n} A_r^{(s)}(y)\,[2u(x) - u(x+y) - u(x-y)]\,dy$$
$$= r^{2s} \int_{|y|>r} \frac{2u(x) - u(x+y) - u(x-y)}{(|y|^2 - r^2)^s |y|^n} dy.$$

This formula gives

$$\lim_{r\to 0^+} \int_{|y|>r} \frac{2u(x) - u(x+y) - u(x-y)}{(|y|^2 - r^2)^s |y|^n} dy = 0.$$

On the other hand (15.9) in Lemma 15.5 gives

$$\lim_{r\to 0^+} \int_{|y|>r} \frac{2u(x) - u(x+y) - u(x-y)}{(|y|^2 - r^2)^s |y|^n} dy = \int_{\mathbb{R}^n} \frac{2u(x) - u(x+y) - u(x-y)}{|y|^{n+2s}} dy$$
$$= \frac{2}{\gamma(n,s)}(-\Delta)^s u(x).$$

We conclude that $(-\Delta)^s u(x) = 0$.

□

COROLLARY 15.8 (Nonlocal Köebe theorem). *Let $\Omega \subset \mathbb{R}^n$ be an open set, and suppose that $u \in \mathscr{L}_s(\mathbb{R}^n) \cap C^{2s+\varepsilon}_{loc}$ for some $\varepsilon > 0$. If for every $x \in \Omega$ and every $0 < r < \operatorname{dist}(x, \partial\Omega)$ we have $u(x) = A_r^{(s)} \star u(x)$, then $(-\Delta)^s u = 0$ in Ω, and therefore by Theorem 12.17 we also have $u \in C^\infty(\Omega)$.*

We next want to establish a converse to Corollary 15.8, namely that if $(-\Delta)^s u = 0$ in an open set $\Omega \subset \mathbb{R}^n$, then $u(x) = A_r^{(s)} \star u(x)$ for all $x \in \Omega$ and for sufficiently small $r > 0$. In the local case, under the assumption that $u \in C^2(\Omega)$, this can be easily achieved by using (ii) of Proposition 12.2. An alternative, less direct way of proving this in the local case is to argue as follows. Suppose without loss of generality that $x = 0 \in \Omega$, and consider the ball $B_r = B_r(0) \subset \overline{B}_r \subset \Omega$. Denote by $P_r(x,y)$ and $G_r(x,y)$ respectively the Poisson kernel and the Green function for the ball B_r. The function

$$v(x) = -\int_{B_r} G_r(x,y)\Delta u(y)dy,$$

solves the problem

$$\begin{cases} \Delta v = \Delta u & \text{in } B_r, \\ v = 0 & \text{on } S_r, \end{cases}$$

whereas the function

$$w(x) = \int_{S_r} P_r(x,y)u(y)d\sigma(y),$$

solves the problem

$$\begin{cases} \Delta w = 0 & \text{in } B_r, \\ w = u & \text{on } S_r. \end{cases}$$

The maximum principle implies that $u = v + w$ in B_r. In particular, we have $u(0) = v(0) + w(0)$. This gives

$$u(0) = \int_{S_r} P_r(0,y)u(y)d\sigma(y) - \int_{B_r} G_r(0,y)\Delta u(y)dy \tag{15.12}$$
$$= \mathscr{M}_r u(0) + \int_{B_r} G_r(0,y)(-\Delta)u(y)dy.$$

From the identity (15.12) it is thus clear that if $\Delta u = 0$ in Ω, then for every $x \in \Omega$ we must have $u(x) = \mathscr{M}_r u(x)$ for all $r > 0$ such that $\overline{B}_r \subset \Omega$. These considerations can be repeated in the nonlocal case, if we use M. Riesz' Poisson kernel and Green function for the ball. Before we turn to this we note that

$$G_r(x,y) = E(x,y) - \int_{S_r} P_r(y,\xi)E(x,\xi)d\sigma(\xi),$$

and therefore, for $n \geq 3$, we have with $c_n = \frac{1}{(n-2)\sigma_{n-1}}$,

$$G_r(0,y) = E(0,y) - \int_{S_r} P_r(y,\xi)E(0,\xi)d\sigma(\xi) = E(0,y) - \frac{c_n}{r^{n-2}} \int_{S_r} P_r(y,\xi)d\sigma(\xi)$$
$$= E(0,y) - \frac{c_n}{r^{n-2}},$$

which, by translation invariance, gives the well-known formula

$$u(x) = \mathscr{M}_r u(x) + \int_{B_r(x)} \left\{ \frac{c_n}{|y-x|^{n-2}} - \frac{c_n}{r^{n-2}} \right\} (-\Delta)u(y)dy.$$

PROPOSITION 15.9. *Let $\Omega \subset \mathbb{R}^n$ be an open set, with $n \geq 2$, and suppose that $u \in \mathscr{L}_s(\mathbb{R}^n) \cap C^{2s+\varepsilon}_{loc}$ for some $\varepsilon > 0$. Assume that $(-\Delta)^s u = 0$ in the open set $\Omega \subset \mathbb{R}^n$. Then, for every $x \in \Omega$ and every $r > 0$ such that $\overline{B}_r \subset \Omega$, we have*

$$u(x) = \mathscr{A}_r^{(s)} u(x) = A_r^{(s)} \star u(x).$$

PROOF. We assume without restriction that $x = 0$, with $B_r \subset \overline{B}_r \subset \Omega$. Let $u \in \mathscr{L}_s(\mathbb{R}^n) \cap C^{2s+\varepsilon}_{loc}$ for some $\varepsilon > 0$ (not necessarily a solution of $(-\Delta)^s u = 0$ in Ω), and consider the function v defined by

$$v(z) = \begin{cases} \int_{\mathbb{R}^n \setminus B_r} P_r^{(s)}(z,y)u(y)dy, & z \in B_r, \\ u(z), & z \in \mathbb{R}^n \setminus B_r. \end{cases} \tag{15.13}$$

Then, by Theorem 15.2 v is the unique solution to the Dirichlet problem for the ball B_r

$$\begin{cases} (-\Delta)^s v = 0 & \text{in } B_r, \\ v = u & \text{in } \mathbb{R}^n \setminus B_r. \end{cases} \tag{15.14}$$

Consider the function $w = u - v$. It solves the problem

$$\begin{cases} (-\Delta)^s w = (-\Delta)^s u & \text{in } B_r, \\ w = 0 & \text{in } \mathbb{R}^n \setminus B_r. \end{cases} \tag{15.15}$$

Then, (see e.g. Theorem 3.2 in [**Bu16**]) the function w is represented by the formula

$$w(z) = \int_{B_r} G_r^{(s)}(z,y)(-\Delta)^s u(y)dy,$$

where for every $z \in B_r$ the function

$$G_r^{(s)}(z,y) = E_s(z,y) - h_z^{(s)}(y) > 0$$

is the Green function for $(-\Delta)^s$ for the ball B_r, see [**R38**], the appendix in [**La72**], and also [**Bu16**]. We notice explicitly that the function $h_z^{(s)}$ represents the solution to the Dirichlet problem

$$\begin{cases} (-\Delta)^s h_z^{(s)} = 0 & \text{in } B_r, \\ h_z^{(s)} = E_s(z,\cdot) & \text{in } \mathbb{R}^n \setminus B_r, \end{cases} \tag{15.16}$$

and therefore it is given by

$$h_z^{(s)}(y) = \begin{cases} \int_{\mathbb{R}^n\setminus B_r} P_r^{(s)}(y,\xi)E_s(z,\xi)d\xi & y \in B_r, \\ E_s(z,y) & y \in \mathbb{R}^n \setminus B_r. \end{cases}$$

We conclude that for every $z \in B_r$ the value of u at z can be written as

$$u(z) = v(z) + w(z).$$

In particular, this gives

$$u(0) = v(0) + w(0) = A_r^{(s)} \star u(0) + \int_{B_r} G_r^{(s)}(0,y)(-\Delta)^s u(y)dy.$$

It is clear from this formula that if $(-\Delta)^s u = 0$ in Ω, then $u(x) = A_r^{(s)} \star u(x)$, provided that $r > 0$ is such that $\overline{B}_r \subset \Omega$.

□

16. The heat semigroup $P_t^{(s)} = e^{t(-\Delta)^s}$

As it is well-known the heat operator $H = \partial_t - \Delta$ plays a fundamental role in almost all areas of mathematics. Since the focus of this note is the fractional Laplacean, it is only natural that we also discuss the nonlocal heat operator $\partial_t + (-\Delta)^s$ and the semigroup associated with it. The present section, as well as the following five ones are devoted to analyze some of the most elementary properties of such semigroup.

A first comment is that this is not such a simple task since, unlike what happens for the classical heat equation, the relevant heat kernel is only explicitly known on the Fourier transform side, a fact that rules out the possibility of explicit elegant and simple computations which characterize the analysis of the classical case. With this state of affairs, even a fundamental fact such as the positivity of the heat kernel is not a priori obvious.

Before we begin our discussion however, we mention that the first pioneering results about the nonlocal heat equation $\partial_t + (-\Delta)^s$ go back to the work of Bochner, see [**B55**] and [**Y78**]. More recently, several authors have analyzed properties such as estimates of the heat kernel, the weak Harnack inequality and the Hölder continuity of solutions, interior and at the boundary, for more general nonlocal parabolic equations, results of Fujita type, a Widder type theorem, Nash-type inequalities, eigenvalue estimates, nonlocal porous medium equation etc., see e.g. [**Ko95**], [**BLW05**], [**BJ07**], [**BM07**], [**BGR10**], [**CKK11**], [**DQRV12**], [**FK13**], [**BP18**], [**BPSV14**], [**AMPP16**], [**BSV17**], [**Fr17**], [**GI17**], [**V17**], [**V17'**] but this list is by no means exhaustive. The regularity theory for very general nonlocal parabolic operators has been developed in the paper [**CCV11**]. Further regularity properties of solutions have been extensively studied in [**CR13**], where the authors study fractional nonlinear parabolic equations, and in [**CF13**] in the context of the nonlocal obstacle problem

$$\begin{cases} \min\{u-\psi, -u_t + (-\Delta)^s u\} = 0, & \text{in } \mathbb{R}^n \times [0,T], \\ u(T) = \psi, \end{cases}$$

where the function ψ represents the obstacle.

We begin by discussing the Cauchy problem: given $0 < s < 1$ and a function $\varphi \in \mathscr{S}(\mathbb{R}^n)$, find a solution to the problem

$$\begin{cases} H_s u = \frac{\partial u}{\partial t} + (-\Delta)^s u = 0 & \text{in } \mathbb{R}^{n+1}_+, \\ \\ u(x,0) = \varphi(x), & x \in \mathbb{R}^n. \end{cases} \tag{16.1}$$

An observation that follows immediately from the definition of H_s and from (2.10) is that the natural scaling for the fractional heat operator is given by the nonisotropic dilations

$$\tilde{\delta}_\lambda(x,t) = (\lambda x, \lambda^{2s} t). \tag{16.2}$$

By this we mean that for every function $u(x,t)$ we have

$$H_s(\tilde{\delta}_\lambda u)(x,t) = \lambda^{2s} \tilde{\delta}_\lambda (H_s u)(x,t). \tag{16.3}$$

Thus, H_s is an operator of fractional "order" $2s$ with respect to the dilations (16.2).

As in the case when $s = 1$, we can solve (16.1) by formally taking a partial Fourier transform with respect to the space variable x. If we let

$$\hat{u}(\xi,t) = \int_{\mathbb{R}^n} e^{-2\pi i <\xi,x>} u(x,t) dx,$$

then (5.2) in Proposition 5.1 gives

$$\begin{cases} \frac{\partial \hat{u}}{\partial t}(\xi,t) + (2\pi|\xi|)^{2s} \hat{u}(\xi,t) = 0 & \text{in } \mathbb{R}^{n+1}_+, \\ \\ \hat{u}(\xi,0) = \hat{\varphi}(\xi), & \xi \in \mathbb{R}^n. \end{cases} \tag{16.4}$$

For every $\xi \in \mathbb{R}^n$ the solution to (16.4) is given by

$$\hat{u}(\xi,t) = \hat{\varphi}(\xi) e^{-t(2\pi|\xi|)^{2s}}.$$

We now define a function $G_s(x,t)$ by the equation

$$\mathscr{F}_{x\to\xi}(G_s(\cdot,t)) = e^{-t(2\pi|\xi|)^{2s}}, \tag{16.5}$$

or, equivalently,

$$G^{(s)}(x,t) = \int_{\mathbb{R}^n} e^{-2\pi i <x,\xi>} e^{-t(2\pi|\xi|)^{2s}} d\xi. \tag{16.6}$$

With this definition it is clear that a solution to (16.4) is given by the formula

$$P_t^{(s)} \varphi(x) \overset{def}{=} u(x,t) = \int_{\mathbb{R}^n} G^{(s)}(x-y,t) \varphi(y) dy. \tag{16.7}$$

We observe that since in view of (16.5) the function $\mathscr{F}_{x\to\xi}(G^{(s)}(\cdot,t))$ decays rapidly, by (4.20) above we conclude that $G^{(s)}(\cdot,t) \in C^\infty(\mathbb{R}^n)$ (this in fact could also be proved directly from (16.6)).

DEFINITION 16.1. The *fractional heat semigroup* $P_t^{(s)}$ is the operator defined by (16.7). When $s = 1$ we indicate with P_t the standard heat semigroup defined by $P_t u(x) = G(\cdot, t) \star u(x)$, where $G(x,t) = (4\pi t)^{-\frac{n}{2}} e^{-\frac{|x|^2}{4t}}$.

Using the Fourier transform it is immediate to verify that $P_t^{(s)}$ is in fact a semigroup, i.e., for every $t, \tau > 0$ the following property holds

$$P_{t+\tau}^{(s)} = P_t^{(s)} \circ P_\tau^{(s)}. \tag{16.8}$$

One immediate important property of $G^{(s)}$ is the following scale invariance

$$G^{(s)}(x,t) = t^{-\frac{n}{2s}} G^{(s)}(\frac{x}{t^{1/2s}}, 1). \tag{16.9}$$

This can be verified by writing (16.6) as follows

$$G^{(s)}(x,t) = \mathscr{F}(\delta_{t^{1/2s}} e^{-(2\pi|\cdot|)^{2s}})(x) = t^{-\frac{n}{2s}} \mathscr{F}(e^{-(2\pi|\cdot|)^{2s}})(\frac{x}{t^{1/2s}}),$$

where we have used (4.16) above. Therefore, if we set

$$\Phi_s(x) = \mathscr{F}(e^{-(2\pi|\cdot|)^{2s}})(x) = G^{(s)}(x,1), \tag{16.10}$$

then (16.9) can be recast in the following self-similar expression

$$G^{(s)}(x,t) = t^{-\frac{n}{2s}} \Phi_s(\frac{x}{t^{1/2s}}). \tag{16.11}$$

Notice that (16.11) implies that $G^{(s)}$ is homogeneous of degree $-n$ with respect to the dilations (16.2), i.e.,

$$G^{(s)}(\lambda x, \lambda^{2s} t) = \lambda^{-n} G^{(s)}(x,t). \tag{16.12}$$

If we denote by Z_s the infinitesimal generator of the dilations (16.2), we thus have

$$Z_s G^{(s)} =< x, \nabla_x G^{(s)} > +2st\frac{\partial G^{(s)}}{\partial t} = -nG^{(s)}.$$

Equivalently, if we introduce the *nonlocal entropy* $\log G^{(s)}$, then we have

$$\frac{\partial(\log G^{(s)})}{\partial t} = -\frac{n}{2s}\frac{1}{t} - \frac{1}{2s} < \frac{x}{t}, \nabla_x(\log G^{(s)}) > . \tag{16.13}$$

The equation (16.13) is a first form of *nonlocal Li-Yau inequality* for solutions of (16.1), and we emphasize that it has been deduced exclusively from the scaling properties of the kernel $G^{(s)}$. We will return to it in Section 21.

Before proceeding we note that, using Theorem 4.4, for $x \neq 0$ we obtain from (16.10)

$$\begin{aligned} \Phi_s(x) &= \frac{2\pi}{|x|^{\frac{n}{2}-1}} \int_0^\infty e^{-(2\pi r)^{2s}} r^{\frac{n}{2}} J_{\frac{n}{2}-1}(2\pi|x|r)dr \\ &= \frac{(2\pi)^{-\frac{n}{2}}}{|x|^n} \int_0^\infty e^{-(\frac{u}{|x|})^{2s}} u^{\frac{n}{2}} J_{\frac{n}{2}-1}(u)du. \end{aligned} \tag{16.14}$$

The equation (5.3.5) on p. 103 in [**Le72**], gives

$$\frac{d}{du}\left[u^{\frac{n}{2}} J_{\frac{n}{2}}(u)\right] = u^{\frac{n}{2}} J_{\frac{n}{2}-1}(u).$$

Substituting this information in (16.14), and integrating by parts (using (4.27) and (4.29)), we find

$$\Phi_s(x) = \frac{2s(2\pi)^{-\frac{n}{2}}}{|x|^{n+2s}} \int_0^\infty e^{-(\frac{u}{|x|})^{2s}} u^{\frac{n}{2}+2s-1} J_{\frac{n}{2}}(u)du. \tag{16.15}$$

When $s = 1/2$, the underlying process is a Poisson process and the integral in the right-hand side of (16.15) can be computed explicitly, see Proposition 16.4 below. However, when $0 < s < 1$ and $s \neq 1/2$ the analysis of such integral is more delicate, see Theorem 16.6 below and the comments that follow.

When $s = 1$ one basic property of the classical heat semigroup intimately connected to the maximum principle is the strict positivity of its kernel $G(x,t) = (4\pi t)^{-\frac{n}{2}} e^{-\frac{|x|^2}{4t}}$ which is of course a trivial consequence of its explicit expression. Since there exists no analogue of such explicit formula for the kernel $G_s(x,t)$, its positivity is far from obvious. The next result establishes this fact. It was certainly obtained by Bochner based on the positivity of the subordination function, see Proposition 2 on p. 261 in [**Y78**] and Theorem 18.3 below, but it was perhaps known earlier to Paul Levy. We have adapted the beautiful proof that follows, which uses a typical Abelian-Tauberian argument, from the one-dimensional presentation in [**DGV03**].

PROPOSITION 16.2. *For every $0 < s < 1$ and for every $(x,t) \in \mathbb{R}^{n+1}_+$, we have*

$$G^{(s)}(x,t) \geq 0.$$

PROOF. In view of (16.9) it suffices to show that for one $T > 0$

$$G^{(s)}(x,T) \geq 0, \tag{16.16}$$

for every $x \in \mathbb{R}^n$. If this holds, in fact, then for every $t > 0$ and every $x \in \mathbb{R}^n$ we have

$$G^{(s)}(x,t) = \left(\frac{T}{t}\right)^{\frac{n}{2s}} G^{(s)}\left(\left(\frac{T}{t}\right)^{1/2s} x, T\right) \geq 0,$$

and we are done. To prove (16.16) consider the spherically symmetric function in $L^1(\mathbb{R}^n)$

$$f = \frac{A}{|\cdot|^{n+2s}} \mathbf{1}_{B(0,1)^c},$$

where the constant $A > 0$ is chosen so that $\hat{f}(0) = \int_{\mathbb{R}^n} f(x)dx = 1$. Using such normalization, we obtain for every $\xi \neq 0$

$$\begin{aligned}
\hat{f}(\xi) &= 1 + \hat{f}(\xi) - \int_{\mathbb{R}^n} f(x)dx \\
&= 1 - \int_{\mathbb{R}^n} [1 - e^{-2\pi i <\xi,x>}] f(x)dx = 1 - A \int_{|x|\geq 1} \frac{1-\cos(2\pi <\xi,x>)}{|x|^{n+2s}} dx \\
&= 1 - A|\xi|^{2s} \int_{|y|\geq|\xi|} \frac{1-\cos(2\pi <\xi/|\xi|, y>)}{|y|^{n+2s}} dy \\
&= 1 - A|\xi|^{2s} \int_{|y|\geq|\xi|} \frac{1-\cos(2\pi y_n)}{|y|^{n+2s}} dy,
\end{aligned}$$

where we have first changed the variable to $y = |\xi|x$, and then used the invariance of the function

$$\xi \longrightarrow \int_{|x|\geq|\xi|} \frac{1-\cos(2\pi < \xi/|\xi|, x >)}{|x|^{n+2s}} dx$$

with respect to orthogonal transformations in $\mathbb{R}^n$. If we denote by

$$\Psi(\xi) = \int_{|x|\geq|\xi|} \frac{1-\cos(2\pi x_n)}{|x|^{n+2s}} dx,$$

then by Lebesgue dominated convergence we see that $\Psi(\xi) \to \alpha > 0$ as $\xi \to 0$, where

$$\alpha = \int_{\mathbb{R}^n} \frac{1-\cos(2\pi x_n)}{|x|^{n+2s}} dx = (2\pi)^{2s} \int_{\mathbb{R}^n} \frac{1-\cos(z_n)}{|z|^{n+2s}} dz = \frac{(2\pi)^{2s}}{\gamma(n,s)},$$

with $\gamma(n,s)$ as in the proof of (5.1) in Proposition 5.1. It follows that we can write

$$\hat{f}(\xi) = 1 - A\alpha|\xi|^{2s}(1+\omega(\xi)),$$

where $\omega(\xi) = O(|\xi|^{2(1-s)}) = o(1)$ as $\xi \to 0$. We will prove that (16.16) holds with

$$T = \frac{A}{\gamma(n,s)}.$$

This will complete the proof.

With this objective in mind, for every positive integer k consider the function

$$f_k(x) = k^{\frac{n}{2s}}(f \star \ldots \star f)(k^{1/2s}x),$$

where the convolution is repeated k times. By (4.16) we have

$$\hat{f}_k(\xi) = \mathscr{F}(f\star\ldots\star f)(k^{-1/2s}\xi) = (\hat{f}(k^{-1/2s}\xi))^k = \left(1 - \frac{A\alpha|\xi|^{2s}(1+\omega(k^{-1/2s}\xi))}{k}\right)^k.$$

This shows the crucial fact that for every $\xi \in \mathbb{R}^n$

$$\lim_{k\to\infty} \hat{f}_k(\xi) = e^{-A\alpha|\xi|^{2s}}. \tag{16.17}$$

Since for every $k \in \mathbb{N}$ we have

$$||\hat{f}_k||_{L^\infty(\mathbb{R}^n)} \leq ||f_k||_{L^1(\mathbb{R}^n)} \leq ||f||^k_{L^1(\mathbb{R}^n)} = 1,$$

we conclude that (16.17) also holds in $\mathscr{S}'(\mathbb{R}^n)$. We thus have for every $\varphi \in \mathscr{S}(\mathbb{R}^n)$

$$< f_k, \hat{\varphi} > = < \hat{f}_k, \varphi > \longrightarrow < e^{-A\alpha|\cdot|^{2s}}, \varphi > = < \mathscr{F}(e^{-A\alpha|\cdot|^{2s}}), \hat{\varphi} > .$$

This is equivalent to saying that in $\mathscr{S}'(\mathbb{R}^n)$

$$\lim_{k\to\infty} f_k = \mathscr{F}(e^{-A\alpha|\cdot|^{2s}}). \tag{16.18}$$

Since $f_k \geq 0$ for every $k \in \mathbb{N}$, we conclude from (16.5) that $\mathscr{F}(e^{-A\alpha|\cdot|^{2s}})(x) = G^{(s)}(x,T) \geq 0$ for every $x \in \mathbb{R}^n$, with $T = \alpha A(2\pi)^{-2s} = \frac{A}{\gamma(n,s)}$. This proves (16.16).

□

PROPOSITION 16.3. *For every $0 < s < 1$ and for every $(x,t) \in \mathbb{R}^{n+1}_+$, we have*

$$G^{(s)}(x,t) > 0.$$

PROOF. From Proposition 16.2 we know that $G^{(s)}(x,t) \geq 0$ globally. Then, its strict positivity follows from the strong maximum principle, or the weak Harnack inequality, see Theorem 1.1 in [**FK13**].

□

The case $s = 1/2$ has a special interest, and since it is quite surprising we state it in a proposition.

PROPOSITION 16.4. *When $s = 1/2$ the heat kernel $G^{(1/2)}(x,t)$ for $\partial_t + (-\Delta)^{1/2}$ is given by the Poisson kernel for the Laplacean for the half-space $\mathbb{R}^n \times \mathbb{R}_+$, i.e.,*

$$G^{(1/2)}(x,t) = P(x,t) = \frac{\Gamma(\frac{n+1}{2})}{\pi^{\frac{n+1}{2}}} \frac{t}{(t^2+|x|^2)^{\frac{n+1}{2}}}. \tag{16.19}$$

PROOF. The equation (16.19) follows from the well-known formula for the Poisson kernel, see e.g. Proposition 5, Sec. 2 in Chap. 3 of [**St70**],

$$\mathscr{F}_{x\to\xi}(e^{-2\pi t|\cdot|}) = P_{1/2}(x,t),$$

see also Remark 10.3 above.

□

In particular, (16.19) says that the decay of the heat kernel for $\partial_t + (-\Delta)^{1/2}$ is not exponential, but polynomial. The next result states that such behavior is shared by all nonlocal semigroups $P_t^{(s)}$, $0 < s < 1$.

PROPOSITION 16.5. *For every $0 < s < 1$ let $\Phi_s(x)$ be as in* (16.10). *Then, there exists a constant $\beta(n,s) > 0$ such that for every $x \in \mathbb{R}^n$*

$$\frac{\beta(n,s)}{1+|x|^{n+2s}} \leq \Phi_s(x) \leq \frac{\beta^{-1}(n,s)}{1+|x|^{n+2s}}.$$

The proof of Proposition 16.5 follows immediately from the following result.

THEOREM 16.6. *For every $0 < s < 1$ one has*

$$\lim_{|x|\to\infty} |x|^{n+2s}\Phi_s(x) = \gamma(n,s) > 0,$$

where $\gamma(n,s)$ is the constant in (5.10) *in Proposition* 5.6.

We note that, using (16.15) one obtains

$$|x|^{n+2s}\Phi_s(x) = 2s(2\pi)^{-\frac{n}{2}} \int_0^\infty e^{-\left(\frac{u}{|x|}\right)^{2s}} u^{\frac{n}{2}+2s-1} J_{\frac{n}{2}}(u) du. \tag{16.20}$$

When $n = 1$ Theorem 16.6 was first proved by G. Pólya in [**Po96**]. It is not easy to find this reference, but a detailed presentation of Pólya's proof can be found in Lemma 3.4 in [**CS15**]. Pólya's argument was generalized to any dimension by Blumenthal and Getoor, see Theorem 2.1 in their paper [**BG60**]. However, the delicate part of the proof is not presented there and the authors refer to [**Po96**]. Integrals like that in the right-hand side of (16.20) are studied in [**PT69**]. A different proof of Theorem 16.6 based on the subordination formula (18.2) below was given by Bendikov in [**Be94**]. One should also see the paper [**Ko00**] by Kolokoltsov for the proof of a more general result. Several heat kernel estimates are contained in the recent paper [**BSV17**], see also [**V17**].

In the local case $s = 1$ a basic property of the heat kernel $G(x,t)$ is that for $n \neq 2$ and for every $x \neq 0$ one has

$$\int_0^\infty G(x,t)dt = \frac{|x|^{2-n}}{(n-2)\sigma_{n-1}}.$$

We recall that the right-hand side is the fundamental solution of $-\Delta$ with pole at $x = 0$. The next result expresses a similar property of $G^{(s)}(x,t)$.

PROPOSITION 16.7. *Let $n \geq 2$. Then, for every $x \neq 0$ one has*

$$\int_0^\infty G^{(s)}(x,t)dt = E_s(x), \tag{16.21}$$

where $E_s(x)$ is the fundamental solution of $(-\Delta)^s$ with singularity at $x = 0$ in Theorem 8.4.

There is more than one way of proving Proposition 16.7. A quick one resorts to the following classical result for which we refer the reader to chapter 5 in [**St70**].

THEOREM 16.8. *For any $0 < \alpha < n$ we have in $\mathscr{S}'(\mathbb{R}^n)$*

$$\mathscr{F}(|\cdot|^{-\alpha}) = \pi^{\alpha-\frac{n}{2}} \frac{\Gamma\left(\frac{n-\alpha}{2}\right)}{\Gamma\left(\frac{\alpha}{2}\right)} |\cdot|^{\alpha-n}.$$

PROOF OF PROPOSITION 16.7. We proceed formally since, similarly to Theorem 16.8, for a rigorous proof we should verify the following steps in $\mathscr{S}'(\mathbb{R}^n)$, and not in the pointwise sense. The interested reader can easily provide the missing details. We notice that, in view of (8.3) and (8.4), proving (16.21) is equivalent to showing

$$\int_0^\infty \mathscr{F}(G^{(s)})(x,t)dt = \hat{E}_s(x) = \frac{\Gamma(\frac{n}{2}-s)}{2^{2s}\pi^{\frac{n}{2}}\Gamma(s)}\mathscr{F}(|\cdot|^{2s-n})(x).$$

In light of (16.5) and Theorem 16.8 (which we can apply with $\alpha = 2s$ since the hypothesis $0 < s < 1$ and $n \geq 2$ automatically guarantee that $0 < \alpha < n$), this is in turn equivalent to showing that

$$\int_0^\infty e^{-t(2\pi|x|)^{2s}}dt = \pi^{\frac{n}{2}-2s}\frac{\Gamma(s)}{\Gamma\left(\frac{n-2s}{2}\right)}\frac{\Gamma(\frac{n}{2}-s)}{2^{2s}\pi^{\frac{n}{2}}\Gamma(s)}(2\pi)^{2s}(2\pi|x|)^{-2s} = (2\pi|x|)^{-2s}.$$

That the integral in the left-hand side is equal to the right-hand side of the above chain of equalities follows from a standard change of variable in the integral.

□

PROPOSITION 16.9. *For every $t > 0$ we have*

$$P_t^{(s)}1(x) = \int_{\mathbb{R}^n} G^{(s)}(x,t)dx = 1.$$

Thus the semigroup $P_t^{(s)}$ is stochastically complete.

PROOF. We have from (16.11)

$$\int_{\mathbb{R}^n} G^{(s)}(x,t)dx = \int_{\mathbb{R}^n} \Phi_s(x)dx = \mathscr{F}^{-1}(\Phi_s)(0).$$

Keeping in mind that $\Phi_s(\xi) = \mathscr{F}(e^{-(2\pi|\cdot|)^{2s}})(\xi)$, we see that $\mathscr{F}^{-1}(\Phi_s)(\xi) = e^{-(2\pi|\xi|)^{2s}}$, and thus $\mathscr{F}^{-1}(\Phi_s)(0) = 1$.

□

Combining Propositions 16.2 and 16.9 we obtain the following maximum principle.

PROPOSITION 16.10. *Let $\varphi \in \mathscr{S}(\mathbb{R}^n)$. Then, for every $x \in \mathbb{R}^n$ and $t > 0$ one has*

$$\inf_{\mathbb{R}^n} \varphi \le P_t^{(s)}\varphi(x) \le \sup_{\mathbb{R}^n} \varphi.$$

17. Bochner's subordination: from P_t to $(-\Delta)^s$

In this section we take a momentary pause from the previous one, to discuss the important fact that, using the standard heat semigroup P_t, we can recover $(-\Delta)^s$. This result is based on Bochner's principle of subordination (for this see Chapter 4 in [**B55**]) and the outcome of it is yet another expression of the fractional Laplacean, see formula (17.1) in Theorem 17.2 below, that is alternative to the ones that we know so far, namely (2.8), (2.11), (10.9) in Theorem 10.1 and (15.11) in Proposition 15.6. We will present two proofs of such result. We begin with a preliminary observation that connects the heat semigroup $P_t = e^{-t\Delta}$ to the spherical mean-value operator $\mathscr{M}_r u(x)$.

LEMMA 17.1. *Let $u \in \mathscr{S}(\mathbb{R}^n)$. Then, for every $0 < t < \infty$ one has*

$$P_t u(x) - u(x) = \sigma_{n-1}(4\pi t)^{-\frac{n}{2}} \int_0^\infty e^{-\frac{r^2}{4t}} r^{n-1}[\mathscr{M}_r u(x) - u(x)]dr.$$

PROOF. This is a simple consequence of Cavalieri's principle, the spherical symmetry of $G(x,t)$, and of the fact that $P_t 1 = 1$. We have

$$\begin{aligned} P_t u(x) - u(x) &= \int_{\mathbb{R}^n} G(x-y,t)[u(y) - u(x)]dy \\ &= \int_0^\infty \int_{S(x,r)} G(x-y,t)[u(y) - u(x)]d\sigma(y)dr \\ &= (4\pi t)^{-\frac{n}{2}} \int_0^\infty e^{-\frac{r^2}{4t}} \int_{S(x,r)} [u(y) - u(x)]d\sigma(y)dr \\ &= \sigma_{n-1}(4\pi t)^{-\frac{n}{2}} \int_0^\infty e^{-\frac{r^2}{4t}} r^{n-1}[\mathscr{M}_r u(x) - u(x)]dr. \end{aligned}$$

□

We are now in a position to establish the main result about Bochner's subordination for the nonlocal operator $(-\Delta)^s$.

THEOREM 17.2. *Let $0 < s < 1$. For any $u \in \mathrm{Dom}(-\Delta)$, hence in particular, for any $u \in \mathscr{S}(\mathbb{R}^n)$, one has*

$$(17.1) \qquad \begin{aligned} (-\Delta)^s u(x) &= \frac{1}{\Gamma(-s)} \int_0^\infty t^{-s-1} \left[P_t u(x) - u(x)\right] dt \\ &= -\frac{s}{\Gamma(1-s)} \int_0^\infty t^{-s-1} \left[P_t u(x) - u(x)\right] dt. \end{aligned}$$

FIRST PROOF. Let $\alpha > 0$. Using Lemma 17.1, we find

$$\begin{aligned}\int_0^\infty & t^{-\alpha-1}[P_t u(x) - u(x)]dt \\ &= \sigma_{n-1}(4\pi)^{-\frac{n}{2}} \int_0^\infty t^{-\alpha-\frac{n}{2}} \int_0^\infty e^{-\frac{r^2}{4t}} r^{n-1}[\mathscr{M}_r u(x) - u(x)]dr \frac{dt}{t} \\ &= \sigma_{n-1}(4\pi)^{-\frac{n}{2}} \int_0^\infty \left(\int_0^\infty t^{-\alpha-\frac{n}{2}} e^{-\frac{r^2}{4t}} \frac{dt}{t} \right) r^{n-1}[\mathscr{M}_r u(x) - u(x)]dr,\end{aligned}$$

assuming that we can exchange the order of integration. Now,

$$\int_0^\infty t^{-\alpha-\frac{n}{2}} e^{-\frac{r^2}{4t}} \frac{dt}{t} = 2^{2\alpha+n}\Gamma(\frac{n}{2}+\alpha) r^{-2\alpha-n}.$$

Substituting in the above formula we find

$$\int_0^\infty t^{-\alpha-1}[P_t u(x) - u(x)]dt = \sigma_{n-1}\pi^{-\frac{n}{2}} 2^{2\alpha}\Gamma(\frac{n}{2}+\alpha) \int_0^\infty r^{-2\alpha-1}[\mathscr{M}_r u(x) - u(x)]dr.$$

Comparing the right-hand side with that of the equation in Proposition 2.11, it is now clear that, in order for the former to provide a multiple of $(-\Delta)^s u(x)$ we must have $\alpha = s$. With such choice we obtain

$$\begin{aligned}- \int_0^\infty & t^{-s-1}[P_t u(x) - u(x)]dt \\ &= -\sigma_{n-1}\pi^{-\frac{n}{2}} 2^{2s}\Gamma(\frac{n}{2}+s) \int_0^\infty r^{-2s-1}[\mathscr{M}_r u(x) - u(x)]dr.\end{aligned}$$

Since by Proposition 2.11 and Proposition 5.6 we find

$$- \int_{\mathbb{R}^n} r^{-1-2s}[\mathscr{M}_r u(x) - u(x)]dr = \frac{(-\Delta)^s u(x)}{\sigma_{n-1}\gamma(n,s)} = \frac{(-\Delta)^s u(x)}{\sigma_{n-1}} \frac{\pi^{\frac{n}{2}}\Gamma(1-s)}{s 2^{2s}\Gamma(\frac{n}{2}+s)},$$

the desired conclusion (17.1) follows.

□

SECOND PROOF. The proof that the right-hand side of (17.1) is in fact equal to $(-\Delta)^s u(x)$ can also be accomplished using the Fourier transform. Thanks to (5.2) in Proposition 5.1, we see that proving (17.1) is equivalent to showing

$$(2\pi|\xi|)^{2s}\hat{u}(\xi) = -\frac{s}{\Gamma(1-s)} \int_0^\infty t^{-s-1} \left(\widehat{P_t u}(\xi) - \hat{u}(\xi) \right) dt.$$

Since $\widehat{P_t u}(\xi) = \mathscr{F}(G(\cdot,t) \star u)(\xi) = \hat{G}(\xi,t)\hat{u}(\xi) = e^{-t(2\pi|\xi|)^2}\hat{u}(\xi)$, we thus see that (17.1) is equivalent to

$$(2\pi|\xi|)^{2s}\hat{u}(\xi) = \frac{s}{\Gamma(1-s)} \int_0^\infty t^{-s-1} \left(1 - e^{-t(2\pi|\xi|)^2}\right) dt\ \hat{u}(\xi).$$

This identity holds if and only if it is true that

$$\begin{aligned}(2\pi|\xi|)^{2s} &= \frac{s}{\Gamma(1-s)} \int_0^\infty t^{-s} \left(1 - e^{-t(2\pi|\xi|)^2}\right) \frac{dt}{t} \\ &= (2\pi|\xi|)^{2s} \frac{s}{\Gamma(1-s)} \int_0^\infty u^{-s-1}\left(1 - e^{-u}\right) du.\end{aligned}$$

The validity of this equation immediately follows from (4.8) above. This completes the proof.

□

Theorem 17.2 extends to a very general framework, and covers situations in which the Euclidean $-\Delta$ is replaced by an operator $-L$ that neither necessarily generates a semigroup, nor its domain is necessarily dense in the relevant Banach space. This result is due to A. V. Balakrishnan, see formula (2.1) in [**Bal60**], or also (4) and (5) p. 260 in [**Y78**].

18. More subordination: from P_t to $P_t^{(s)}$

"*The 'stable laws' $\{e^{-t|\alpha|^{2p}}\}$, $0<p<1$, are each subordinate to the Gaussian law $\{e^{-t|\alpha|^2}\}$, and quite generally if $\{e^{-t\psi(\alpha)}\}$ is any subdivisible process then so is $\{e^{-t\psi(\alpha)^p}\}$ for any $0<p<1$*".

This quote is from p. 93 in [**B55**]. This section is devoted to further illustrating Bochner's beautiful subordination idea with a twofold purpose. On one hand, it leads to an explicit representation of the nonlocal semigroup $P_t^{(s)}$ in terms of the standard heat semigroup, see Theorem 18.3 below. On the other hand, as we have already mentioned in the closing of the previous section, the principle of subordination allows for far-reaching generalizations. These developments were already envisioned by Bochner himself, when he said on p. 95 of [**B55**]:

"*Now, our 'subordination' can also be introduced on spaces in general provided we shift the emphasis from Fourier transformation (which may not even be definable) to (generalizations of) the distributions $F(u;A)$...*"

In what follows we will discuss material from [**B55**], [**P52**], [**Bal60**] and p. 259-268 in [**Y78**]. We begin with a definition.

DEFINITION 18.1 (Bochner's subordinator). Let $0<s<1$. For every fixed $t>0$ we introduce the *subordinator* function as the inverse Laplace transform of $z\to e^{-tz^s}$, $\Re z>0$,

$$f_s(t;\tau)=\begin{cases}\frac{1}{2\pi i}\int_{\varepsilon-i\infty}^{\varepsilon+i\infty} e^{z\tau-tz^s}dz & \tau\geq 0,\\ 0 & \tau<0.\end{cases}\tag{18.1}$$

In (18.1) the parameter $\varepsilon>0$ is fixed and z^s denotes the branch such that $\Re(z^s)>0$ when $\Re z>0$. In this way, z^s is a one-valued holomorphic function in the z-plane cut along the negative real axis. It is clear that, thanks to Cauchy's integral formula, the value of the integral is independent of $\varepsilon>0$. If $\tau>0$ we have

$$f_s(t;\tau)=\frac{e^{\varepsilon\tau}}{2\pi}\int_{-\infty}^{\infty} e^{iy\tau}e^{-t(\varepsilon+iy)^s}dy.$$

Thus, the convergence of the integral in (18.1) is guaranteed by the decay of the factor e^{-tz^s}, and we have $f_s(t;\tau)\cong e^{\varepsilon\tau}$. The following simple, yet crucial formula, is key to the subordination principle. It expresses the fact that the Laplace transform of $f_s(t;\cdot)$ is the function $z\to e^{-tz^s}$.

LEMMA 18.2. *For every $t>0$ and $a>0$ one has*

$$e^{-ta^s}=\int_0^\infty f_s(t;\tau)e^{-\tau a}d\tau.$$

PROOF. Consider the function

$$g(z)=\frac{e^{-tz^s}}{z-a}.$$

It is a holomorphic function in $\{z \in \mathbb{C} \mid \Re z > 0, z \neq a\}$, with a simple pole in $z = a$. Its residue is given by $\operatorname{Res}(g, a) = e^{-ta^s}$. Having observed this, for any $a > 0$ consider the line $\varepsilon + iy$, with $0 < \varepsilon < a$. We have from (18.1), after an exchange of the order of integration,

$$\int_0^\infty e^{-\tau a} f_s(t;\tau)d\tau = \frac{1}{2\pi i}\int_{\varepsilon - i\infty}^{\varepsilon + i\infty} e^{-tz^s}\int_0^\infty e^{-(a-z)\tau}d\tau dz$$
$$= \frac{1}{2\pi i}\int_{\varepsilon - i\infty}^{\varepsilon + i\infty} e^{-tz^s}\left[\frac{e^{-(a-z)\tau}}{z-a}\right]_0^\infty dz$$
$$= -\frac{1}{2\pi i}\int_{\varepsilon - i\infty}^{\varepsilon + i\infty} \frac{e^{-tz^s}}{z-a}dz = -\frac{1}{2\pi i}\int_{\varepsilon - i\infty}^{\varepsilon + i\infty} g(z)dz.$$

We now apply Cauchy's residue theorem to the function $g(z)$ and to the curve Γ_R, composed of a vertical piece $\varepsilon + iy$, with $|y| \leq R$ and by a half-circle C_R of points $z = \varepsilon + Re^{i\vartheta}$, for large R. From what observed above, we obtain

$$\int_{\Gamma_R} g(z)dz = 2\pi i e^{-ta^s}.$$

Since $\int_{C_R} g(z)dz \to 0$ as $R \to \infty$ we conclude that

$$-\frac{1}{2\pi i}\int_{\varepsilon - i\infty}^{\varepsilon + i\infty} g(z)dz = e^{-ta^s}.$$

This completes the proof. □

A remarkable consequence of Lemma 18.2 is the following result that connects the fractional semigroup $P_t^{(s)}$ to the standard heat semigroup P_t.

THEOREM 18.3. *Let $0 < s < 1$. Then, for every $x \in \mathbb{R}^n$ and $t > 0$ one has*

$$G^{(s)}(x,t) = \int_0^\infty f_s(t;\tau)G(x,\tau)d\tau. \tag{18.2}$$

As a consequence, for any $u \in \mathscr{S}(\mathbb{R}^n)$ one has

$$P_t^{(s)}u(x) = \int_0^\infty f_s(t;\tau)P_\tau u(x)d\tau. \tag{18.3}$$

PROOF. For every $t > 0$ and $\xi \in \mathbb{R}^n$ we have using Fubini's theorem

$$\mathscr{F}\left(\int_0^\infty f_s(t;\tau)P_\tau u(\cdot)d\tau\right)(\xi) = \int_0^\infty f_s(t;\tau)\mathscr{F}(P_\tau u(\cdot))(\xi)d\tau$$
$$= \int_0^\infty f_s(t;\tau)\mathscr{F}(G(\cdot,\tau)\star u)(\xi)d\tau = \int_0^\infty f_s(t;\tau)\mathscr{F}(G(\cdot,\tau))(\xi)\hat{u}(\xi)d\tau$$
$$= \hat{u}(\xi)\int_0^\infty f_s(t;\tau)e^{-\tau(2\pi|\xi|)^2}d\tau = \hat{u}(\xi)e^{-t(2\pi|\xi|)^{2s}}$$
$$= \hat{u}(\xi)\mathscr{F}(G^{(s)}(\cdot,t))(\xi) = \mathscr{F}(P_t^{(s)}u)(\xi),$$

where in the third to the last equality we have used Lemma 18.2 with $a = (2\pi|\xi|)^2$. This proves (18.3). The proof of (18.2) is done in a similar, but simpler, way. □

We can now use Theorem 18.3 to draw two basic properties of the subordination function $f_s(t;\tau)$. We only prove one of them, (18.4) below, and refer to Proposition 3 on p.262 in [**Y78**] for a proof of (18.5).

PROPOSITION 18.4. *For any fixed* $t, t' > 0$ *one has*

$$\int_0^\infty f_s(t;\tau)d\tau = 1, \tag{18.4}$$

and

$$f_s(t+t';\tau) = \int_0^\infty f_s(t;\tau-\sigma)f_s(t';\sigma)d\sigma. \tag{18.5}$$

PROOF. Using Proposition 16.9 and the identity (18.2), we find

$$\begin{aligned} 1 &= \int_{\mathbb{R}^n} G^{(s)}(x,t)dx = \int_{\mathbb{R}^n}\int_0^\infty f_s(t;\tau)G(x,\tau)d\tau dx \\ &= \int_0^\infty f_s(t;\tau)\int_{\mathbb{R}^n} G(x,\tau)dx d\tau = \int_0^\infty f_s(t;\tau)d\tau, \end{aligned}$$

which proves (18.4). □

19. A chain rule for $(-\Delta)^s$

In [**CC03**] the authors proved a basic pointwise inequality for the fractional Laplacean. Such inequality, which can be seen as a form of nonlocal chain-rule, plays a remarkable role in many problems from the applied sciences involving the nonlocal operator $(-\Delta)^s$, see for instance the beautiful papers [**CC04**] and [**CV10**], respectively on the two-dimensional quasi-geostrophic equation, and nonlinear evolution equations with fractional diffusion.

We will present two accounts of the chain rule, the former from [**CC03**], the latter from an interesting generalization given in [**CM15**]. Let us begin with a simple observation. If $u \in C^2(\mathbb{R}^n)$ and $\varphi \in C^2(\mathbb{R})$ the standard chain rule gives

$$\Delta\varphi(u) = \varphi''(u)|\nabla u|^2 + \varphi'(u)\Delta u.$$

If we assume that φ is also convex, then $\varphi'' \geq 0$, and we obtain in a trivial way

$$(-\Delta)\varphi(u) \leq \varphi'(u)(-\Delta)u.$$

The next result generalizes to the nonlocal setting this observation.

THEOREM 19.1 (Chain rule for $(-\Delta)^s$). *Let* $0 < s \leq 1$ *and* $\varphi \in C^1(\mathbb{R})$ *be a convex function. Then, for any* $u \in \mathscr{S}(\mathbb{R}^n)$ *one has*

$$(-\Delta)^s\varphi(u) \leq \varphi'(u(x))(-\Delta)^s u.$$

FIRST PROOF. Since the function $\varphi \in C^1(\mathbb{R})$ and is convex, we have for any $\tau, \sigma \in \mathbb{R}$

$$\varphi'(\sigma)(\tau-\sigma) \leq \phi(\tau) - \phi(\sigma).$$

This inequality easily gives for $u \in \mathscr{S}(\mathbb{R}^n)$ and for every $x, y \in \mathbb{R}^n$

$$2\varphi(u(x)) - \varphi(u(x+y)) - \varphi(u(x-y)) \leq \varphi'(u(x))(2u(x) - u(x+y) - u(x-y)).$$

Dividing the latter inequality by $|y|^{n+2s}$ and integrating in $y \in \mathbb{R}^n$ we immediately obtain the desired conclusion keeping in mind the definition (2.8) of $(-\Delta)^s$. □

SECOND PROOF. The second proof we present is taken from [**CM15**] and it has the advantage of carrying over to a situation where $\mathbb{R}^n$ is replaced by a compact n-dimensional manifold M, in which case the representation (2.8) is no longer available. Consider the Cauchy problems

$$\begin{cases} \frac{\partial U}{\partial t} + (-\Delta)^s U = 0 & \text{in } \mathbb{R}^{n+1}_+, \\ U(x,0) = u(x), & x \in \mathbb{R}^n. \end{cases} \tag{19.1}$$

and

$$\begin{cases} \frac{\partial V}{\partial t} + (-\Delta)^s V = 0 & \text{in } \mathbb{R}^{n+1}_+, \\ V(x,0) = \varphi(u)(x), & x \in \mathbb{R}^n. \end{cases} \tag{19.2}$$

Their solutions are respectively given by $U(x,t) = P_t^{(s)} u(x)$ and $V(x,t) = P_t^{(s)} \varphi(u)(x)$. Now, by Jensen inequality, Proposition 16.2 and 16.9, we obtain

$$\varphi(U(x,t)) = \varphi\left(\int_{\mathbb{R}^n} G^{(s)}(x-y,t)u(y)dy\right) \le \int_{\mathbb{R}^n} \varphi(u)(y)G^{(s)}(x-y,t)dy = V(x,t).$$

This shows that if, for a fixed $x \in \mathbb{R}^n$, we consider the function

$$\Psi(t) = V(x,t) - \varphi(U(x,t)),$$

then we have $\Psi(t) \ge 0$ for every $t > 0$. Since we clearly have

$$\Psi(0) = V(x,0) - \varphi(U(x,0)) = \varphi(u(x)) - \varphi(u(x)) = 0,$$

we conclude that it must be $\Psi'(0) \ge 0$.

We now have

$$\begin{aligned} \Psi'(t) &= \frac{\partial P_t \varphi(u)}{\partial t}(x) - \varphi'(P_t u(x)) \frac{\partial P_t u}{\partial t}(x) \\ &= -(-\Delta)^s P_t(\varphi(u))(x) + \varphi'(P_t u(x))(-\Delta)^s P_t u(x). \end{aligned}$$

This formula gives

$$0 \le \Psi'(0) = -(-\Delta)^s \varphi(u)(x) + \varphi'(u(x))(-\Delta)^s u(x),$$

which gives the desired conclusion.

□

20. The Gamma calculus for $(-\Delta)^s$

In the applications of pde's to geometry there is a remarkable tool that allows to connect the heat semigroup to the geometry of the underlying manifold. This tool is the so-called Bakry-Emery Gamma calculus, for which we refer the reader to the beautiful recent book [**BGL14**] and the references therein. At the heart of this calculus there is the so-called *curvature-dimension inequality* $\mathrm{CD}(\kappa, n)$ that we introduce in (20.11) and Definition 20.7 below. It is a remarkable fact that, on a n-dimensional Riemannian manifold $\mathbb{M}$, such inequality on functions is in fact equivalent to the lower bound $\mathrm{Ric} \ge \kappa$ on the Ricci tensor. More importantly, a stunning aspect of the gamma calculus is that the curvature-dimension inequality alone, in combination with properties of the heat semigroup, suffices to develop a wide program that connects the geometry of $\mathbb{M}$ to various global properties of the manifold itself, such as:

- global volume bounds for the geodesic balls,
- the Bonnet-Myers compactness theorem,
- global Poincaré inequalities,
- parabolic scale invariant Harnack inequalities,
- Gaussian upper and lower bounds,
- Liouville theorems
- De Giorgi-Nash-Moser estimates...and much more.

Traditionally, the majority of these fundamental results from Riemannian geometry rest on two pillars:

(i) the celebrated *Bochner identity*;
(ii) the *Laplacean comparison theorem.*

Whereas (i) is a purely pointwise identity on functions (see (20.7) below), (ii) relies on deeper aspects which are more genuinely Riemannian, such as the fact that the exponential map is locally a diffeomorphism and the theory of Jacobi fields. The gamma calculus allows to remove from the equation the Laplacean comparison theorem, and in a way it elevates the Bochner identity to a preeminent role in the development of the Li-Yau theory. This is especially important in situations where the above mentioned Riemannian tools are lacking. In this connection one should see the works [**BaG11**], [**BaG17**], [**BaG13**], [**BBG14**], [**BBGM14**].

Inspired by the above discussion, in this section we propose that a gamma calculus be developed in the context of nonlocal operators such as $(-\Delta)^s$. We stress that although this would have an interest in its own even in the setting of flat $\mathbb{R}^n$, a nonlocal gamma calculus would also be instrumental to considerable developments both in analysis and geometry. In what follows for the sake of simplicity we write $L = -(-\Delta)^s$, for a given $0 < s < 1$.

DEFINITION 20.1 (Nonlocal *carré du champ*). Given $u, v \in \mathscr{S}(\mathbb{R}^n)$ we define

$$\Gamma^{(s)}(u, v) = \frac{1}{2}[L(uv) - uLv - vLu]. \tag{20.1}$$

When $u = v$ in (20.1), we simply write

$$\Gamma^{(s)}(u) \overset{def}{=} \Gamma^{(s)}(u, u).$$

It is obvious that $\Gamma^{(s)}(u, v) = \Gamma^{(s)}(v, u)$. We have the following result.

LEMMA 20.2. *For $u, v \in \mathscr{S}(\mathbb{R}^n)$ one has*

$$\Gamma^{(s)}(u, v)(x) = \frac{\gamma(n, s)}{2} \int_{\mathbb{R}^n} \frac{(u(x) - u(y))(v(x) - v(y))}{|x - y|^{n+2s}} dy.$$

PROOF. It is easier to adopt the alternative expression (2.11) of $(-\Delta)^s$. Keeping in mind our sign convention for L and using (20.1) we find

$$\begin{aligned}
&\Gamma^{(s)}(u, v)\\
&= -\frac{\gamma(n, s)}{2} \mathrm{PV} \int_{\mathbb{R}^n} \frac{u(x)v(x) - u(y)v(y) - u(x)(v(x) - v(y)) - v(x)(u(x) - u(y))}{|x - y|^{n+2s}} dy\\
&= -\frac{\gamma(n, s)}{2} \mathrm{PV} \int_{\mathbb{R}^n} \frac{u(x)v(y) + v(x)u(y) - u(x)v(x) - u(y)v(y)}{|x - y|^{n+2s}} dy\\
&= \frac{\gamma(n, s)}{2} \int_{\mathbb{R}^n} \frac{(u(x) - u(y))(v(x) - v(y))}{|x - y|^{n+2s}} dy.
\end{aligned}$$

Notice that we have dropped the principal value sign in front of the last integral since, thanks to a cancellation that has occurred, the numerator in the last integral is $O(|x-y|^2)$ near x, so that the integrand is in fact locally in L^1.

□

When $u = v$ we obtain from Lemma 20.2 for every $x \in \mathbb{R}^n$

$$\Gamma^{(s)}(u)(x) = \frac{\gamma(n,s)}{2}\int_{\mathbb{R}^n} \frac{(u(x)-u(y))^2}{|x-y|^{n+2s}}dy \geq 0. \tag{20.2}$$

REMARK 20.3. We note that the positivity of $\Gamma^{(s)}(u)$ could have also been deduced immediately from its definition

$$-\Gamma^{(s)}(u) = \frac{1}{2}\left[(-\Delta)^s(u^2) - 2u(-\Delta)^s u\right], \tag{20.3}$$

and the chain rule in Theorem 19.1. The latter gives in fact

$$(-\Delta)^s(u^2) \leq 2u(-\Delta)^s u, \tag{20.4}$$

which shows $-\Gamma^{(s)}(u) \leq 0$.

DEFINITION 20.4 (Nonlocal energy). Given a function $u \in \mathscr{S}(\mathbb{R}^n)$ we define its $s-$*energy* as follows

$$\mathscr{E}_{(s)}(u) = \frac{1}{2}\int_{\mathbb{R}^n} \Gamma^{(s)}(u)(x)dx = \frac{\gamma(n,s)}{4}\int_{\mathbb{R}^n}\int_{\mathbb{R}^n} \frac{(u(x)-u(y))^2}{|x-y|^{n+2s}}dydx.$$

The reader should note that the energy $\mathscr{E}_{(s)}(u)$ is precisely the one that enters in the definition of the fractional Sobolev space

$$W^{s,2}(\mathbb{R}^n) = \{u \in L^2(\mathbb{R}^n) \mid \mathscr{E}_{(s)}(u) < \infty\}. \tag{20.5}$$

This is a Hilbert space if we endow it with the norm

$$||u||_{W^{s,2}(\mathbb{R}^n)} = \left(||u||^2_{L^2(\mathbb{R}^n)} + \mathscr{E}_{(s)}(u)\right)^{1/2},$$

see [**Ad75**], and also [**DPV12**]. The space $W^{s,2}(\mathbb{R}^n)$ can also be characterized using the Fourier transform. Consider in fact the Sobolev space $H^{s,2}(\mathbb{R}^n)$ recalled in (9.5) above. Using the Fourier transform it is easy to show that $W^{2,s}(\mathbb{R}^n) \cong H^{s,2}(\mathbb{R}^n)$. In this connection we note that if $u \in \mathscr{S}(\mathbb{R}^n)$, then by applying in this order Corollary 5.3, Lemma 5.4 and Plancherel's theorem, we find

$$\int_{\mathbb{R}^n} u\,(-\Delta)^s u\,dx = \int_{\mathbb{R}^n} \left((-\Delta)^{s/2}u\right)^2 dx = \int_{\mathbb{R}^n}\left(\mathscr{F}\left((-\Delta)^{s/2}u\right)\right)^2 dx$$
$$= \int_{\mathbb{R}^n} (2\pi|\xi|)^{2s}\,|\hat{u}|^2 dx.$$

We have already discussed related questions in Section 6 above.

Returning to Definition 20.4, we have the following result that shows that $(-\Delta)^s$ is the first variation of the energy $\mathscr{E}_{(s)}$, and thus such operator also has a nice variational structure.

PROPOSITION 20.5. *The fractional Laplacean is the Euler-Lagrange equation of the functional $u \to \mathscr{E}_{(s)}(u)$. Given $u \in \mathscr{S}(\mathbb{R}^n)$, we have in fact for every $\varphi \in \mathscr{S}(\mathbb{R}^n)$*

$$\frac{d}{dt}\mathscr{E}_{(s)}(u+t\varphi)\big|_{t=0} = \int_{\mathbb{R}^n} (-\Delta)^s u(x)\varphi(x)dx. \tag{20.6}$$

This shows that u is a critical point of $\mathscr{E}_{(s)}$ if and only if $(-\Delta)^s u = 0$.

PROOF. Given $u, \varphi \in \mathscr{S}(\mathbb{R}^n)$ we consider the function $t \to \mathscr{E}_{(s)}(u + t\varphi)$, and take its derivative at $t = 0$. After some elementary computations we obtain

$$\begin{aligned}
\frac{d}{dt}\mathscr{E}_{(s)}(u+t\varphi)\big|_{t=0} &= \frac{\gamma(n,s)}{2}\int_{\mathbb{R}^n} PV\int_{\mathbb{R}^n} \frac{u(x)\varphi(x)-u(y)\varphi(y)}{|x-y|^{n+2s}}dydx \\
&= \frac{\gamma(n,s)}{2}\int_{\mathbb{R}^n} PV\int_{\mathbb{R}^n} \frac{u(x)(\varphi(x)-\varphi(y))+(u(x)-u(y))\varphi(y)}{|x-y|^{n+2s}}dydx \\
&= \frac{\gamma(n,s)}{2}\Big\{\int_{\mathbb{R}^n} u(x)PV\int_{\mathbb{R}^n} \frac{\varphi(x)-\varphi(y)}{|x-y|^{n+2s}}dydx \\
&\qquad + \int_{\mathbb{R}^n} \varphi(y)PV\int_{\mathbb{R}^n} \frac{u(x)-u(y)}{|x-y|^{n+2s}}dxdy\Big\} \\
&= \frac{1}{2}\int_{\mathbb{R}^n} u(x)(-\Delta)^s\varphi(x)dx + \frac{1}{2}\int_{\mathbb{R}^n} \varphi(x)(-\Delta)^s u(x)dx \\
&= \int_{\mathbb{R}^n} \varphi(x)(-\Delta)^s u(x)dx,
\end{aligned}$$

where in the second to the last equality we have used (2.11), while in the last equality we have used Lemma 5.4.

□

In connection with the variational structure of $(-\Delta)^s$ we mention that in the existing literature the notion of weak solution of the problem (13.11) above is formulated by saying the $u \in H^{s,2}(\mathbb{R}^n)$, $u = 0$ a.e. in $\mathbb{R}^n \setminus \Omega$, and

$$\int_{\mathbb{R}^n} (-\Delta)^{s/2}u(-\Delta)^{s/2}\varphi dx = \int_\Omega f\varphi dx,$$

for every $\varphi \in H^{s,2}(\mathbb{R}^n)$, such that $\varphi = 0$ a.e. in $\mathbb{R}^n \setminus \Omega$. It is worth observing here that such notion is equivalent to requesting that

$$\int_{\mathbb{R}^n} \Gamma^{(s)}(u,\varphi)(x)dx = \int_\Omega f\varphi dx,$$

for all φ as above.

Using Theorem 17.2 we can obtain an alternative expression of $\Gamma^{(s)}(u)$ based on the heat semigroup. This observation is important since it allows to introduce a notion of nonlocal energy in non-Euclidean situations. We leave the proof of the following Proposition 20.6 to the interested reader.

PROPOSITION 20.6. *Let $u \in \mathscr{S}(\mathbb{R}^n)$, then*

$$\Gamma^{(s)}(u)(x) = \frac{s}{2\Gamma(1-s)}\int_0^\infty t^{-s-1}\left(P_t u^2(x) - 2u(x)P_t u(x) + u^2(x)\right)dt.$$

Notice that since by Jensen's inequality, or simply Cauchy-Schwarz, we have $P_t u^2(x) \geq (P_t u(x))^2$, we obtain

$$\Gamma^{(s)}(u)(x) \geq \frac{s}{2\Gamma(1-s)}\int_0^\infty t^{-s-1}\left(P_t u(x) - u(x)\right)^2 dt \geq 0,$$

which confirms the positivity of $\Gamma^{(s)}(u)$, see (20.2) and Remark 20.3 above.

Suppose we are on a n-dimensional Riemannian manifold M, with Laplacean L. In the local case $s = 1$, let us simply denote $\Gamma^{(1)}(u)$ by $\Gamma(u)$. We easily obtain from (20.3) that $\Gamma(u) = |\nabla u|^2$. In such situation, the celebrated Bochner's identity gives

$$L(\Gamma(u)) = 2||\nabla^2 u||^2 + 2\Gamma(u, Lu) + 2\,\mathrm{Ric}(\nabla u, \nabla u), \tag{20.7}$$

where Ric indicates the Ricci tensor on M, and we have denoted by ∇^2 the Hessian on M. The central idea of the Bakry-Emery's gamma calculus is to reverse the recipe and use (20.7) as an analytical definition of Ricci tensor. Such intuition is implemented through Definition 20.7 below, see [**BGL14**]. First, we introduce the functional

$$\Gamma_2(u) = \frac{1}{2}\left[L(\Gamma(u)) - 2\Gamma(u, Lu)\right]. \tag{20.8}$$

With this definition it is clear that we can rewrite (20.7) as

$$\Gamma_2(u) = ||\nabla^2 u||^2 + \mathrm{Ric}(\nabla u, \nabla u). \tag{20.9}$$

Suppose now that M satisfies the Ricci lower bound Ric$\geq \kappa$. Then, $\mathrm{Ric}(\nabla u, \nabla u) \geq \kappa|\nabla|^2 = \kappa\Gamma(u)$, and we obtain from (20.9)

$$\Gamma_2(u) \geq ||\nabla^2 u||^2 + \kappa\Gamma(u), \tag{20.10}$$

for every $f \in C^\infty(M)$. On the other hand, Newton's inequality gives

$$||\nabla^2 u||^2 \geq \frac{1}{n}(Lu)^2,$$

and we thus find from (20.10) that for every $f \in C^\infty(M)$ one has

$$\Gamma_2(u) \geq \frac{1}{n}(Lu)^2 + \kappa\Gamma(u). \tag{20.11}$$

DEFINITION 20.7 (Bakry-Emery). The manifold M and the operator L are said to satisfy the *curvature-dimension inequality* $CD(\kappa, n)$ if (20.11) holds true for every $u \in C^\infty(M)$.

We have shown above that

$$\mathrm{Ric}\ \geq \kappa \implies \mathrm{CD}(\kappa, n).$$

It is a remarkable fact that the inequality $\mathrm{CD}(\kappa, n)$, which is a condition on functions, does in fact completely characterize Ricci lower bounds on M, in the sense that

$$\mathrm{Ric}\ \geq \kappa \iff \mathrm{CD}(\kappa, n).$$

For the implication $\Longleftarrow$ the reader should see Proposition 6.2 in [**B94**].

In view of the above discussion, and since as we have shown in (20.2) above there exists a natural nonlocal *carré du champ*, we next introduce a nonlocal counterpart of the form Γ_2.

DEFINITION 20.8 (Nonlocal $\Gamma_2^{(s)}$). Given $u, v \in \mathscr{S}(\mathbb{R}^n)$ we define

$$\Gamma_2^{(s)}(u, v) = \frac{1}{2}\left[L\Gamma_s(u, v) - \Gamma_s(u, Lv) - \Gamma_s(v, Lu)\right]. \tag{20.12}$$

Again, it is obvious that $\Gamma_2^{(s)}(u, v) = \Gamma_2^{(s)}(v, u)$. As for $\Gamma^{(s)}$ we set

$$\Gamma_2^{(s)}(u) \overset{def}{=} \Gamma_2^{(s)}(u, u) = \frac{1}{2}\left[L\Gamma^{(s)}(u) - 2\Gamma^{(s)}(u, Lu)\right]. \tag{20.13}$$

Open problem: Is there a number $d > 0$ such that

$$\Gamma_2^{(s)}(u) \geq \frac{1}{d}((-\Delta)^s u)^2 \ ? \tag{20.14}$$

Is $d = n$? If true, the inequality (20.14) would be quite relevant in adapting to the nonlocal setting the Bakry-Emery gamma calculus, and for instance obtain Riemannian results in the spirit of [**BaG11**], or the extensions to some sub-Riemannian spaces as in [**BaG17**], [**BaG13**], [**BBG14**] and [**BBGM14**]. Understanding the question (20.14) is inextricably connected to understanding

$$(-\Delta)^s \Gamma^{(s)}(u) = ...?$$

i.e., a nonlocal analogue of the celebrated identity of Bochner (20.7) above. We plan to come back to these questions in future works.

21. Are there nonlocal Li-Yau inequalities?

The celebrated Li-Yau inequality states that if M is a n-dimensional boundariless complete Riemannian manifold with nonnegative Ricci tensor, then for any positive solution $f(x,t)$ of the heat equation on $M \times (0,\infty)$, with $u = \log f$ one has

$$|\nabla u|^2 - u_t \leq \frac{n}{2t}. \tag{21.1}$$

To be precise, (21.1) is only one case of the more general inequality of Li and Yau, see [**LY86**]. One fundamental consequence of (21.1) is the following scale invariant *Harnack inequality*: under the above hypothesis on M, let $f > 0$ be a solution of the heat equation on $M \times (0,\infty)$. Then, for every $x, y \in M$ and any $0 < \tau < t < \infty$ one has

$$f(x,\tau) \leq f(y,t) \left(\frac{t}{\tau}\right)^{\frac{n}{2}} \exp\left(\frac{d(x,y)^2}{4(t-\tau)}\right). \tag{21.2}$$

This section is devoted to setting forth some interesting conjectures concerning the heat kernel $G_s(x,t)$ defined in (16.6). But before we do that, we would like to provide some motivation. The first observation is that the standard heat kernel in flat $\mathbb{R}^n$ satisfies (21.1) above with equality.

LEMMA 21.1. *Let* $G(x,t) = (4\pi t)^{-\frac{n}{2}} e^{-\frac{|x|^2}{4t}}$ *be the Gauss-Weierstrass kernel, and define the* entropy $E(x,t) = \log G(x,t)$. *Then, for every* $(x,t) \in \mathbb{R}^{n+1}_+$ *one has*

$$|\nabla_x E(x,t)|^2 - E_t(x,t) = \frac{n}{2t}. \tag{21.3}$$

PROOF. The proof is a simple computation based on the observation that

$$E(x,t) = -\frac{n}{2}\log(4\pi t) - \frac{|x|^2}{4t}.$$

This gives

$$\nabla_x E(x,t) = -\frac{x}{2t}, \qquad E_t(x,t) = -\frac{n}{2t} + \frac{|x|^2}{4t^2},$$

and the result immediately follows. □

REMARK 21.2. It is interesting to observe that Lemma 21.1 can also be derived by the self-similar equation (16.12) which, for $s = 1$, we rewrite

$$- < \frac{x}{2t}, \nabla_x E(x,t) > - E_t(x,t) = \frac{n}{2t}. \tag{21.4}$$

Since, as one easily recognizes,

$$- < \frac{x}{2t}, \nabla_x E(x,t) >= |\nabla_x E(x,t)|^2, \tag{21.5}$$

we immediately obtain (21.3) from the self-similar equation (21.4).

A remarkable consequence of Lemma 21.1 is the following inequality of Li-Yau type for the heat semigroup in $\mathbb{R}^n$.

THEOREM 21.3. *Let $\varphi \in C(\mathbb{R}^n) \cap L^\infty(\mathbb{R}^n)$, $\varphi \geq 0$, and consider the function $f(x,t) = P_t\varphi(x)$. If $u(x,t) = \log f(x,t)$, then*

$$|\nabla_x u(x,t)|^2 - u_t(x,t) \leq \frac{n}{2t}.$$

PROOF. We begin by observing that the sought for conclusion can be reformulated in the following way

$$\frac{|\nabla_x f(x,t)|^2}{f(x,t)} \leq f_t(x,t) + \frac{n}{2t} f(x,t). \tag{21.6}$$

Similarly, the conclusion in Lemma 21.1 can be written as

$$\frac{|\nabla_x G(x,t)|^2}{G(x,t)} = G_t(x,t) + \frac{n}{2t} G(x,t). \tag{21.7}$$

Since

$$f(x,t) = P_t\varphi(x) = \int_{\mathbb{R}^n} G(x-y,t)\varphi(y)dy,$$

denoting $D_i = \frac{\partial}{\partial x_i}$, we now have

$$\begin{aligned} D_i f(x,t) &= \int_{\mathbb{R}^n} D_i G(x-y,t)\varphi(y)dy \\ &= \int_{\mathbb{R}^n} \frac{D_i G(x-y,t)}{G(x-y,t)^{1/2}} \varphi(y)^{1/2} G(x-y,t)^{1/2} \varphi(y)^{1/2} dy \\ &\leq \left(\int_{\mathbb{R}^n} \frac{D_i G(x-y,t)^2}{G(x-y,t)} \varphi(y)dy \right)^{1/2} \left(\int_{\mathbb{R}^n} G(x-y,t)\varphi(y)dy \right)^{1/2}. \end{aligned}$$

This gives

$$\begin{aligned} |\nabla_x f(x,t)|^2 &\leq \sum_{i=1}^n \int_{\mathbb{R}^n} \frac{D_i G(x-y,t)^2}{G(x-y,t)} \varphi(y)dy \int_{\mathbb{R}^n} G(x-y,t)\varphi(y)dy \\ &= f(x,t) \int_{\mathbb{R}^n} \frac{|\nabla_x G(x-y,t)|^2}{G(x-y,t)} \varphi(y)dy \\ &= f(x,t) \int_{\mathbb{R}^n} \left(G_t(x-y,t) + \frac{n}{2t} G(x-y,t) \right) \varphi(y)dy, \end{aligned}$$

where in the last equality we have used (21.7). From this estimate we infer

$$\begin{aligned}\frac{|\nabla_x f(x,t)|^2}{f(x,t)} &\le \int_{\mathbb{R}^n} G_t(x-y,t)\varphi(y)dy + \frac{n}{2t}\int_{\mathbb{R}^n} G(x-y,t)\varphi(y)dy \\ &= f_t(x,t) + \frac{n}{2t} f(x,t),\end{aligned}$$

which is the desired inequality (21.6).

□

We are now ready to prove the following fundamental result that was first independently obtained by B. Pini [**P54**] and J. Hadamard [**H54**] in the plane $\mathbb{R}_x \times \mathbb{R}_t$. It extends to the heat equation (with some fundamental differences) the celebrated inequality first proved for the Laplacean by Axel Harnack, see [**Har87**].

THEOREM 21.4 (The Pini-Hadamard scale invariant Harnack inequality). *Let $\varphi \in C(\mathbb{R}^n) \cap L^\infty(\mathbb{R}^n)$, $\varphi \ge 0$, and consider the function $f(x,t) = P_t\varphi(x)$. For every $x, y \in \mathbb{R}^n$, and every $0 < s < t < \infty$ one has*

$$f(x,s) \le f(y,t) \left(\frac{t}{s}\right)^{\frac{n}{2}} \exp\left(\frac{|x-y|^2}{4(t-s)}\right).$$

PROOF. Consider the function

$$\psi(\tau) = \log f(\gamma(\tau)), \qquad 0 \le \tau \le 1,$$

where $\gamma(\tau)$ is a straight line (geodesic) starting at the "upper" point (y,t) and ending at the "lower" point (x,s), i.e.,

$$\gamma(\tau) = (y + \tau(x-y), t + \tau(s-t)), \qquad 0 \le \tau \le 1.$$

We clearly have

$$\begin{aligned}\log\frac{f(x,s)}{f(y,t)} &= \log f(\gamma(1)) - \log f(\gamma(0)) = \psi(1) - \psi(0) = \int_0^1 \psi'(\tau)d\tau \\ &= \int_0^1 < \frac{\nabla_x f(\gamma(\tau))}{f(\gamma(\tau))}, x-y > d\tau - (t-s)\int_0^1 \frac{f_t(\gamma(\tau))}{f(\gamma(\tau))}d\tau \\ &\le |x-y|\int_0^1 \frac{|\nabla_x f(\gamma(\tau))|}{f(\gamma(\tau))}d\tau + (t-s)\int_0^1 \frac{n}{2(t+\tau(s-t))}d\tau \\ &\quad - (t-s)\int_0^1 \frac{|\nabla_x f(\gamma(\tau))|^2}{f(\gamma(\tau))^2}d\tau,\end{aligned}$$

where in the last line we have used the Li-Yau inequality for f in Theorem 21.3, see also (21.6). An easy argument now gives

$$\int_0^1 \frac{n}{2(t+\tau(s-t))}d\tau = \frac{n}{2(t-s)}\int_s^t \frac{dr}{r} = \frac{1}{t-s}\log\left(\frac{t}{s}\right)^{\frac{n}{2}}.$$

On the other hand, for every $\varepsilon > 0$ we have

$$\begin{aligned}|x-y|\int_0^1 \frac{|\nabla_x f(\gamma(\tau))|}{f(\gamma(\tau))}d\tau &\le |x-y|\left(\int_0^1 \frac{|\nabla_x f(\gamma(\tau))|^2}{f(\gamma(\tau))^2}d\tau\right)^{1/2} \\ &\le \frac{\varepsilon}{2}\int_0^1 \frac{|\nabla_x f(\gamma(\tau))|^2}{f(\gamma(\tau))^2}d\tau + \frac{1}{2\varepsilon}|x-y|^2.\end{aligned}$$

We thus find for every $\varepsilon > 0$

$$\log \frac{f(x,s)}{f(y,t)} \le \frac{\varepsilon}{2} \int_0^1 \frac{|\nabla_x f(\gamma(\tau))|^2}{f(\gamma(\tau))^2} d\tau - (t-s) \int_0^1 \frac{|\nabla_x f(\gamma(\tau))|^2}{f(\gamma(\tau))^2} d\tau$$
$$+ \frac{1}{2\varepsilon}|x-y|^2 + \log\left(\frac{t}{s}\right)^{\frac{n}{2}}.$$

Choosing $\varepsilon = 2(t-s)$ in the latter inequality, we obtain

$$\log \frac{f(x,s)}{f(y,t)} \le \log\left(\frac{t}{s}\right)^{\frac{n}{2}} + \frac{|x-y|^2}{4(t-s)}.$$

Exponentiating, we reach the desired conclusion.

□

In view of Lemma 21.1, Theorem 21.3 and Theorem 21.4 it is natural to wonder whether such results have nonlocal analogues. As we have indicated in Section 20, besides having an interest in its own right, such question is also relevant to the development of a nonlocal Li-Yau theory. We are thus led to formulating the following:

Conjecture 1: With $\Gamma^{(s)}$ defined as in (20.2) above, is it true that for every $x \in \mathbb{R}^n$ and $t > 0$ one has

$$\frac{\Gamma^{(s)}(G^{(s)})(x,t)}{G^{(s)}(x,t)^2} - \frac{\frac{\partial G^{(s)}}{\partial t}(x,t)}{G^{(s)}(x,t)} \le \frac{n}{2s}\frac{1}{t}\ ? \tag{21.8}$$

Equivalently, we can write this conjecture in the following way

$$\Gamma^{(s)}(G^{(s)})(x,t) - \frac{\partial G^{(s)}}{\partial t}(x,t) G^{(s)}(x,t) \le \frac{n}{2s}\frac{1}{t} G^{(s)}(x,t)^2\ ? \tag{21.9}$$

Recall that we have observed in (16.13) above that

$$-\frac{\partial G^{(s)}}{\partial t} = \frac{n}{2s}\frac{1}{t}G^{(s)}(x,t) + \frac{1}{2s} < \frac{x}{t}, \nabla_x G^{(s)} > .$$

It follows that (21.9) is true if and only if:

Conjecture 2: Is it true that

$$\Gamma^{(s)}(G^{(s)}(\cdot,t))(x) + \frac{1}{2s} < \frac{x}{t}, \nabla_x G^{(s)}(x,t) > G^{(s)}(x,t)\ \le\ 0\ ? \tag{21.10}$$

We observe explicitly that (21.10) represents the nonlocal counterpart of the local identity (21.5) noted above. We also note that from the formula (18.2) in Theorem 18.3 above we have for every $x \in \mathbb{R}^n$ and $t > 0$

$$\begin{aligned} < \frac{x}{2}, \nabla_x G^{(s)}(x,t) > &= \int_0^\infty f_s(t;\tau) < \frac{x}{2}, \nabla_x G(x,\tau) > d\tau \\ &= -\pi|x|^2 \int_0^\infty f_s(t;\tau)\tilde{G}(x,\tau) d\tau, \end{aligned} \tag{21.11}$$

where we have denoted by $\tilde{G}(x,\tau) = (4\pi\tau)^{-\frac{n+2}{2}} e^{-\frac{|x|^2}{4\tau}}$.

22. A Li-Yau inequality for Bessel operators

The discussion of the extension problem (10.5) in Section 10 has evidenced the key role of the Bessel operator

$$\mathscr{B}_a = \frac{\partial^2}{\partial y^2} + \frac{a}{y}\frac{\partial}{\partial y}, \qquad a = 1 - 2s, \tag{22.1}$$

in the analysis of the fractional Laplacean $(-\Delta)^s$. But $\mathscr{B}_a$ plays an equally important role in the study of other nonlocal operators such as, for instance, the fractional heat operator $(\partial_t - \Delta)^s$, see [**NS16**], [**ST17**] and also [**BG17**]. Because of the ubiquitous presence of (22.1) in the fractional world, and since this topic is perhaps more frequented by workers in probability than analysts and geometers, in this section we provide a purely analytical construction of the fundamental solution of the heat semigroup associated with $\mathscr{B}_a$, see Proposition 22.3 below. After that result, we recall a proposition which states that the Neumann heat semigroup associated with $\mathscr{B}_a$ satisfies a *curvature-dimension inequality.*

If with $-1 < a < 1$ we define the parameter ν by the equation

$$a = 2\nu + 1,$$

then we can write (22.1) in the following way

$$\mathscr{B}_\nu = \frac{\partial^2}{\partial y^2} + \frac{2\nu+1}{y}\frac{\partial}{\partial y}, \qquad -1 < \nu < 0. \tag{22.2}$$

We intend to study the Cauchy problem for the heat equation associated with (22.2),

$$\begin{cases} \frac{\partial^2 f}{\partial y^2} + \frac{2\nu+1}{y}\frac{\partial f}{\partial y} = \frac{\partial f}{\partial t}, & y > 0, t > 0, \\ f(y,0) = \varphi(y). \end{cases} \tag{22.3}$$

We remark that the case $a = 0$ (which in the extension problem corresponds to the critical exponent $s = 1/2$) corresponds to the value $\nu = -1/2$.

To solve (22.3) we introduce the *modified Hankel transform* of a function f

$$\mathcal{H}_\nu(f)(x) = \int_0^\infty f(y) G_\nu(xy) y^{2\nu+1} dy, \tag{22.4}$$

see [**MS65**], where we have let

$$G_\nu(z) = z^{-\nu} J_\nu(z). \tag{22.5}$$

We recall, see e.g. 5.3.5 on p. 103 in [**Le72**], that

$$G_\nu'(z) = -z G_{\nu+1}(z). \tag{22.6}$$

Since for $z \in \mathbb{C}$ such that $|\arg z| < \pi$ we have

$$\begin{cases} G_\nu(z) \to \frac{2^{-\nu}}{\Gamma(\nu+1)}, & \text{as } z \to 0, \\ |G_\nu(z)| \le \frac{C}{|z|^{\nu+\frac{1}{2}}}, & \text{as } |z| \to \infty, \end{cases} \tag{22.7}$$

the integral defining (22.4) is finite if $f \in \mathcal{C}_\nu(0,\infty)$, where

$$\mathcal{C}_\nu(0,\infty) = \{f \in C(0,\infty) \mid \forall R > 0$$
$$\text{one has } \int_0^R |f(y)| y^{2\nu+1} dy < \infty, \ \int_R^\infty |f(y)| y^{\nu+\frac{1}{2}} dy < \infty\}.$$

Note that $f \in \mathcal{C}_\nu(0,\infty)$ implies, in particular, that

$$\liminf_{y\to 0^+} y^{2\nu+2}|f(y)| = 0, \qquad \liminf_{y\to\infty} y^{\nu+\frac{3}{2}}|f(y)| = 0.$$

We will work with functions f in such class. We will also need the following class

$$\mathcal{C}^1_\nu(0,\infty) = \{f \in C^1(0,\infty) \mid f, \frac{1}{y}f' \in \mathcal{C}_\nu(0,\infty)\}.$$

We notice that membership in $\mathcal{C}^1_\nu(0,\infty)$ imposes, in particular, the weak *Neumann condition*

(22.8)
$$\liminf_{y\to 0^+} y^{2\nu+1}|f'(y)| = 0.$$

LEMMA 22.1. *Let* $f \in C^1_\nu(0,\infty)$. *Then,*

$$\mathcal{H}_\nu(\frac{1}{y}\frac{df}{dy})(x) = -\mathcal{H}_{\nu-1}(f)(x).$$

PROOF. Under the given assumptions on f we can integrate by parts in the following integral and omit the boundary contributions since they vanish. Using (22.6), we find

$$\begin{aligned}\mathcal{H}_\nu(\frac{1}{y}\frac{df}{dy})(x) &= \int_0^\infty f'(y)G_\nu(xy)y^{2\nu}dy = -\int_0^\infty f(y)xG'_\nu(xy)y^{2\nu}dy \\ &\quad - 2\nu\int_0^\infty f(y)G_\nu(xy)y^{2\nu-1}dy \\ &= \int_0^\infty f(y)\left[(xy)^2G_{\nu+1}(xy) - 2\nu G_\nu(xy)\right]y^{2\nu-1}dy.\end{aligned}$$

If we now use formula 5.3.6 on p. 103 in [**Le72**]

$$J_{\nu+1}(z) = \frac{2\nu}{z}J_\nu(z) - J_{\nu-1}(z),$$

we find

(22.9)
$$z^2G_{\nu+1}(z) = 2\nu G_\nu(z) - G_{\nu-1}(z).$$

Using this identity in the above integral we finally obtain

$$\mathcal{H}_\nu(\frac{1}{y}\frac{df}{dy})(x) = -\int_0^\infty f(y)G_\nu(xy)y^{2\nu-1}dy = -\mathcal{H}_{\nu-1}(f)(x).$$

□

LEMMA 22.2. *Suppose now that* $f \in C^2(0,\infty)$ *and that* $f, f'' \in \mathcal{C}_\nu(0,\infty)$. *Then,*

$$\mathcal{H}_\nu(f'')(x) = (2\nu+1)\mathcal{H}_{\nu-1}(f)(x) - x^2\mathcal{H}_\nu(f)(x).$$

PROOF. Again, we can integrate by parts omitting the boundary terms in the integral defining $\mathcal{H}_\nu(f'')(x)$. This gives

$$\begin{aligned}&\mathcal{H}_\nu(f'')(x) \\ &= \int_0^\infty f(y)\left[(xy)^2G''_\nu(xy) + 2(2\nu+1)xyG'_\nu(xy) + 2\nu(2\nu+1)G_\nu(xy)\right]y^{2\nu-1}dy\end{aligned}$$

Next we use the fact that J_ν satisfies Bessel's differential equation (4.22),

$$z^2J''_\nu(z) + zJ'_\nu(z) + (z^2-\nu^2)J_\nu(z) = 0,$$

to deduce that

$$z^2 G_\nu''(z) + (2\nu+1) z G_\nu'(z) + z^2 G_\nu(z) = 0. \tag{22.10}$$

Using (22.10) we find, after some simplification,

$$\begin{aligned}&(xy)^2 G_\nu''(xy) + 2(2\nu+1) xy G_\nu'(xy) + 2\nu(2\nu+1) G_\nu(xy)\\ &= -(2\nu+1)(xy)^2 G_{\nu+1}(xy) + 2\nu(2\nu+1) G_\nu(xy) - (xy)^2 G_\nu(xy),\end{aligned}$$

where in the last equality we have used (22.6). Next, we use (22.9) to conclude

$$\begin{aligned}&(xy)^2 G_\nu''(xy) + 2(2\nu+1) xy G_\nu'(xy) + 2\nu(2\nu+1) G_\nu(xy)\\ &= (2\nu+1) G_{\nu-1}(xy) - (xy)^2 G_\nu(xy).\end{aligned}$$

Substituting in the above integral we finally obtain

$$\begin{aligned}\mathcal{H}_\nu(f'')(x) &= \int_0^\infty f(y) \left[(2\nu+1) G_{\nu-1}(xy) - (xy)^2 G_\nu(xy)\right] y^{2\nu-1} dy\\ &= (2\nu+1)\mathcal{H}_{\nu-1}(f)(x) - x^2 \mathcal{H}_\nu(f)(x).\end{aligned}$$

This completes the proof. □

With Lemmas 22.1 and 22.2 in hands, we return to the Cauchy problem with the purpose of finding a representation formula of the solution. We have the following result. For a different probabilistic approach see [**BS02**], but one should keep in mind that the probabilist's generator is $\frac{1}{2}\mathscr{B}_a$, with $\mathscr{B}_a$ as in (22.2).

PROPOSITION 22.3. *Let $\nu > -1$ and consider the heat semigroup $P_t^\nu = e^{-t\mathscr{B}_\nu}$ on $(\mathbb{R}_+, y^{2\nu+1}dy)$ with generator $\mathscr{B}_\nu$ as in (22.2). Then, the Neumann heat kernel associated with $e^{-t\mathscr{B}_\nu}$ is given by*

$$p_\nu^N(x,y,t) = p_\nu^N(y,x,t) = (2t)^{-(\nu+1)} \left(\frac{xy}{2t}\right)^{-\nu} I_\nu\left(\frac{xy}{2t}\right) e^{-\frac{x^2+y^2}{4t}}, \tag{22.11}$$

where we have denoted by $I_\nu(z)$ the modified Bessel function of the first kind defined by (4.31). This means that for any given function $\varphi \in \mathcal{C}_\nu^1(0,\infty)$ the solution of the Cauchy problem (22.3) is given by

$$P_t^\nu \varphi(x) = \int_0^\infty \varphi(y) p_\nu^N(x,y,t) y^{2\nu+1} dy. \tag{22.12}$$

PROOF. We assume that f be a solution to (22.3). We formally apply to (22.3) the Hankel transform $\mathcal{H}_\nu$ with respect to the variable y. I.e., we let

$$\mathcal{H}_\nu(f)(x,t) = \int_0^\infty f(y,t) G_\nu(xy) y^{2\nu+1} dy.$$

Remarkably, if we use Lemma 22.2 and Lemma 22.1, the term $(2\nu+1)\mathcal{H}_{\nu-1}(f)(x)$ magically drops, and the Cauchy problem (22.3) is converted into the following one

$$\begin{cases} \frac{\partial \mathcal{H}_\nu(f)}{\partial t}(x,t) = -x^2 \mathcal{H}_\nu(f)(x,t), & x>0, t>0,\\ \mathcal{H}_\nu(f)(x,0) = \mathcal{H}_\nu(\varphi)(x), \end{cases} \tag{22.13}$$

whose unique solution is

$$\mathcal{H}_\nu(f)(x,t) = \mathcal{H}_\nu(\varphi)(x) e^{-tx^2}. \tag{22.14}$$

We next apply formally $\mathcal{H}_\nu$ to (22.14), and use Hankel's inversion formula

$$\mathcal{H}_\nu(\mathcal{H}_\nu(f)) = f$$

to find

$$\begin{aligned} f(x,t) &= \int_0^\infty (xz)^{-\nu} e^{-tz^2} \left(\int_0^\infty (zy)^{-\nu} J_\nu(zy)\varphi(y) y^{2\nu+1} dy \right) J_\nu(xz) z^{2\nu+1} dz \\ &= x^{-\nu} \int_0^\infty \varphi(y) y^{\nu+1} \left(\int_0^\infty z e^{-tz^2} J_\nu(yz) J_\nu(xz) dz \right) dy. \end{aligned}$$

At this point we appeal to formula 3. on p. 223 in [**PBM88**] that gives

$$\int_0^\infty z e^{-tz^2} J_\nu(yz) J_\nu(xz) dz = \frac{1}{2t} I_\nu\left(\frac{xy}{2t}\right) e^{-\frac{x^2+y^2}{4t}}, \tag{22.15}$$

provided that $\Re\nu > -1$. Since $\nu \in \mathbb{R}$ and $\nu > -1$, we can use (22.15) and finally obtain

$$f(x,t) = \int_0^\infty \varphi(y) p_\nu^N(x,y,t) y^{2\nu+1} dy, \tag{22.16}$$

where

$$p_\nu^N(x,y,t) = (2t)^{-(\nu+1)} \left(\frac{xy}{2t}\right)^{-\nu} I_\nu\left(\frac{xy}{2t}\right) e^{-\frac{x^2+y^2}{4t}}.$$

This establishes (22.11), (22.12), thus completing the proof.

□

REMARK 22.4. We note explicitly that for every $y > 0, t > 0$ one has

$$p_\nu^N(0,y,t) = \frac{1}{2^{2\nu+1}\Gamma(\nu+1)} t^{-(\nu+1)} e^{-\frac{y^2}{4t}}. \tag{22.17}$$

This can be seen by the following power series representation

$$z^{-\nu} I_\nu(z) = 2^{-\nu} \sum_{k=0}^\infty \frac{(z/2)^{2k}}{\Gamma(k+1)\Gamma(k+\nu+1)}, \tag{22.18}$$

valid for $|\arg z| < \pi$. From (22.18) we immediately recognize that, similarly to (4.27), we have

$$z^{-\nu} I_\nu(z) \cong \frac{2^{-\nu}}{\Gamma(\nu+1)}, \quad \text{as } z \to 0, \quad |\arg z| < \pi. \tag{22.19}$$

The desired conclusion (22.17) immediately follows from (22.11) and (22.19). Note that when $\nu = -\frac{1}{2}$ we obtain from (22.17)

$$p^N_{-\frac{1}{2}}(0, y, t) = 2(4\pi t)^{-1/2} e^{-\frac{y^2}{4t}}.$$

Our next result shows that the Neumann heat kernel associated with the Bessel operator $\mathscr{B}_a$ satisfies an inequality of Li-Yau type reminiscent of that in Lemma 21.1 above (notice however that, unlike the classical heat kernel, we presently have an inequality, not an equality).

PROPOSITION 22.5 (Li-Yau inequality). *Let* $-\frac{1}{2} \leq \nu < 0$, *and denote by* $E_\nu(x, y, t) = \log p^N_\nu(x, y, t)$, *where* $p^N_\nu(x, y, t)$ *is as in* (22.11) *in Proposition* 22.3 *above. Then, the following Li-Yau type inequality holds*

$$(D_y \log E_\nu)^2 - D_t \log E_\nu \leq \frac{\nu + 1}{t}. \tag{22.20}$$

This result is derived from Proposition 22.3 and we omit the relevant details. Similarly to Theorems 21.3 and 21.4, Proposition 22.5 leads to a related Li-Yau inequality and to a Harnack inequality for positive solutions of the heat equation $\partial_t - \mathscr{B}_\nu$, where $\mathscr{B}_\nu$ is given in (22.2) above.

23. The fractional p-Laplacean

We cannot close this fractional note without a brief discussion of a nonlinear nonlocal operator which has been attracting a great deal of attention over the past few years, and whose analysis poses remarkable challenges and open questions.

We have seen in Proposition 20.5 that the fractional Laplacean has a variational structure, in the sense that it also arises as the Euler-Lagrange equation of the energy functional $\mathscr{E}_{(s)}(u)$ in Definition 20.4. In the local case $s = 1$ the corresponding energy $\mathscr{E}(u) = \frac{1}{2}\int_{\mathbb{R}^n} |\nabla u|^2 dx$ is only one in the infinite scale of exponents

$$\mathscr{E}_p(u) = \frac{1}{p}\int_{\mathbb{R}^n} |\nabla u|^p dx, \qquad 1 < p < \infty,$$

(we leave out the end-point cases $p = 1$ and $p = \infty$ since their discussion would deserve a book in its own). It is well-known that the Euler-Lagrange equation of the functional $u \to \mathscr{E}_p(u)$ is the so-called *p-Laplace equation*

$$\Delta_p u = \operatorname{div}(|\nabla u|^{p-2}\nabla u) = 0, \qquad 1 < p < \infty.$$

This operator is of course nonlinear and degenerate elliptic and, despite some similarities with its linear ancestor, the Laplacean, its analysis is much harder and not yet completely understood. The most fundamental open problem in dimension $n \geq 3$ remains to present day the *unique continuation property*: it is disheartening that we do not know whether a nontrivial solution of $\Delta_p u = 0$ in a connected open set can have a zero of infinite order, or vanish in an open subset. When $n = 2$ the strong unique continuation does hold as a consequence of the results of Bojarski and Iwaniec [**BI87**] ($p > 2$), Alessandrini [**Al87**] ($1 < p < \infty$), and Manfredi [**Ma88**] ($1 < p < \infty$).

It has long been known, however, that weak solutions of $\Delta_p u = 0$ have at best a locally Hölder continuous gradient. For instance, the function $u(x) = |x|^{p/(p-1)}$ satisfies the equation $\Delta_p u = c(n,p)$, and clearly we have $u \in C^{1,\alpha}_{loc}$, but $u \notin C^2$, at least when $p > 2$. The fundamental $C^{1,\alpha}$ regularity result was first proved in the late 60's by N. Ural'tseva when $p \geq 2$, and subsequently independently generalized to all $1 < p < \infty$ (and to more general quasilinear equations) by J. Lewis [**Le83**], Di Benedetto [**DB83**] and Tolksdorff [**To84**].

Because of its variational structure the p-Laplacean presents itself in connection with the case $p \neq 2$ of the Sobolev embedding theorem

$$W^{1,p}(\mathbb{R}^n) \hookrightarrow L^q(\mathbb{R}^n), \qquad \frac{1}{p} - \frac{1}{q} = \frac{1}{n}.$$

But Δ_p plays an important role also in the applied sciences, for instance in the study of non-Newtonian fluids.

In view of what has been said so far it seems natural to consider for any $0 < s < 1$ the following fractional energy

$$\mathscr{E}_{(s),p}(u) = \frac{1}{p} \int_{\mathbb{R}^n} \int_{\mathbb{R}^n} \frac{|u(x) - u(y)|^p}{|x-y|^{n+ps}} dxdy, \qquad 1 < p < \infty. \tag{23.1}$$

When $\mathbb{R}^n$ is replaced by the boundary of a bounded open set $\Omega \subset \mathbb{R}^n$, such energy was introduced independently by Gagliardo [**Ga57**] and Slobodeckji [**Slo58**] in connection with the characterization of the traces on $\partial\Omega$ of functions in the Sobolev space $W^{1,p}(\Omega)$. For a general geometric approach to the characterization of traces we refer the reader to the Memoir of the AMS [**DGN06**].

For any $0 < s < 1$ the *fractional p-Laplace operator* $(-\Delta_p)^s$ is defined as the Euler-Lagrange equation of the energy functional $u \to \mathscr{E}_{(s),p}(u)$. It is an easy exercise to show that a function u in the fractional Sobolev space

$$W^{s,p}(\mathbb{R}^n) = \{u \in L^p(\mathbb{R}^n) \mid \mathscr{E}_{(s),p}(u) < \infty\},$$

endowed with the natural norm

$$||u||_{W^{s,p}(\mathbb{R}^n)} = \left(||u||^p_{L^p(\mathbb{R}^n)} + \mathscr{E}_{(s),p}(u)\right)^{1/p},$$

is a weak solution of $(-\Delta_p)^s u = 0$ if for any $\varphi \in W^{s,p}(\mathbb{R}^n)$ having compact support one has

$$\int_{\mathbb{R}^n} \int_{\mathbb{R}^n} \frac{|u(x) - u(y)|^{p-2}(u(x) - u(y))(\varphi(x) - \varphi(y))}{|x-y|^{n+ps}} dxdy = 0. \tag{23.2}$$

The equation $(-\Delta_p)^s u = 0$ was first independently introduced in the papers [**AMRT09**] and [**IN10**].

There presently exists a large literature on the fractional p-Laplacean. Unfortunately, in this brief section we cannot go into a detailed discussion of all the interesting work that has been done. We will only quote some papers, referring the interested reader to those sources and the references therein. The Perron method for $(-\Delta_p)^s$ has been studied in [**LL17**]. The analogue of Serrin's 1964 $C^{0,\alpha}_{loc}$ regularity for weak solutions of $(-\Delta_p)^s u = 0$ is known, and it has been proved in [**DKP16**]. The same authors established the Harnack inequality in [**DKP14**]. The Hölder continuity up to the boundary for the Dirichlet problem in $C^{1,1}$ domains was proved in [**IMS16**]. One should also see the preprint [**Co16**] which contains related results for minimizers of nonlocal functionals of the calculus of variations. Regularity estimates for solutions with measure data were established in [**KMS15**].

There are of course many basic open questions, and the reader could derive some of them from the discussion of the nonlocal linear case $p = 2$ in this note. But at present a fundamental open problem concerning the operator $(-\Delta_p)^s$ is the nonlocal counterpart of the above cited $C^{1,\alpha}$ regularity theorem for the local case. In this connection, an interesting new contribution has been recently given in [**BL17**], where the authors establish the nonlocal counterpart of a famous theorem of K. Uhlenbeck stating that when $p \geq 2$ weak solutions in $W^{1,p}_{loc}$ of the p-Laplacean system have in fact

$$|\nabla u|^{\frac{p-2}{2}} \nabla u \in W^{1,2}_{loc}.$$

We note in passing that this fact implies that

$$\nabla u \in W^{s,p}_{loc}, \qquad 0 < s < \frac{2}{p}.$$

The main result in [**BL17**], which is Theorem 1.5, provides a nonlocal analogue of this latter conclusion. Since the precise statement is somewhat involved, we refer the reader to their paper.

Added in proof: In connection with the problem of the optimal regularity discussed above we mention that, after the posting of the present paper, we have received from L. Brasco his joint preprint [**BLS17**] with E. Lindgren and A. Schikorra. In such work the authors establish the optimal $C^{s+\alpha}$ Hölder regularity result for solutions to the equation $(-\Delta_p)^s u = f$ in the range $p \geq 2$. Thus, only the range $1 < p < 2$ presently remains *terra incognita.*

Acknowledgments

I thank A. Banerjee, D. M. Nhieu, A. Petrosyan, C. Pop and X. Ros-Oton for so graciously taking the time and effort to read the manuscript at different stages of its preparation and for providing me with much valuable feedback.

I also want to thank the anonymous reviewers for their several valuable suggestions which have contributed to improve the presentation.

References

[Ab15] N. Abatangelo, *Large S-harmonic functions and boundary blow-up solutions for the fractional Laplacian*, Discrete Contin. Dyn. Syst. **35** (2015), no. 12, 5555–5607, DOI 10.3934/dcds.2015.35.5555. MR3393247

[Ab17] N. Abatangelo, *Very large solutions for the fractional Laplacian: towards a fractional Keller-Osserman condition*, Adv. Nonlinear Anal. **6** (2017), no. 4, 383–405, DOI 10.1515/anona-2015-0150. MR3719289

[AV17] N. Abatangelo & E. Valdinoci, *Getting acquainted with the fractional Laplacean*, arXiv:1710.1156 (2017).

[AMPP16] B. Abdellaoui, M. Medina, I. Peral, and A. Primo, *Optimal results for the fractional heat equation involving the Hardy potential*, Nonlinear Anal. **140** (2016), 166–207, DOI 10.1016/j.na.2016.03.013. MR3492734

[Ad75] R. A. Adams, *Sobolev spaces*, Academic Press [A subsidiary of Harcourt Brace Jovanovich, Publishers], New York-London, 1975. Pure and Applied Mathematics, Vol. 65. MR0450957

[ABG15] H. Aimar, G. Beltritti, and I. Gómez, *Improvement of Besov regularity for solutions of the fractional Laplacian*, Constr. Approx. **41** (2015), no. 2, 219–229, DOI 10.1007/s00365-014-9256-0. MR3315676

[Al87] G. Alessandrini, *Critical points of solutions to the p-Laplace equation in dimension two* (English, with Italian summary), Boll. Un. Mat. Ital. A (7) **1** (1987), no. 2, 239–246. MR898285

[A79] F. J. Almgren Jr., *Dirichlet's problem for multiple valued functions and the regularity of mass minimizing integral currents*, Minimal submanifolds and geodesics (Proc. Japan-United States Sem., Tokyo, 1977), North-Holland, Amsterdam-New York, 1979, pp. 1–6. MR574247

[AT91] C. J. Amick and J. F. Toland, *Uniqueness and related analytic properties for the Benjamin-Ono equation—a nonlinear Neumann problem in the plane*, Acta Math. **167** (1991), no. 1-2, 107–126, DOI 10.1007/BF02392447. MR1111746

[AMRT09] F. Andreu, J. M. Mazón, J. D. Rossi, and J. Toledo, *A nonlocal p-Laplacian evolution equation with nonhomogeneous Dirichlet boundary conditions*, SIAM J. Math. Anal. **40** (2008/09), no. 5, 1815–1851, DOI 10.1137/080720991. MR2471902

[AC04] I. Athanasopoulos and L. A. Caffarelli, *Optimal regularity of lower dimensional obstacle problems* (English, with English and Russian summaries), Zap. Nauchn. Sem. S.-Peterburg. Otdel. Mat. Inst. Steklov. (POMI) **310** (2004), no. Kraev. Zadachi Mat. Fiz. i Smezh. Vopr. Teor. Funkts. 35 [34], 49–66, 226, DOI 10.1007/s10958-005-0496-1; English transl., J. Math. Sci. (N.Y.) **132** (2006), no. 3, 274–284. MR2120184

[ACS08] I. Athanasopoulos, L. A. Caffarelli, and S. Salsa, *The structure of the free boundary for lower dimensional obstacle problems*, Amer. J. Math. **130** (2008), no. 2, 485–498, DOI 10.1353/ajm.2008.0016. MR2405165

[ACDH04] P. Auscher, T. Coulhon, X. T. Duong, and S. Hofmann, *Riesz transform on manifolds and heat kernel regularity* (English, with English and French summaries), Ann. Sci. École Norm. Sup. (4) **37** (2004), no. 6, 911–957, DOI 10.1016/j.ansens.2004.10.003. MR2119242

[Bak87] D. Bakry, *Étude des transformations de Riesz dans les variétés riemanniennes à courbure de Ricci minorée* (French), Séminaire de Probabilités, XXI, Lecture Notes in Math., vol. 1247, Springer, Berlin, 1987, pp. 137–172, DOI 10.1007/BFb0077631. MR941980

[B94] D. Bakry, *L'hypercontractivité et son utilisation en théorie des semigroupes* (French), Lectures on probability theory (Saint-Flour, 1992), Lecture Notes in Math., vol. 1581, Springer, Berlin, 1994, pp. 1–114, DOI 10.1007/BFb0073872. MR1307413

[BGL14] D. Bakry, I. Gentil, and M. Ledoux, *Analysis and geometry of Markov diffusion operators*, Grundlehren der Mathematischen Wissenschaften [Fundamental Principles of Mathematical Sciences], vol. 348, Springer, Cham, 2014. MR3155209

[Bal60] A. V. Balakrishnan, *Fractional powers of closed operators and the semigroups generated by them*, Pacific J. Math. **10** (1960), 419–437. MR0115096

[BG17] A. Banerjee and N. Garofalo, *Monotonicity of generalized frequencies and the strong unique continuation property for fractional parabolic equations*, Adv. Math. **336** (2018), 149–241, DOI 10.1016/j.aim.2018.07.021. MR3846151

[BDS15] V. Banica, M. d. M. González, and M. Sáez, *Some constructions for the fractional Laplacian on noncompact manifolds*, Rev. Mat. Iberoam. **31** (2015), no. 2, 681–712, DOI 10.4171/RMI/850. MR3375868

[Ba67] M. S. Baouendi, *Sur une classe d'opérateurs elliptiques dégénérés* (French), Bull. Soc. Math. France **95** (1967), 45–87. MR0228819

[BPSV14] B. Barrios, I. Peral, F. Soria, and E. Valdinoci, *A Widder's type theorem for the heat equation with nonlocal diffusion*, Arch. Ration. Mech. Anal. **213** (2014), no. 2, 629–650, DOI 10.1007/s00205-014-0733-1. MR3211862

[BBG14] F. Baudoin, M. Bonnefont, and N. Garofalo, *A sub-Riemannian curvature-dimension inequality, volume doubling property and the Poincaré inequality*, Math. Ann. **358** (2014), no. 3-4, 833–860, DOI 10.1007/s00208-013-0961-y. MR3175142

[BBGM14] F. Baudoin, M. Bonnefont, N. Garofalo, and I. H. Munive, *Volume and distance comparison theorems for sub-Riemannian manifolds*, J. Funct. Anal. **267** (2014), no. 7, 2005–2027, DOI 10.1016/j.jfa.2014.07.030. MR3250358

[BaG11] F. Baudoin and N. Garofalo, *Perelman's entropy and doubling property on Riemannian manifolds*, J. Geom. Anal. **21** (2011), no. 4, 1119–1131, DOI 10.1007/s12220-010-9180-x. MR2836593

[BaG13] F. Baudoin and N. Garofalo, *A note on the boundedness of Riesz transform for some subelliptic operators*, Int. Math. Res. Not. IMRN **2** (2013), 398–421, DOI 10.1093/imrn/rnr271. MR3010694

[BaG17] F. Baudoin and N. Garofalo, *Curvature-dimension inequalities and Ricci lower bounds for sub-Riemannian manifolds with transverse symmetries*, J. Eur. Math. Soc. (JEMS) **19** (2017), no. 1, 151–219, DOI 10.4171/JEMS/663. MR3584561

[Be94] A. Bendikov, *Asymptotic formulas for symmetric stable semigroups*, Exposition. Math. **12** (1994), no. 4, 381–384. MR1297844

[BM07] A. Bendikov and P. Maheux, *Nash type inequalities for fractional powers of non-negative self-adjoint operators*, Trans. Amer. Math. Soc. **359** (2007), no. 7, 3085–3097, DOI 10.1090/S0002-9947-07-04020-2. MR2299447

[BMST16] A. Bernardis, F. J. Martín-Reyes, P. R. Stinga, and J. L. Torrea, *Maximum principles, extension problem and inversion for nonlocal one-sided equations*, J. Differential Equations **260** (2016), no. 7, 6333–6362, DOI 10.1016/j.jde.2015.12.042. MR3456835

[BIK15] P. Biler, C. Imbert, and G. Karch, *The nonlocal porous medium equation: Barenblatt profiles and other weak solutions*, Arch. Ration. Mech. Anal. **215** (2015), no. 2, 497–529, DOI 10.1007/s00205-014-0786-1. MR3294409

[BLW05] M. Birkner, J. A. López-Mimbela, and A. Wakolbinger, *Comparison results and steady states for the Fujita equation with fractional Laplacian* (English, with English and French summaries), Ann. Inst. H. Poincaré Anal. Non Linéaire **22** (2005), no. 1, 83–97, DOI 10.1016/j.anihpc.2004.05.002. MR2114412

[BG60] R. M. Blumenthal and R. K. Getoor, *Some theorems on stable processes*, Trans. Amer. Math. Soc. **95** (1960), 263–273, DOI 10.2307/1993291. MR0119247

[B55] S. Bochner, *Harmonic analysis and the theory of probability*, University of California Press, Berkeley and Los Angeles, 1955. MR0072370

[B59] S. Bochner, *Lectures on Fourier integrals. With an author's supplement on monotonic functions, Stieltjes integrals, and harmonic analysis*, Translated by Morris Tenenbaum and Harry Pollard. Annals of Mathematics Studies, No. 42, Princeton University Press, Princeton, N.J., 1959. MR0107124

[BC49] S. Bochner and K. Chandrasekharan, *Fourier Transforms*, Annals of Mathematics Studies, no. 19, Princeton University Press, Princeton, N. J.; Oxford University Press, London, 1949. MR0031582

[Bo97] K. Bogdan, *The boundary Harnack principle for the fractional Laplacian*, Studia Math. **123** (1997), no. 1, 43–80, DOI 10.4064/sm-123-1-43-80. MR1438304

[Bo99] K. Bogdan, *Representation of α-harmonic functions in Lipschitz domains*, Hiroshima Math. J. **29** (1999), no. 2, 227–243. MR1704245

[BB99] K. Bogdan and T. Byczkowski, *Probabilistic proof of boundary Harnack principle for α-harmonic functions*, Potential Anal. **11** (1999), no. 2, 135–156, DOI 10.1023/A:1008637918784. MR1703823

[BB99'] K. Bogdan and T. Byczkowski, *Potential theory for the α-stable Schrödinger operator on bounded Lipschitz domains*, Studia Math. **133** (1999), no. 1, 53–92, DOI 10.4064/sm-133-1-53-92. MR1671973

[BGR10] K. Bogdan, T. Grzywny, and M. Ryznar, *Heat kernel estimates for the fractional Laplacian with Dirichlet conditions*, Ann. Probab. **38** (2010), no. 5, 1901–1923, DOI 10.1214/10-AOP532. MR2722789

[BJ07] K. Bogdan and T. Jakubowski, *Estimates of heat kernel of fractional Laplacian perturbed by gradient operators*, Comm. Math. Phys. **271** (2007), no. 1, 179–198, DOI 10.1007/s00220-006-0178-y. MR2283957

[BKN02] K. Bogdan, T. Kulczycki, and A. Nowak, *Gradient estimates for harmonic and q-harmonic functions of symmetric stable processes*, Illinois J. Math. **46** (2002), no. 2, 541–556. MR1936936

[BI87] B. Bojarski and T. Iwaniec, *p-harmonic equation and quasiregular mappings*, Partial differential equations (Warsaw, 1984), Banach Center Publ., vol. 19, PWN, Warsaw, 1987, pp. 25–38. MR1055157

[BSV17] M. Bonforte, Y. Sire, and J. L. Vázquez, *Optimal existence and uniqueness theory for the fractional heat equation*, Nonlinear Anal. **153** (2017), 142–168, DOI 10.1016/j.na.2016.08.027. MR3614666

[BS02] A. N. Borodin and P. Salminen, *Handbook of Brownian motion—facts and formulae*, 2nd ed., Probability and its Applications, Birkhäuser Verlag, Basel, 2002. MR1912205

[BCPS13] C. Brändle, E. Colorado, A. de Pablo, and U. Sánchez, *A concave-convex elliptic problem involving the fractional Laplacian*, Proc. Roy. Soc. Edinburgh Sect. A **143** (2013), no. 1, 39–71, DOI 10.1017/S0308210511000175. MR3023003

[BP18] C. Brändle and A. de Pablo, *Nonlocal heat equations: regularizing effect, decay estimates and Nash inequalities*, Commun. Pure Appl. Anal. **17** (2018), no. 3, 1161–1178, DOI 10.3934/cpaa.2018056. MR3809118

[BL17] L. Brasco and E. Lindgren, *Higher Sobolev regularity for the fractional p-Laplace equation in the superquadratic case*, Adv. Math. **304** (2017), 300–354, DOI 10.1016/j.aim.2016.03.039. MR3558212

[BLS17] L. Brasco, E. Lindgren & A. Schikorra, *Higher Hölder regularity for the fractional p-Laplacian in the superquadratic case*, arXiv:1711.09835 (2017).

[Bu16] C. Bucur, *Some observations on the Green function for the ball in the fractional Laplace framework*, Commun. Pure Appl. Anal. **15** (2016), no. 2, 657–699, DOI 10.3934/cpaa.2016.15.657. MR3461641

[BuV16] C. Bucur and E. Valdinoci, *Nonlocal diffusion and applications*, Lecture Notes of the Unione Matematica Italiana, vol. 20, Springer, [Cham]; Unione Matematica Italiana, Bologna, 2016. MR3469920

[BuSq18] C. Bucur and M. Squassina, *Asymptotic mean value properties for fractional anisotropic operators*, J. Math. Anal. Appl. **466** (2018), no. 1, 107–126, DOI 10.1016/j.jmaa.2018.05.063. MR3818107

[Ca08] X. Cabré, *Elliptic PDE's in probability and geometry: symmetry and regularity of solutions*, Discrete Contin. Dyn. Syst. **20** (2008), no. 3, 425–457, DOI 10.3934/dcds.2008.20.425. MR2373200

[Ca17] X. Cabré, *Isoperimetric, Sobolev, and eigenvalue inequalities via the Alexandroff-Bakelman-Pucci method: a survey*, Chin. Ann. Math. Ser. B **38** (2017), no. 1, 201–214, DOI 10.1007/s11401-016-1067-0. MR3592160

[CR13] X. Cabré and J.-M. Roquejoffre, *The influence of fractional diffusion in Fisher-KPP equations*, Comm. Math. Phys. **320** (2013), no. 3, 679–722, DOI 10.1007/s00220-013-1682-5. MR3057187

[CS14] X. Cabré and Y. Sire, *Nonlinear equations for fractional Laplacians, I: Regularity, maximum principles, and Hamiltonian estimates*, Ann. Inst. H. Poincaré Anal. Non Linéaire **31** (2014), no. 1, 23–53, DOI 10.1016/j.anihpc.2013.02.001. MR3165278

[CS15] X. Cabré and Y. Sire, *Nonlinear equations for fractional Laplacians II: Existence, uniqueness, and qualitative properties of solutions*, Trans. Amer. Math. Soc. **367** (2015), no. 2, 911–941, DOI 10.1090/S0002-9947-2014-05906-0. MR3280032

[C37] R. Caccioppoli, *Sui teoremi d'esistenza di Riemann* (Italian), Ann. Scuola Norm. Sup. Pisa Cl. Sci. (2) **7** (1938), no. 2, 177–187. MR1556804

[CCV11] L. Caffarelli, C. H. Chan, and A. Vasseur, *Regularity theory for parabolic nonlinear integral operators*, J. Amer. Math. Soc. **24** (2011), no. 3, 849–869, DOI 10.1090/S0894-0347-2011-00698-X. MR2784330

[CDS16] L. Caffarelli, D. De Silva, and O. Savin, *Obstacle-type problems for minimal surfaces*, Comm. Partial Differential Equations **41** (2016), no. 8, 1303–1323, DOI 10.1080/03605302.2016.1192646. MR3532394

[CFMS81] L. Caffarelli, E. Fabes, S. Mortola, and S. Salsa, *Boundary behavior of nonnegative solutions of elliptic operators in divergence form*, Indiana Univ. Math. J. **30** (1981), no. 4, 621–640, DOI 10.1512/iumj.1981.30.30049. MR620271

[CF13] L. Caffarelli and A. Figalli, *Regularity of solutions to the parabolic fractional obstacle problem*, J. Reine Angew. Math. **680** (2013), 191–233, DOI 10.1515/crelle.2012.036. MR3100955

[CRS10] L. Caffarelli, J.-M. Roquejoffre, and O. Savin, *Nonlocal minimal surfaces*, Comm. Pure Appl. Math. **63** (2010), no. 9, 1111–1144, DOI 10.1002/cpa.20331. MR2675483

[CSS08] L. A. Caffarelli, S. Salsa, and L. Silvestre, *Regularity estimates for the solution and the free boundary of the obstacle problem for the fractional Laplacian*, Invent. Math. **171** (2008), no. 2, 425–461, DOI 10.1007/s00222-007-0086-6. MR2367025

[CSV15] L. Caffarelli, O. Savin, and E. Valdinoci, *Minimization of a fractional perimeter-Dirichlet integral functional*, Ann. Inst. H. Poincaré Anal. Non Linéaire **32** (2015), no. 4, 901–924, DOI 10.1016/j.anihpc.2014.04.004. MR3390089

[CS07] L. Caffarelli and L. Silvestre, *An extension problem related to the fractional Laplacian*, Comm. Partial Differential Equations **32** (2007), no. 7-9, 1245–1260, DOI 10.1080/03605300600987306. MR2354493

[CSo10] L. A. Caffarelli and P. E. Souganidis, *Convergence of nonlocal threshold dynamics approximations to front propagation*, Arch. Ration. Mech. Anal. **195** (2010), no. 1, 1–23, DOI 10.1007/s00205-008-0181-x. MR2564467

[CS16] L. A. Caffarelli and P. R. Stinga, *Fractional elliptic equations, Caccioppoli estimates and regularity*, Ann. Inst. H. Poincaré Anal. Non Linéaire **33** (2016), no. 3, 767–807, DOI 10.1016/j.anihpc.2015.01.004. MR3489634

[CV10] L. A. Caffarelli and A. Vasseur, *Drift diffusion equations with fractional diffusion and the quasi-geostrophic equation*, Ann. of Math. (2) **171** (2010), no. 3, 1903–1930, DOI 10.4007/annals.2010.171.1903. MR2680400

[CDPT] L. Capogna, D. Danielli, S. D. Pauls, and J. T. Tyson, *An introduction to the Heisenberg group and the sub-Riemannian isoperimetric problem*, Progress in Mathematics, vol. 259, Birkhäuser Verlag, Basel, 2007. MR2312336

[CG11] S.-Y. A. Chang and M. d. M. González, *Fractional Laplacian in conformal geometry*, Adv. Math. **226** (2011), no. 2, 1410–1432, DOI 10.1016/j.aim.2010.07.016. MR2737789

[CDL15] W. Chen, L. D'Ambrosio, and Y. Li, *Some Liouville theorems for the fractional Laplacian*, Nonlinear Anal. **121** (2015), 370–381, DOI 10.1016/j.na.2014.11.003. MR3348929

[CLL17] W. Chen, C. Li, and Y. Li, *A direct method of moving planes for the fractional Laplacian*, Adv. Math. **308** (2017), 404–437, DOI 10.1016/j.aim.2016.11.038. MR3600062

[CLO06] W. Chen, C. Li, and B. Ou, *Classification of solutions for an integral equation*, Comm. Pure Appl. Math. **59** (2006), no. 3, 330–343, DOI 10.1002/cpa.20116. MR2200258

[CLZ16] Y.-H. Chen, C. Liu, and Y. Zheng, *Existence results for the fractional Nirenberg problem*, J. Funct. Anal. **270** (2016), no. 11, 4043–4086, DOI 10.1016/j.jfa.2016.03.013. MR3484963

[CKK11] Z.-Q. Chen, P. Kim, and T. Kumagai, *Global heat kernel estimates for symmetric jump processes*, Trans. Amer. Math. Soc. **363** (2011), no. 9, 5021–5055, DOI 10.1090/S0002-9947-2011-05408-5. MR2806700

[Chu92] S.-K. Chua, *Extension theorems on weighted Sobolev spaces*, Indiana Univ. Math. J. **41** (1992), no. 4, 1027–1076, DOI 10.1512/iumj.1992.41.41053. MR1206339

[Ci38] G. Cimmino, *Sulle equazioni lineari alle derivate parziali del secondo ordine di tipo ellittico sopra una superficie chiusa* (Italian), Ann. Scuola Norm. Sup. Pisa Cl. Sci. (2) **7** (1938), no. 1, 73–96. MR1556798

[Ci38'] G. Cimmino, *Nuovo tipo di condizioni al contorno e nuovo metodo di trattazione per il problema generalizzato di Dirichlet*, Rend. Circ. Mat. Palermo **61** (1938), 177-221.

[CF74] R. R. Coifman and C. Fefferman, *Weighted norm inequalities for maximal functions and singular integrals*, Studia Math. **51** (1974), 241–250, DOI 10.4064/sm-51-3-241-250. MR0358205

[CC03] A. Córdoba and D. Córdoba, *A pointwise estimate for fractionary derivatives with applications to partial differential equations*, Proc. Natl. Acad. Sci. USA **100** (2003), no. 26, 15316–15317, DOI 10.1073/pnas.2036515100. MR2032097

[CC04] A. Córdoba and D. Córdoba, *A maximum principle applied to quasi-geostrophic equations*, Comm. Math. Phys. **249** (2004), no. 3, 511–528, DOI 10.1007/s00220-004-1055-1. MR2084005

[CM15] A. Córdoba and Á. D. Martínez, *A pointwise inequality for fractional Laplacians*, Adv. Math. **280** (2015), 79–85, DOI 10.1016/j.aim.2015.02.018. MR3350213

[CT04] A. Cotsiolis and N. K. Tavoularis, *Best constants for Sobolev inequalities for higher order fractional derivatives*, J. Math. Anal. Appl. **295** (2004), no. 1, 225–236, DOI 10.1016/j.jmaa.2004.03.034. MR2064421

[CD99] T. Coulhon and X. T. Duong, *Riesz transforms for $1 \leq p \leq 2$*, Trans. Amer. Math. Soc. **351** (1999), no. 3, 1151–1169, DOI 10.1090/S0002-9947-99-02090-5. MR1458299

[Co16] M. Cozzi, *Regularity results and Harnack inequalities for minimizers and solutions of nonlocal problems: a unified approach via fractional De Giorgi classes*, J. Funct. Anal. **272** (2017), no. 11, 4762–4837, DOI 10.1016/j.jfa.2017.02.016. MR3630640

[DGN06] D. Danielli, N. Garofalo, and D.-M. Nhieu, *Non-doubling Ahlfors measures, perimeter measures, and the characterization of the trace spaces of Sobolev functions in Carnot-Carathéodory spaces*, Mem. Amer. Math. Soc. **182** (2006), no. 857, x+119, DOI 10.1090/memo/0857. MR2229731

[DGPT17] D. Danielli, N. Garofalo, A. Petrosyan, and T. To, *Optimal regularity and the free boundary in the parabolic Signorini problem*, Mem. Amer. Math. Soc. **249** (2017), no. 1181, v + 103, DOI 10.1090/memo/1181. MR3709717

[DS17] D. Danielli & S. Salsa, *Obstacle Problems Involving The Fractional Laplacean*, Recent Developments in Nonlocal Theory Ed. by G. Palatucci & T. Kuusi, to appear.

[DDGW17] A. DelaTorre, M. del Pino, M. d. M. González, and J. Wei, *Delaunay-type singular solutions for the fractional Yamabe problem*, Math. Ann. **369** (2017), no. 1-2, 597–626, DOI 10.1007/s00208-016-1483-1. MR3694655

[DQRV12] A. de Pablo, F. Quirós, A. Rodríguez, and J. L. Vázquez, *A general fractional porous medium equation*, Comm. Pure Appl. Math. **65** (2012), no. 9, 1242–1284, DOI 10.1002/cpa.21408. MR2954615

[DS16] D. De Silva and O. Savin, *Boundary Harnack estimates in slit domains and applications to thin free boundary problems*, Rev. Mat. Iberoam. **32** (2016), no. 3, 891–912, DOI 10.4171/RMI/902. MR3556055

[DS15] D. De Silva and O. Savin, *C^∞ regularity of certain thin free boundaries*, Indiana Univ. Math. J. **64** (2015), no. 5, 1575–1608, DOI 10.1512/iumj.2015.64.5632. MR3418452

[DB83] E. DiBenedetto, *$C^{1+\alpha}$ local regularity of weak solutions of degenerate elliptic equations*, Nonlinear Anal. **7** (1983), no. 8, 827–850, DOI 10.1016/0362-546X(83)90061-5. MR709038

[DKP14] A. Di Castro, T. Kuusi, and G. Palatucci, *Nonlocal Harnack inequalities*, J. Funct. Anal. **267** (2014), no. 6, 1807–1836, DOI 10.1016/j.jfa.2014.05.023. MR3237774

[DKP16] A. Di Castro, T. Kuusi, and G. Palatucci, *Local behavior of fractional p-minimizers*, Ann. Inst. H. Poincaré Anal. Non Linéaire **33** (2016), no. 5, 1279–1299, DOI 10.1016/j.anihpc.2015.04.003. MR3542614

[DMV17] S. Dipierro, M. Medina, and E. Valdinoci, *Fractional elliptic problems with critical growth in the whole of $\mathbb{R}^n$*, Appunti. Scuola Normale Superiore di Pisa (Nuova Serie) [Lecture Notes. Scuola Normale Superiore di Pisa (New Series)], vol. 15, Edizioni della Normale, Pisa, 2017. MR3617721

[DMPS16] S. Dipierro, L. Montoro, I. Peral, and B. Sciunzi, *Qualitative properties of positive solutions to nonlocal critical problems involving the Hardy-Leray potential*, Calc. Var. Partial Differential Equations **55** (2016), no. 4, Art. 99, 29, DOI 10.1007/s00526-016-1032-5. MR3528440

[DSV16] S. Dipierro, O. Savin, and E. Valdinoci, *Graph properties for nonlocal minimal surfaces*, Calc. Var. Partial Differential Equations **55** (2016), no. 4, Art. 86, 25, DOI 10.1007/s00526-016-1020-9. MR3516886

[DSV16'] S. Dipierro, O. Savin & E. Valdinoci, *Definition of fractional Laplacian for functions with polynomial growth*, arXiv:1610.04663 (2016).

[DSV17] S. Dipierro, O. Savin, and E. Valdinoci, *All functions are locally s-harmonic up to a small error*, J. Eur. Math. Soc. (JEMS) **19** (2017), no. 4, 957–966, DOI 10.4171/JEMS/684. MR3626547

[DSV17'] S. Dipierro, O. Savin, and E. Valdinoci, *Boundary behavior of nonlocal minimal surfaces*, J. Funct. Anal. **272** (2017), no. 5, 1791–1851, DOI 10.1016/j.jfa.2016.11.016. MR3596708

[DPV12] E. Di Nezza, G. Palatucci, and E. Valdinoci, *Hitchhiker's guide to the fractional Sobolev spaces*, Bull. Sci. Math. **136** (2012), no. 5, 521–573, DOI 10.1016/j.bulsci.2011.12.004. MR2944369

[DGV03] J. Droniou, T. Gallouet, and J. Vovelle, *Global solution and smoothing effect for a non-local regularization of a hyperbolic equation*, J. Evol. Equ. **3** (2003), no. 3, 499–521, DOI 10.1007/s00028-003-0503-1. Dedicated to Philippe Bénilan. MR2019032

[DP70] N. du Plessis, *An introduction to potential theory*, Hafner Publishing Co., Darien, Conn.; Oliver and Boyd, Edinburgh, 1970. University Mathematical Monographs, No. 7. MR0435422

[DL69] G. Duvaut and J.-L. Lions, *Inequalities in mechanics and physics*, Springer-Verlag, Berlin-New York, 1976. Translated from the French by C. W. John; Grundlehren der Mathematischen Wissenschaften, 219. MR0521262

[D12] B. Dyda, *Fractional calculus for power functions and eigenvalues of the fractional Laplacian*, Fract. Calc. Appl. Anal. **15** (2012), no. 4, 536–555, DOI 10.2478/s13540-012-0038-8. MR2974318

[DKK17] B. Dyda, A. Kuznetsov, and M. Kwaśnicki, *Eigenvalues of the fractional Laplace operator in the unit ball*, J. Lond. Math. Soc. (2) **95** (2017), no. 2, 500–518, DOI 10.1112/jlms.12024. MR3656279

[Dy65] E. B. Dynkin, *Markov processes. Vols. I, II*, Translated with the authorization and assistance of the author by J. Fabius, V. Greenberg, A. Maitra, G. Majone. Die Grundlehren der Mathematischen Wissenschaften, Bände 121, vol. 122, Academic Press Inc., Publishers, New York; Springer-Verlag, Berlin-Göttingen-Heidelberg, 1965. MR0193671

[EP16] C. L. Epstein and C. A. Pop, *Regularity for the supercritical fractional Laplacian with drift*, J. Geom. Anal. **26** (2016), no. 2, 1231–1268, DOI 10.1007/s12220-015-9590-x. MR3472835

[EMOT53] A. Erdélyi, W. Magnus, F. Oberhettinger, and F. G. Tricomi, *Higher transcendental functions. Vols. I, II*, McGraw-Hill Book Company, Inc., New York-Toronto-London, 1953. Based, in part, on notes left by Harry Bateman. MR0058756

[EG15] L. C. Evans and R. F. Gariepy, *Measure theory and fine properties of functions*, Revised edition, Textbooks in Mathematics, CRC Press, Boca Raton, FL, 2015. MR3409135

[FJK82] E. Fabes, D. Jerison, and C. Kenig, *The Wiener test for degenerate elliptic equations* (English, with French summary), Ann. Inst. Fourier (Grenoble) **32** (1982), no. 3, vi, 151–182. MR688024

[FKS81] E. B. Fabes, C. E. Kenig, and R. P. Serapioni, *The local regularity of solutions of degenerate elliptic equations*, Comm. Partial Differential Equations **7** (1982), no. 1, 77–116, DOI 10.1080/03605308208820218. MR643158

[Fa16] M. M. Fall, *Entire s-harmonic functions are affine*, Proc. Amer. Math. Soc. **144** (2016), no. 6, 2587–2592, DOI 10.1090/proc/13021. MR3477075

[FF14] M. M. Fall and V. Felli, *Unique continuation property and local asymptotics of solutions to fractional elliptic equations*, Comm. Partial Differential Equations **39** (2014), no. 2, 354–397, DOI 10.1080/03605302.2013.825918. MR3169789

[FJ15] M. M. Fall and S. Jarohs, *Overdetermined problems with fractional Laplacian*, ESAIM Control Optim. Calc. Var. **21** (2015), no. 4, 924–938, DOI 10.1051/cocv/2014048. MR3395749

[FV17] A. Farina and E. Valdinoci, *Regularity and rigidity theorems for a class of anisotropic nonlocal operators*, Manuscripta Math. **153** (2017), no. 1-2, 53–70, DOI 10.1007/s00229-016-0875-6. MR3635973

[FV17'] A. Farina and E. Valdinoci, *A rigidity result for non-local semilinear equations*, Proc. Roy. Soc. Edinburgh Sect. A **147** (2017), no. 5, 1009–1018, DOI 10.1017/S0308210516000391. MR3705316

[FdlL86] C. Fefferman and R. de la Llave, *Relativistic stability of matter. I*, Rev. Mat. Iberoamericana **2** (1986), no. 1-2, 119–213, DOI 10.4171/RMI/30. MR864658

[FK13] M. Felsinger and M. Kassmann, *Local regularity for parabolic nonlocal operators*, Comm. Partial Differential Equations **38** (2013), no. 9, 1539–1573, DOI 10.1080/03605302.2013.808211. MR3169755

[FKV15] M. Felsinger, M. Kassmann, and P. Voigt, *The Dirichlet problem for nonlocal operators*, Math. Z. **279** (2015), no. 3-4, 779–809, DOI 10.1007/s00209-014-1394-3. MR3318251

[FR17] X. Fernández-Real and X. Ros-Oton, *Regularity theory for general stable operators: parabolic equations*, J. Funct. Anal. **272** (2017), no. 10, 4165–4221, DOI 10.1016/j.jfa.2017.02.015. MR3626038

[FF15] F. Ferrari and B. Franchi, *Harnack inequality for fractional sub-Laplacians in Carnot groups*, Math. Z. **279** (2015), no. 1-2, 435–458, DOI 10.1007/s00209-014-1376-5. MR3299862

[FV12] F. Ferrari and I. E. Verbitsky, *Radial fractional Laplace operators and Hessian inequalities*, J. Differential Equations **253** (2012), no. 1, 244–272, DOI 10.1016/j.jde.2012.03.024. MR2917408

[FoS17] M. Focardi and E. Spadaro, *On the Measure and the Structure of the Free Boundary of the Lower Dimensional Obstacle Problem*, Arch. Ration. Mech. Anal. **230** (2018), no. 1, 125–184, DOI 10.1007/s00205-018-1242-4. MR3840912

[Fo89] G. B. Folland, *Harmonic analysis in phase space*, Annals of Mathematics Studies, vol. 122, Princeton University Press, Princeton, NJ, 1989. MR983366

[FL82] B. Franchi and E. Lanconelli, *De Giorgi's theorem for a class of strongly degenerate elliptic equations* (English, with Italian summary), Atti Accad. Naz. Lincei Rend. Cl. Sci. Fis. Mat. Natur. (8) **72** (1982), no. 5, 273–277 (1983). MR728257

[FL83] B. Franchi and E. Lanconelli, *Une métrique associée à une classe d'opérateurs elliptiques dégénérés* (French), Rend. Sem. Mat. Univ. Politec. Torino **Special Issue** (1983), 105–114 (1984). Conference on linear partial and pseudodifferential operators (Torino, 1982). MR745979

[FL83'] B. Franchi and E. Lanconelli, *Hölder regularity theorem for a class of linear nonuniformly elliptic operators with measurable coefficients*, Ann. Scuola Norm. Sup. Pisa Cl. Sci. (4) **10** (1983), no. 4, 523–541. MR753153

[FL84] B. Franchi and E. Lanconelli, *An embedding theorem for Sobolev spaces related to nonsmooth vector fields and Harnack inequality*, Comm. Partial Differential Equations **9** (1984), no. 13, 1237–1264, DOI 10.1080/03605308408820362. MR764663

[FL85] B. Franchi and E. Lanconelli, *Une condition géométrique pour l'inégalité de Harnack* (French), J. Math. Pures Appl. (9) **64** (1985), no. 3, 237–256. MR823405

[FS87] B. Franchi and R. Serapioni, *Pointwise estimates for a class of strongly degenerate elliptic operators: a geometrical approach*, Ann. Scuola Norm. Sup. Pisa Cl. Sci. (4) **14** (1987), no. 4, 527–568 (1988). MR963489

[Fr17] R. L. Frank, *Eigenvalue bounds for the fractional Laplacian: A review*, arXiv:1603.09736v2 (2017).

[FGMT15] R. L. Frank, M. d. M. González, D. D. Monticelli, and J. Tan, *An extension problem for the CR fractional Laplacian*, Adv. Math. **270** (2015), 97–137, DOI 10.1016/j.aim.2014.09.026. MR3286532

[FLe13] R. L. Frank and E. Lenzmann, *Uniqueness of non-linear ground states for fractional Laplacians in* $\mathbb{R}$, Acta Math. **210** (2013), no. 2, 261–318, DOI 10.1007/s11511-013-0095-9. MR3070568

[FLeS16] R. L. Frank, E. Lenzmann, and L. Silvestre, *Uniqueness of radial solutions for the fractional Laplacian*, Comm. Pure Appl. Math. **69** (2016), no. 9, 1671–1726, DOI 10.1002/cpa.21591. MR3530361

[Ga57] E. Gagliardo, *Caratterizzazioni delle tracce sulla frontiera relative ad alcune classi di funzioni in n variabili* (Italian), Rend. Sem. Mat. Univ. Padova **27** (1957), 284–305. MR0102739

[Ga70] L. Gårding, *Marcel Riesz in memoriam*, Acta Math. **124** (1970), no. nos. 1-2, x-xi, DOI 10.1007/BF02394565. MR0256837

[G84] N. Garofalo, *Second order parabolic equations in nonvariational forms: boundary Harnack principle and comparison theorems for nonnegative solutions*, Ann. Mat. Pura Appl. (4) **138** (1984), 267–296, DOI 10.1007/BF01762548. MR779547

[G93] N. Garofalo, *Unique continuation for a class of elliptic operators which degenerate on a manifold of arbitrary codimension*, J. Differential Equations **104** (1993), no. 1, 117–146, DOI 10.1006/jdeq.1993.1065. MR1224123

[G16] N. Garofalo, *Hypoelliptic operators and some aspects of analysis and geometry of sub-Riemannian spaces*, Geometry, analysis and dynamics on sub-Riemannian manifolds. Vol. 1, EMS Ser. Lect. Math., Eur. Math. Soc., Zürich, 2016, pp. 123–257. MR3587667

[GLe89] N. Garofalo and J. L. Lewis, *A symmetry result related to some overdetermined boundary value problems*, Amer. J. Math. **111** (1989), no. 1, 9–33, DOI 10.2307/2374477. MR980297

[GL86] N. Garofalo and F.-H. Lin, *Monotonicity properties of variational integrals,* A_p *weights and unique continuation*, Indiana Univ. Math. J. **35** (1986), no. 2, 245–268, DOI 10.1512/iumj.1986.35.35015. MR833393

[GL87] N. Garofalo and F.-H. Lin, *Unique continuation for elliptic operators: a geometric-variational approach*, Comm. Pure Appl. Math. **40** (1987), no. 3, 347–366, DOI 10.1002/cpa.3160400305. MR882069

[GP09] N. Garofalo and A. Petrosyan, *Some new monotonicity formulas and the singular set in the lower dimensional obstacle problem*, Invent. Math. **177** (2009), no. 2, 415–461, DOI 10.1007/s00222-009-0188-4. MR2511747

[GPPS17] N. Garofalo, A. Petrosyan, C. A. Pop, and M. Smit Vega Garcia, *Regularity of the free boundary for the obstacle problem for the fractional Laplacian with drift*, Ann. Inst. H. Poincaré Anal. Non Linéaire **34** (2017), no. 3, 533–570, DOI 10.1016/j.anihpc.2016.03.001. MR3633735

[GPS16] N. Garofalo, A. Petrosyan, and M. Smit Vega Garcia, *An epiperimetric inequality approach to the regularity of the free boundary in the Signorini problem with variable coefficients* (English, with English and French summaries), J. Math. Pures Appl. (9) **105** (2016), no. 6, 745–787, DOI 10.1016/j.matpur.2015.11.013. MR3491531

[GRO17] N. Garofalo & X. Ros-Oton, *Structure and regularity of the singular set in the obstacle problem for the fractional Laplacean*, Revista Matematica Iberoamericana, to appear.

[GS14] N. Garofalo and M. Smit Vega Garcia, *New monotonicity formulas and the optimal regularity in the Signorini problem with variable coefficients*, Adv. Math. **262** (2014), 682–750, DOI 10.1016/j.aim.2014.05.021. MR3228440

[GS64] I. M. Gel'fand and G. E. Shilov, *Generalized functions. Vol. I: Properties and operations*, Translated by Eugene Saletan, Academic Press, New York-London, 1964. MR0166596

[Ge61] R. K. Getoor, *First passage times for symmetric stable processes in space*, Trans. Amer. Math. Soc. **101** (1961), 75–90, DOI 10.2307/1993412. MR0137148

[Go16] M. del Mar González, *Recent progress on the fractional Laplacean in conformal geometry*, arXiv:1609.08988 (2016).

[GSS14] M. d. M. González, M. Sáez, and Y. Sire, *Layer solutions for the fractional Laplacian on hyperbolic space: existence, uniqueness and qualitative properties*, Ann. Mat. Pura Appl. (4) **193** (2014), no. 6, 1823–1850, DOI 10.1007/s10231-013-0358-2. MR3275264

[GR80] H. van Haeringen and L. P. Kok, *Table errata: Table of integrals, series, and products [corrected and enlarged edition, Academic Press, New York, 1980; MR 81g:33001] by I. S. Gradshteyn [I. S. Gradshteĭn] and I. M. Ryzhik*, Math. Comp. **39** (1982), no. 160, 747–757. MR669666

[GZ03] C. R. Graham and M. Zworski, *Scattering matrix in conformal geometry*, Invent. Math. **152** (2003), no. 1, 89–118, DOI 10.1007/s00222-002-0268-1. MR1965361

[GI17] A. Greco and A. Iannizzotto, *Existence and convexity of solutions of the fractional heat equation*, Commun. Pure Appl. Anal. **16** (2017), no. 6, 2201–2226, DOI 10.3934/cpaa.2017109. MR3693879

[Gr14] G. Grubb, *Local and nonlocal boundary conditions for μ-transmission and fractional elliptic pseudodifferential operators*, Anal. PDE **7** (2014), no. 7, 1649–1682, DOI 10.2140/apde.2014.7.1649. MR3293447

[Gr15] G. Grubb, *Fractional Laplacians on domains, a development of Hörmander's theory of μ-transmission pseudodifferential operators*, Adv. Math. **268** (2015), 478–528, DOI 10.1016/j.aim.2014.09.018. MR3276603

[Gr16] G. Grubb, *Integration by parts and Pohozaev identities for space-dependent fractional-order operators*, J. Differential Equations **261** (2016), no. 3, 1835–1879, DOI 10.1016/j.jde.2016.04.017. MR3501834

[Gru70] V. V. Grušin, *A certain class of hypoelliptic operators* (Russian), Mat. Sb. (N.S.) **83 (125)** (1970), 456–473. MR0279436

[Gru71] V. V. Grušin, *A certain class of elliptic pseudodifferential operators that are degenerate on a submanifold* (Russian), Mat. Sb. (N.S.) **84 (126)** (1971), 163–195. MR0283630

[Gui09] N. Guillen, *Optimal regularity for the Signorini problem*, Calc. Var. Partial Differential Equations **36** (2009), no. 4, 533–546, DOI 10.1007/s00526-009-0242-5. MR2558329

[H54] J. Hadamard, *Extension à l'équation de la chaleur d'un théorème de A. Harnack* (French), Rend. Circ. Mat. Palermo (2) **3** (1954), 337–346 (1955), DOI 10.1007/BF02849264. MR0068713

[Har87] C. G. Axel Harnack, *Die Grundlagen der Theorie des logarithmischen Potentiales und der eindeutigen Potentialfunktion in der Ebene*, Teubner, Leipzig, 1887.

[He69] L. L. Helms, *Introduction to potential theory*, Pure and Applied Mathematics, Vol. XXII, Wiley-Interscience A Division of John Wiley & Sons, New York-London-Sydney, 1969. MR0261018

[Ho66] L. Hörmander, *Pseudo-differential operators and hypoelliptic equations*, Singular integrals (Proc. Sympos. Pure Math., Vol. X, Chicago, Ill., 1966), Amer. Math. Soc., Providence, R.I., 1967, pp. 138–183. MR0383152

[Ho67] L. Hörmander, *Hypoelliptic second order differential equations*, Acta Math. **119** (1967), 147–171, DOI 10.1007/BF02392081. MR0222474

[IMS16] A. Iannizzotto, S. Mosconi, and M. Squassina, *Global Hölder regularity for the fractional p-Laplacian*, Rev. Mat. Iberoam. **32** (2016), no. 4, 1353–1392, DOI 10.4171/RMI/921. MR3593528

[IN10] H. Ishii and G. Nakamura, *A class of integral equations and approximation of p-Laplace equations*, Calc. Var. Partial Differential Equations **37** (2010), no. 3-4, 485–522, DOI 10.1007/s00526-009-0274-x. MR2592984

[JKS17] S. Jarohs, T. Kulczycki & P. Salani, *Starshape of the superlevel sets of solutions to equations involving the fractional Laplacian in starshaped rings*, arXiv:1701.02367 (2017).

[JW14] S. Jarohs and T. Weth, *Asymptotic symmetry for a class of nonlinear fractional reaction-diffusion equations*, Discrete Contin. Dyn. Syst. **34** (2014), no. 6, 2581–2615, DOI 10.3934/dcds.2014.34.2581. MR3177650

[Je81] D. S. Jerison, *The Dirichlet problem for the Kohn Laplacian on the Heisenberg group. I*, J. Funct. Anal. **43** (1981), no. 1, 97–142, DOI 10.1016/0022-1236(81)90040-9. MR639800

[JK82] D. S. Jerison and C. E. Kenig, *Boundary behavior of harmonic functions in non-tangentially accessible domains*, Adv. in Math. **46** (1982), no. 1, 80–147, DOI 10.1016/0001-8708(82)90055-X. MR676988

[JK85] D. Jerison and C. E. Kenig, *Unique continuation and absence of positive eigenvalues for Schrödinger operators*, Ann. of Math. (2) **121** (1985), no. 3, 463–494, DOI 10.2307/1971205. With an appendix by E. M. Stein. MR794370

[JN17] Y. Jhaveri and R. Neumayer, *Higher regularity of the free boundary in the obstacle problem for the fractional Laplacian*, Adv. Math. **311** (2017), 748–795, DOI 10.1016/j.aim.2017.03.006. MR3628230

[KH73] L. I. Kamynin and B. N. Himčenko, *The analogues of the Giraud theorem for a second order parabolic equation* (Russian), Sibirsk. Mat. Ž. **14** (1973), 86–110, 236. MR0333453

[Ka09] M. Kassmann, *A priori estimates for integro-differential operators with measurable kernels*, Calc. Var. Partial Differential Equations **34** (2009), no. 1, 1–21, DOI 10.1007/s00526-008-0173-6. MR2448308

[Ka11] M. Kassmann, *Harnack inequalities and Hölder regularity estimates for nonlocal operator revisited*, available as SFB 701-preprint no. 11015, 2011, https://www.math.uni-bielefeld.de/kassmann

[Ka11'] M. Kassmann, *A new formulation of Harnack's inequality for nonlocal operators* (English, with English and French summaries), C. R. Math. Acad. Sci. Paris **349** (2011), no. 11-12, 637–640, DOI 10.1016/j.crma.2011.04.014. MR2817382

[KPS15] H. Koch, A. Petrosyan, and W. Shi, *Higher regularity of the free boundary in the elliptic Signorini problem*, Nonlinear Anal. **126** (2015), 3–44, DOI 10.1016/j.na.2015.01.007. MR3388870

[KRS16] H. Koch, A. Rüland, and W. Shi, *The variable coefficient thin obstacle problem: Carleman inequalities*, Adv. Math. **301** (2016), 820–866, DOI 10.1016/j.aim.2016.06.023. MR3539391

[KRS17] H. Koch, A. Rüland, and W. Shi, *The variable coefficient thin obstacle problem: optimal regularity and regularity of the regular free boundary*, Ann. Inst. H. Poincaré Anal. Non Linéaire **34** (2017), no. 4, 845–897, DOI 10.1016/j.anihpc.2016.08.001. MR3661863

[KN65] J. J. Kohn and L. Nirenberg, *An algebra of pseudo-differential operators*, Comm. Pure Appl. Math. **18** (1965), 269–305, DOI 10.1002/cpa.3160180121. MR0176362

[KN65'] J. J. Kohn and L. Nirenberg, *Non-coercive boundary value problems*, Comm. Pure Appl. Math. **18** (1965), 443–492, DOI 10.1002/cpa.3160180305. MR0181815

[Ko00] V. Kolokoltsov, *Symmetric stable laws and stable-like jump-diffusions*, Proc. London Math. Soc. (3) **80** (2000), no. 3, 725–768, DOI 10.1112/S0024611500012314. MR1744782

[K89] T. Kolsrud, *Traces of harmonic functions, capacities, and traces of symmetric Markov processes*, J. Theoret. Probab. **2** (1989), no. 4, 399–418, DOI 10.1007/BF01051874. MR1011195

[Ko95] T. Komatsu, *Uniform estimates for fundamental solutions associated with non-local Dirichlet forms*, Osaka J. Math. **32** (1995), no. 4, 833–860. MR1380729

[KMS15] T. Kuusi, G. Mingione, and Y. Sire, *Nonlocal equations with measure data*, Comm. Math. Phys. **337** (2015), no. 3, 1317–1368, DOI 10.1007/s00220-015-2356-2. MR3339179

[KP17] T. Kuusi & G. Palatucci, *Recent Developments in Nonlocal Theory*, Ed. by G. Palatucci & T. Kuusi, De Gruyter Poland Ltd., Berlin/Boston (2017).

[Kwa17] M. Kwaśnicki, *Ten equivalent definitions of the fractional Laplace operator*, Fract. Calc. Appl. Anal. **20** (2017), no. 1, 7–51, DOI 10.1515/fca-2017-0002. MR3613319

[La72] N. S. Landkof, *Foundations of modern potential theory*, Springer-Verlag, New York-Heidelberg, 1972. Translated from the Russian by A. P. Doohovskoy; Die Grundlehren der mathematischen Wissenschaften, Band 180. MR0350027

[Le72] N. N. Lebedev, *Special functions and their applications*, Dover Publications, Inc., New York, 1972. Revised edition, translated from the Russian and edited by Richard A. Silverman; Unabridged and corrected republication. MR0350075

[Le83] J. L. Lewis, *Regularity of the derivatives of solutions to certain degenerate elliptic equations*, Indiana Univ. Math. J. **32** (1983), no. 6, 849–858, DOI 10.1512/iumj.1983.32.32058. MR721568

[LY86] P. Li and S.-T. Yau, *On the parabolic kernel of the Schrödinger operator*, Acta Math. **156** (1986), no. 3-4, 153–201, DOI 10.1007/BF02399203. MR834612

[LL17] E. Lindgren and P. Lindqvist, *Perron's method and Wiener's theorem for a nonlocal equation*, Potential Anal. **46** (2017), no. 4, 705–737, DOI 10.1007/s11118-016-9603-9. MR3636596

[LM72] J.-L. Lions and E. Magenes, *Non-homogeneous boundary value problems and applications. Vol. I*, Springer-Verlag, New York-Heidelberg, 1972. Translated from the French by P. Kenneth; Die Grundlehren der mathematischen Wissenschaften, Band 181. MR0350177

[Ma88] J. J. Manfredi, *p-harmonic functions in the plane*, Proc. Amer. Math. Soc. **103** (1988), no. 2, 473–479, DOI 10.2307/2047164. MR943069

[Mi74] K. Miller, *Nonunique continuation for uniformly parabolic and elliptic equations in self-adjoint divergence form with Hölder continuous coefficients*, Arch. Rational Mech. Anal. **54** (1974), 105–117, DOI 10.1007/BF00247634. MR0342822

[MO69] S. A. Molčanov and E. Ostrovskiĭ, *Symmetric stable processes as traces of degenerate diffusion processes.* (Russian, with English summary), Teor. Verojatnost. i Primenen. **14** (1969), 127–130. MR0247668

[MRS16] G. Molica Bisci, V. D. Radulescu, and R. Servadei, *Variational methods for nonlocal fractional problems*, Encyclopedia of Mathematics and its Applications, vol. 162, Cambridge University Press, Cambridge, 2016. With a foreword by Jean Mawhin. MR3445279

[MS65] B. Muckenhoupt and E. M. Stein, *Classical expansions and their relation to conjugate harmonic functions*, Trans. Amer. Math. Soc. **118** (1965), 17–92, DOI 10.2307/1993944. MR0199636

[NS16] K. Nyström and O. Sande, *Extension properties and boundary estimates for a fractional heat operator*, Nonlinear Anal. **140** (2016), 29–37, DOI 10.1016/j.na.2016.02.027. MR3492726

[OS74] K. B. Oldham and J. Spanier, *The fractional calculus*, Academic Press [A subsidiary of Harcourt Brace Jovanovich, Publishers], New York-London, 1974. Theory and applications of differentiation and integration to arbitrary order; With an annotated chronological bibliography by Bertram Ross; Mathematics in Science and Engineering, Vol. 111. MR0361633

[PP15] A. Petrosyan and C. A. Pop, *Optimal regularity of solutions to the obstacle problem for the fractional Laplacian with drift*, J. Funct. Anal. **268** (2015), no. 2, 417–472, DOI 10.1016/j.jfa.2014.10.009. MR3283160

[PSU12] A. Petrosyan, H. Shahgholian, and N. Uraltseva, *Regularity of free boundaries in obstacle-type problems*, Graduate Studies in Mathematics, vol. 136, American Mathematical Society, Providence, RI, 2012. MR2962060

[P52] R. S. Phillips, *On the generation of semigroups of linear operators*, Pacific J. Math. **2** (1952), 343–369. MR0050797

[P54] B. Pini, *Sulla soluzione generalizzata di Wiener per il primo problema di valori al contorno nel caso parabolico* (Italian), Rend. Sem. Mat. Univ. Padova **23** (1954), 422–434. MR0065794

[Pl63] A. Pliś, *On non-uniqueness in Cauchy problem for an elliptic second order differential equation*, Bull. Acad. Polon. Sci. Sér. Sci. Math. Astronom. Phys. **11** (1963), 95–100. MR0153959

[Po96] C.-C. Poon, *Unique continuation for parabolic equations*, Comm. Partial Differential Equations **21** (1996), no. 3-4, 521–539, DOI 10.1080/03605309608821195. MR1387458

[PBM88] A. P. Prudnikov, Yu. A. Brychkov, and O. I. Marichev, *Integrals and series. Vol. 2*, 2nd ed., Gordon & Breach Science Publishers, New York, 1988. Special functions; Translated from the Russian by N. M. Queen. MR950173

[PT69] W. E. Pruitt and S. J. Taylor, *The potential kernel and hitting probabilities for the general stable process in R^N*, Trans. Amer. Math. Soc. **146** (1969), 299–321, DOI 10.2307/1995174. MR0250372

[PS16] P. Pucci and S. Saldi, *Critical stationary Kirchhoff equations in $\mathbb{R}^N$ involving nonlocal operators*, Rev. Mat. Iberoam. **32** (2016), no. 1, 1–22, DOI 10.4171/RMI/879. MR3470662

[RS75] M. Reed and B. Simon, *Methods of modern mathematical physics. II. Fourier analysis, self-adjointness*, Academic Press [Harcourt Brace Jovanovich, Publishers], New York-London, 1975. MR0493420

[Re82] R. C. Reilly, *Mean curvature, the Laplacian, and soap bubbles*, Amer. Math. Monthly **89** (1982), no. 3, 180–188, 197–198, DOI 10.2307/2320201. MR645791

[Rel40] F. Rellich, *Darstellung der Eigenwerte von $\Delta u + \lambda u = 0$ durch ein Randintegral* (German), Math. Z. **46** (1940), 635–636, DOI 10.1007/BF01181459. MR0002456

[R38] M. Riesz, *Intégrales de Riemann-Liouville et potentiels*, Acta Sci. Math. Szeged, **9** (1938), 1-42.

[R49] M. Riesz, *L'intégrale de Riemann-Liouville et le problème de Cauchy* (French), Acta Math. **81** (1949), 1–223, DOI 10.1007/BF02395016. MR0030102

[RO15] X. Ros-Oton, *Nonlocal elliptic equations in bounded domains: a survey*, Publ. Mat. **60** (2016), no. 1, 3–26. MR3447732

[RO17] X. Ros-Oton, *Obstacle problems and free boundaries: an overview*, arXiv:1707.00992 (2017).

[RO17'] X. Ros-Oton, *Boundary regularity, Pohozaev identities and nonexistence results*, Recent Developments in Nonlocal Theory Ed. by G. Palatucci & T. Kuusi, to appear.

[ROS14] X. Ros-Oton and J. Serra, *The Dirichlet problem for the fractional Laplacian: regularity up to the boundary* (English, with English and French summaries), J. Math. Pures Appl. (9) **101** (2014), no. 3, 275–302, DOI 10.1016/j.matpur.2013.06.003. MR3168912

[ROS14'] X. Ros-Oton and J. Serra, *The Pohozaev identity for the fractional Laplacian*, Arch. Ration. Mech. Anal. **213** (2014), no. 2, 587–628, DOI 10.1007/s00205-014-0740-2. MR3211861

[Ro77] B. Ross, *The development of fractional calculus 1695–1900* (English, with German and French summaries), Historia Math. **4** (1977), 75–89, DOI 10.1016/0315-0860(77)90039-8. MR0444394

[Ru15] A. Rüland, *Unique continuation for fractional Schrödinger equations with rough potentials*, Comm. Partial Differential Equations **40** (2015), no. 1, 77–114, DOI 10.1080/03605302.2014.905594. MR3268922

[SKM93] S. G. Samko, A. A. Kilbas, and O. I. Marichev, *Fractional integrals and derivatives*, Gordon and Breach Science Publishers, Yverdon, 1993. Theory and applications; Edited and with a foreword by S. M. Nikol′skiĭ; Translated from the 1987 Russian original; Revised by the authors. MR1347689

[SV14] O. Savin and E. Valdinoci, *Density estimates for a variational model driven by the Gagliardo norm* (English, with English and French summaries), J. Math. Pures Appl. (9) **101** (2014), no. 1, 1–26, DOI 10.1016/j.matpur.2013.05.001. MR3133422

[Sch66] L. Schwartz, *Théorie des distributions* (French), Publications de l'Institut de Mathématique de l'Université de Strasbourg, No. IX-X. Nouvelle édition, entiérement corrigée, refondue et augmentée, Hermann, Paris, 1966. MR0209834

[Sec13] S. Secchi, *Ground state solutions for nonlinear fractional Schrödinger equations in* $\mathbb{R}^N$, J. Math. Phys. **54** (2013), no. 3, 031501, 17, DOI 10.1063/1.4793990. MR3059423

[Se71] J. Serrin, *A symmetry problem in potential theory*, Arch. Rational Mech. Anal. **43** (1971), 304–318, DOI 10.1007/BF00250468. MR0333220

[SV15] R. Servadei and E. Valdinoci, *The Brezis-Nirenberg result for the fractional Laplacian*, Trans. Amer. Math. Soc. **367** (2015), no. 1, 67–102, DOI 10.1090/S0002-9947-2014-05884-4. MR3271254

[Si07] L. Silvestre, *Regularity of the obstacle problem for a fractional power of the Laplace operator*, Comm. Pure Appl. Math. **60** (2007), no. 1, 67–112, DOI 10.1002/cpa.20153. MR2270163

[Slo58] L. N. Slobodeckiĭ, *S. L. Sobolev's spaces of fractional order and their application to boundary problems for partial differential equations* (Russian), Dokl. Akad. Nauk SSSR (N.S.) **118** (1958), 243–246. MR0106325

[Sp58] F. Spitzer, *Some theorems concerning 2-dimensional Brownian motion*, Trans. Amer. Math. Soc. **87** (1958), 187–197, DOI 10.2307/1993096. MR0104296

[St70] E. M. Stein, *Singular integrals and differentiability properties of functions*, Princeton Mathematical Series, No. 30, Princeton University Press, Princeton, N.J., 1970. MR0290095

[St76] J. M. Ash (ed.), *Studies in harmonic analysis*, Mathematical Association of America, Washington, D. C., 1976. MAA Studies in Mathematics. Vol. 13. MR0442565

[SW78] E. M. Stein and S. Wainger, *Problems in harmonic analysis related to curvature*, Bull. Amer. Math. Soc. **84** (1978), no. 6, 1239–1295, DOI 10.1090/S0002-9904-1978-14554-6. MR508453

[SW71] E. M. Stein and G. Weiss, *Introduction to Fourier analysis on Euclidean spaces*, Princeton University Press, Princeton, N.J., 1971. Princeton Mathematical Series, No. 32. MR0304972

[ST10] P. R. Stinga and J. L. Torrea, *Extension problem and Harnack's inequality for some fractional operators*, Comm. Partial Differential Equations **35** (2010), no. 11, 2092–2122, DOI 10.1080/03605301003735680. MR2754080

[ST17] P. R. Stinga and J. L. Torrea, *Regularity theory and extension problem for fractional nonlocal parabolic equations and the master equation*, SIAM J. Math. Anal. **49** (2017), no. 5, 3893–3924, DOI 10.1137/16M1104317. MR3709888

[Str83] R. S. Strichartz, *Analysis of the Laplacian on the complete Riemannian manifold*, J. Funct. Anal. **52** (1983), no. 1, 48–79, DOI 10.1016/0022-1236(83)90090-3. MR705991

[Tan11] J. Tan, *The Brezis-Nirenberg type problem involving the square root of the Laplacian*, Calc. Var. Partial Differential Equations **42** (2011), no. 1-2, 21–41, DOI 10.1007/s00526-010-0378-3. MR2819627

[Ta81] M. E. Taylor, *Pseudodifferential operators*, Princeton Mathematical Series, vol. 34, Princeton University Press, Princeton, N.J., 1981. MR618463

[Ta11] M. E. Taylor, *Partial differential equations II. Qualitative studies of linear equations*, 2nd ed., Applied Mathematical Sciences, vol. 116, Springer, New York, 2011. MR2743652

[To84] P. Tolksdorf, *Regularity for a more general class of quasilinear elliptic equations*, J. Differential Equations **51** (1984), no. 1, 126–150, DOI 10.1016/0022-0396(84)90105-0. MR727034

[T67] F. Trèves, *Topological vector spaces, distributions and kernels*, Academic Press, New York-London, 1967. MR0225131

[T75] F. Trèves, *Basic linear partial differential equations*, Academic Press [A subsidiary of Harcourt Brace Jovanovich, Publishers], New York-London, 1975. Pure and Applied Mathematics, Vol. 62. MR0447753

[V17] J. L. Vázquez, *The mathematical theories of diffusion: nonlinear and fractional diffusion*, Nonlocal and nonlinear diffusions and interactions: new methods and directions, Lecture Notes in Math., vol. 2186, Springer, Cham, 2017, pp. 205–278. MR3588125

[V17'] J. L. Vázquez, *Asymptotic behaviour for the fractional heat equation in the Euclidean space*, Complex Var. Elliptic Equ. **63** (2018), no. 7-8, 1216–1231, DOI 10.1080/17476933.2017.1393807. MR3802823

[W62] G. N. Watson, *A treatise in the theory of Bessel functions*, 2nd ed., Cambridge University Press, London, 1962.

[We71] H. F. Weinberger, *Remark on the preceding paper of Serrin*, Arch. Rational Mech. Anal. **43** (1971), 319–320, DOI 10.1007/BF00250469. MR0333221

[W40] H. Weyl, *The method of orthogonal projection in potential theory*, Duke Math. J. **7** (1940), 411–444. MR0003331

[Y78] K. Yosida, *Functional analysis* Fifth edition. Grundlehren der Mathematischen Wissenschaften, Band 123. Springer-Verlag, Berlin-New York, 1978. xii+501 pp.

[Yu17] H. Yu, *Unique continuation for fractional orders of elliptic equations*, Ann. PDE **3** (2017), no. 2, Art. 16, 21, DOI 10.1007/s40818-017-0033-9. MR3695403

Dipartimento d'Ingegneria Civile e Ambientale (DICEA), Università di Padova, Via Marzolo, 9 - 35131 Padova, Italy

Email address: nicola.garofalo@unipd.it

Contemporary Mathematics
Volume **723**, 2019
https://doi.org/10.1090/conm/723/14544

Uniqueness for weak solutions of parabolic equations with a fractional time derivative

Mark Allen

ABSTRACT. We prove uniqueness for weak solutions to abstract parabolic equations with the fractional Marchaud or Caputo time derivative. We consider weak solutions in time for divergence form equations when the fractional derivative is transferred to the test function.

1. Introduction

We prove a uniqueness result for weak-in-time solutions to abstract evolutionary equations. In order to state the equation, we recall notation from [**11**]. Let $\mathscr{V}$ and $\mathscr{H}$ be real separable Hilbert spaces such that $\mathscr{V}$ is densely and continuously embedded into $\mathscr{H}$. If $(\cdot,\cdot)$ is the pairing for $\mathscr{H}$ and $\langle\cdot,\cdot\rangle$ is the pairing for $\mathscr{V}' \times \mathscr{V}$ with $\mathscr{V}'$ the dual of $\mathscr{V}$, then we consider the equation

$$(1.1)\qquad \begin{aligned} &(\partial_t^\alpha u(t), v(t)) + a(t, u(t), v(t)) = \langle f(t), v(t)\rangle, \\ &\text{if } v \in \mathscr{V} \text{ for a.e. } t \in (0,T), \end{aligned}$$

where ∂_t^α is the Caputo or Marchaud derivative, and $a : (0,T) \times \mathscr{V} \times \mathscr{V} \to \mathbb{R}$ is a coercive bilinear form. Zacher [**11**] proved existence and uniqueness for an equation similar to (1.1) but with a different weak formulation than what we will consider and with an even more general fractional derivative. We will consider a different notion of weak solution made precise in (1.14) that naturally accounts for initial data defined on $(-\infty, 0]$ as $v(t)$ rather than only defining initial data at the initial time 0 as $u(0) = v$.

The evolutionary equation (1.1) with the fractional time derivative is analogous to abstract evolutionary equations with the usual time derivative. One of the most notable examples being the equation

$$(1.2)\qquad u_t - (a^{ij}(x,t)u_i)_j = f.$$

The function u is a weak solution to the above equation if for every $t \in (0,T)$ and $v \in W_0^{1,2}(\Omega)$, one has

$$(1.3)\qquad \int_\Omega u_t v + a^{ij}(x,t) u_i v_j = \int_\Omega fv.$$

2010 *Mathematics Subject Classification.* 35K90, 35R09, 35R11, 45N05, 45K05.

Solutions to (1.3) are considered weak solutions that are strong in time. If u is a solution to (1.3), then by integrating in time and transferring the derivative in time to the test function, u is also a solution to

$$\int_\Omega u(x,0)v(x,0) - \int_0^T \int_\Omega uv_t + a^{ij}(x,t)u_i v_j = \int_0^T \int_\Omega f(x,t)v(x,t) \tag{1.4}$$

for all $v \in L^2(0,T;H_0^1(\Omega))$ with $v' \in L^2(0,T;L^2(\Omega))$ and $v(x,T) \equiv 0$. Proving uniqueness for solutions to (1.3) is straightforward since u is a valid test function, so one simply multiplies by u and integrates in time. Proving uniqueness for solutions to (1.4) is not straightforward, however, since a priori it is not known if u is a valid test function. In this paper we say "weak-in-time" to emphasize that via an integration by parts formula, the fractional time derivative is transferred to the test function as in (1.4).

It is advantageous to consider solutions to (1.4) which are weak solutions in time since it is often convenient [**8**] to work in the larger class of solutions to prove existence as well as regularity estimates such as Hölder continuity. This is also true for parabolic equations involving the Caputo derivative. Zacher [**12**] utilized a larger class of solutions to prove Hölder continuity of solutions to parabolic equations involving the Caputo derivative which are local in the spatial variables. Recently in [**3**] the author with Caffarelli and Vasseur proved Hölder continuity for solutions to a nonlocal parabolic equation of divergence form which is a nonlocal analogue to (1.2) and is given by

$$\partial_t^\alpha u - \int [w(y,t) - w(x,t)]K(x,y,t) = f(x,t). \tag{1.5}$$

The authors in [**3**] utilized solutions that are weak-in-time to prove Hölder continuity. The same authors also adapted the methods in [**4**] to prove Hölder continuity for a nonlocal porous medium equation. We also remark that Hölder continuity for nondivergence form parabolic equations involving fractional derivatives was shown in [**2**] and [**1**].

Considering weak-in-time solutions is advantageous for existence and regularity results; however, proving uniqueness becomes more difficult. Uniqueness for weak-in-time solutions to (1.5) was only shown in [**3**] if $K(x,y,t)$ is independent of t. In this paper we show uniqueness for $K(x,y,t)$ with no regularity assumptions on the variable t. As the methods apply to the more general parabolic equation (1.1), we show uniqueness for weak-in-time solutions to (1.1).

In order to formulate weak-in-time solutions, we recall the definition of the Caputo derivative and an integration by parts formula.

1.1. The Marchaud and Caputo Derivatives. The Caputo fractional time derivative for $0 < \alpha < 1$ is defined as

$${}_a\partial_t^\alpha u(t) := \frac{1}{\Gamma(1-\alpha)} \int_a^t \frac{u'(s)}{(t-s)^\alpha}\, ds, \tag{1.6}$$

and a reference on the Caputo derivative is [**7**]. An equivalent formulation of the Caputo derivative is

$${}_a\partial_t^\alpha u(t) = \frac{d}{dt}\frac{1}{\Gamma(1-\alpha)} \int_a^t \frac{u(s)-u(a)}{(t-s)^\alpha}\, ds. \tag{1.7}$$

This is the formulation of the Caputo derivative utilized in [**11**] and [**12**]. If $u \in C^1$, then using integration by parts on (1.6) one also has another equivalent formulation

$$
{}_a\partial_t^\alpha u(t) = \frac{1}{\Gamma(1-\alpha)} \frac{u(t)-u(a)}{(t-a)^\alpha} + \frac{\alpha}{\Gamma(1-\alpha)} \int_a^t \frac{u(t)-u(s)}{(t-s)^{1+\alpha}}\, ds. \tag{1.8}
$$

For simplicity in this paper we multiply ${}_a\partial_t^\alpha$ by the constant $\Gamma(1-\alpha)$. In (1.1) this constant will be absorbed into $a(t,u(t),v(t))$ and the right hand side $f(t)$. If we define $u(t) = u(a)$ for $t < a$, then up to constant ${}_a\partial_t^\alpha$ is also equivalent to

$$
\partial_t^\alpha u(t) = \alpha \int_{-\infty}^t \frac{u(t)-u(s)}{(t-s)^{1+\alpha}}\, ds. \tag{1.9}
$$

The formulation in (1.9) is also known as the Marchaud derivative [**10**] and was recently studied in [**5**]. We will use the Marchaud derivative in (1.9) in this paper. There are three main advantages to considering the Marchaud derivative. First, it is no longer necessary to consider the initial point a of integration. For this reason we have dropped the subscript a in front of ∂_t^α. Second, one may consider more general initial data for a parabolic problem. Rather than $u(0) = u_0 \in \mathscr{H}$, one may consider for $t \leq 0$ the initial data $u(t) = v(t)$ where $v(t)(1-t)^{(-1-\alpha)/2} \in L^2(-\infty, 0; \mathscr{H})$. Third, the formulation in 1.9 allows one to more easily utilize the nonlocal nature of the fractional time derivative to prove regularity results. This was accomplished for divergence form problems in [**3**] and [**4**] and nondivergence problems in [**2**] and [**1**].

For divergence form problems, the nonlocal nature is utilized by considering an integration by parts formula which also gives a weak-in-time formulation of the problem. If $u, \phi \in C^1((-\infty, T])$ and $\phi(t) = 0$ for $t \leq -M$ for some $M > 0$, then

$$
\begin{aligned}
&\int_{-\infty}^T \phi \partial_t^\alpha u \, dt + \int_{-\infty}^T u \partial_t^\alpha \phi \, dt \\
&\quad = \int_{-\infty}^T \partial_t^\alpha (u\phi)\, dt + \alpha \int_{-\infty}^T \int_{-\infty}^t \frac{[u(t)-u(s)][\phi(t)-\phi(s)]}{(t-s)^{1+\alpha}}\, ds\, dt.
\end{aligned} \tag{1.10}
$$

The Marchaud derivative (1.9) looks similar to the one-dimensional fractional Laplacian except the integration occurs from only one side. Because of this the Marchaud derivative retains some features of the directional derivative. However, the Marchaud derivative also behaves similarly to the fractional Laplacian. This is illustrated by the fact that (1.10) seems to be a combination of

$$
\int \phi \partial_t u + \int u \partial_t \phi = \int \partial_t (u\phi)
$$

and

$$
\int \phi (-\Delta)^\sigma u = c_\sigma \int \int \frac{[u(x)-u(y)][\phi(x)-\phi(y)]}{|x-y|^{1+2\sigma}}\, dx\, dy.
$$

The first term in the right hand side of (1.10) can be rewritten as shown in [**3**] to accommodate less regular functions, so that the integration by parts formula becomes

$$
\begin{aligned}
&\int_{-\infty}^T \phi \partial_t^\alpha u \, dt + \int_{-\infty}^T u \partial_t^\alpha \phi \, dt \\
&\quad = \int_{-\infty}^T \frac{u(t)\phi(t)}{(T-t)^\alpha}\, dt + \alpha \int_{-\infty}^T \int_{-\infty}^t \frac{[u(t)-u(s)][\phi(t)-\phi(s)]}{(t-s)^{1+\alpha}}\, ds\, dt.
\end{aligned} \tag{1.11}
$$

If we only integrate from 0 to T, then the integration by parts formula takes the form

$$\begin{aligned}\int_0^T \phi\partial_t^\alpha u = \int_0^T \frac{u(t)\phi(t)}{(T-t)^\alpha}\,dt + \alpha\int_0^T\int_{-\infty}^t \frac{[u(t)-u(s)][\phi(t)-\phi(s)]}{(t-s)^{1+\alpha}}\,ds\,dt \\ - \int_0^T u\partial_t^\alpha\phi\,dt + \int_{-\infty}^0 u(t)\phi(t)\left[\frac{1}{(T-t)^\alpha} - \frac{1}{(0-t)^\alpha}\right]dt\end{aligned} \tag{1.12}$$

1.2. Formulation of the weak solution and Main Result. With an integration by parts formula in hand we may give a definition of a weak solution. We first recall the definition of the fractional Sobolev space

$$H^{\alpha/2}(0,T;\mathscr{H}) := \left\{u \in L^2(0,T;\mathscr{H}) : \int_0^T\int_0^T \frac{\|u(t)-u(s)\|_{\mathscr{H}}^2}{|t-s|^{1+\alpha}} < \infty\right\}. \tag{1.13}$$

See [**6**] for a reference of the fractional Sobolev spaces and their properties. The proper initial condition for u will be that $u(t) = v(t)$ for almost all $t \in (-\infty, 0]$ with $v(t)(1-t)^{(-1-\alpha)/2} \in L^2(-\infty,0;\mathscr{H})$. Notice that this initial condition includes the possibility that $v(t)$ is time independent so that $v(t) = v \in \mathscr{H}$ which corresponds to the usual initial condition for the Caputo derivative. We then say that u is a weak solution with initial condition v, if $u(t) = v(t)$ for almost every $t \leq 0$ and

$$\begin{aligned}&\int_0^T \frac{(u(t),\phi(t))_{\mathscr{H}}}{(T-t)^{1+\alpha}}\,dt + \alpha\int_0^T\int_{-\infty}^t \frac{(u(t)-u(s),\phi(t)-\phi(s))_{\mathscr{H}}}{(t-s)^{1+\alpha}}\,ds\,dt \\ &\quad - \int_0^T (u(t),\partial_t^\alpha\phi(t))_{\mathscr{H}}\,dt + \int_0^T a(t,u(t),\phi(t))\,dt \\ &\quad + \int_{-\infty}^0 (v(t),\phi(t))_{\mathscr{H}}\left[\frac{1}{(T-t)^\alpha} - \frac{1}{(0-t)^\alpha}\right]dt \\ &= \int_0^T \langle f,\phi\rangle_{\mathscr{V}'\times\mathscr{V}}\,dt\end{aligned} \tag{1.14}$$

for all

$$\begin{gathered}\phi \in L^2(0,T;\mathscr{V}) \\ \partial_t^\alpha\phi \in L^2(0,T;\mathscr{H}) \\ \phi(t)(1-t)^{-(1+\alpha)/2} \in L^2(-\infty,0;\mathscr{H}) \\ \phi(t) \equiv 0 \text{ for } t \leq -M \text{ for some } M > 0.\end{gathered}$$

Since for any test function ϕ, we assume $\phi(t) \equiv 0$ for $t \leq -M$ for some $M > 0$, and since $\phi \in L^2(-M,T;\mathscr{H})$, from [**5**] the Marchaud derivative of ϕ can be defined in the distributional sense, and this is how to understand the second requirement above. For the test function we will use later, $\partial_t^\alpha\phi$ will be defined classically everywhere and will be continuous.

Our main theorem is the following uniqueness result.

THEOREM 1.1. *Solutions to* (1.14) *are unique; i.e. if* u_1, u_2 *are both solutions to* (1.14) *with prescribed initial data* $v(t)$ *and right hand side* $f(t)$, *then* $u_1 - u_2 \equiv 0$.

A future question of interest would be to determine necessary and sufficient conditions on the initial data $v(t)$, the bilinear form a, and right hand side f, so that for a solution u of (1.14), the fractional derivative $\partial_t^\alpha u \in L^2(0,T;\mathscr{H})$ and consequently u is a strong solution to (1.14). Another question of interest is proving uniqueness for weak-in-time solutions for fractional derivatives defined by a kernel $K(t,s) \approx (t-s)^{-1-\alpha}$ and satisfying $K(t,t-s) = K(t+s,t)$. Such kernels were considered in [**4**] and the Hölder continuity results in [**3**] will also apply to fractional derivatives defined by such a kernel.

As this paper is concerned with uniqueness, we do not prove existence of solutions to (1.14). However, we do mention that for a specific $\mathscr{V}$ and $\mathscr{H}$ and a restricted class of right-hand side functions, existence was proven in [**3**].

1.3. Outline and Notation. To prove uniqueness we use Steklov averages. In Section 2 we present preliminary results on how the Steklov averages behave with the fractional Marchaud derivative. In Section 3 we prove our main result. Rather than show that any solution u to (1.14) is contained in successively better spaces as is done in the local case (see [**8**]), we utilize the nonlocal nature of the Marchaud derivative to prove uniqueness directly. This is the essence of Lemma 3.1.

We define the notation that will be consistent throughout the paper.

- ∂_t^α - the Marchaud derivative as defined in (1.9).
- α - the order of the Marchaud derivative.
- t, s - will always be variables reserved as time variables.
- $H^{\alpha/2}(0,T,\mathscr{H})$ - the fractional Sobolev space as defined in (1.13)
- $a(t,\cdot,\cdot)$ a bounded $\mathscr{V}$-coercive bilinear form.
- λ, Λ - The coercivity constants for the bilinear form $a(t,u(t),v(t))$, so that for almost all t

$$|a(t,u(t),v(t))| \leq \Lambda \|u(t)\|_{\mathscr{V}} \cdot \|v(t)\|_{\mathscr{V}} \quad \text{and}$$
$$\lambda \|u(t)\|_{\mathscr{V}}^2 \leq |a(t,u(t),u(t))|.$$

- ψ a cut-off function defined in (2.1).
- The pairing $(\cdot,\cdot)$ will refer to $(\cdot,\cdot)_{\mathscr{H}}$.
- The norm $\|\cdot\|$ will refer to $\|\cdot\|_{\mathscr{H}}$ unless otherwise stated.

2. Steklov averages

In order to apply the Steklov averages technique, we will utilize the following cut-off function ψ throughout the paper.

$$\text{(2.1)} \qquad \begin{aligned} &(1) \quad \psi : \mathbb{R} \to \mathbb{R} \text{ and } \psi \in C^\infty(\mathbb{R}) \\ &(2) \quad \psi' \leq 0 \\ &(3) \quad 0 \leq \psi \leq 1 \\ &(4) \quad \psi(t) = 1 \text{ for } t \leq T - 2\epsilon \\ &(5) \quad \psi(t) = 0 \text{ for } t \geq T - \epsilon. \end{aligned}$$

The quantity $\epsilon > 0$ will be made precise later. The following Lemma states what equation $u\psi$ will satisfy. We recall that the pairing $(\cdot,\cdot)$ represents $(\cdot,\cdot)_{\mathscr{H}}$.

LEMMA 2.1. *Let u be a solution to (1.14) with right hand side $f \equiv 0$ and initial data $v \equiv 0$. Then $\psi(t)u(t)$ satisfies the equation*

(2.2)
$$\int_{-\infty}^{T} \frac{(\psi(t)u(t), \phi(t))}{(T-t)^{\alpha}} \, dt + \alpha \int_{-\infty}^{T} \int_{-\infty}^{s} \frac{(\psi(t)u(t) - \psi(s)u(s), \phi(t) - \phi(s))}{(t-s)^{1+\alpha}} \, ds \, dt$$
$$- \int_{-\infty}^{T} (\psi(t)u(t), \partial_t^{\alpha}\phi(t)) \, dt + \int_{0}^{T} a(t, \psi u, \phi) \, dt$$
$$= \alpha \int_{-\infty}^{T} \int_{-\infty}^{t} \frac{(\phi(t), u(s)[\psi(t) - \psi(s)])}{(t-s)^{1+\alpha}} \, ds \, dt.$$

PROOF. We first show how to transfer ψ from ϕ to u. We note that if $g, \phi \in C^1(-\infty, T; \mathscr{H})$ with $g(t) = 0$ for $t \leq 0$, and $\phi(t) = 0$ for $t \leq -M$ for some M, then

$$\int_{-\infty}^{T} \frac{(\psi(t)g(t), \phi(t))}{(T-t)^{\alpha}} \, dt + \alpha \int_{-\infty}^{T} \int_{-\infty}^{t} \frac{(\psi(t)g(t) - \psi(s)g(s), \phi(t) - \phi(s))}{(t-s)^{1+\alpha}} \, ds \, dt$$
$$- \int_{-\infty}^{T} (\psi(t)g(t), \partial_t^{\alpha}\phi(t)) \, dt$$
$$= \int_{-\infty}^{T} (\phi, \partial_t^{\alpha}(g\psi)) \, dt$$
$$= \int_{-\infty}^{T} (\phi\psi, \partial_t^{\alpha} g) \, dt + \alpha \int_{-\infty}^{T} \int_{-\infty}^{t} \frac{(\phi(t), g(s)[\psi(t) - \psi(s)])}{(t-s)^{1+\alpha}} \, ds \, dt$$
$$= \alpha \int_{-\infty}^{T} \int_{-\infty}^{t} \frac{(\phi(t), g(s)[\psi(t) - \psi(s)])}{(t-s)^{1+\alpha}} \, ds \, dt$$
$$+ \int_{-\infty}^{T} \frac{(g(t), \psi(t)\phi(t))}{(T-t)^{\alpha}} \, dt + \alpha \int_{-\infty}^{T} \int_{-\infty}^{t} \frac{(g(t) - g(s), \psi(t)\phi(t) - \psi(s)u(s))}{(t-s)^{1+\alpha}} \, ds \, dt$$
$$- \int_{-\infty}^{T} (g, \partial_t^{\alpha}(\psi\phi)) \, dt.$$

We then have that

$$\int_{-\infty}^{T} \frac{(\psi g, \phi)}{(T-t)^{\alpha}} \, dt + \alpha \int_{-\infty}^{T} \int_{-\infty}^{t} \frac{(g(t)\psi(t) - g(s)\psi(s), \phi(t) - \phi(s))}{(t-s)^{1+\alpha}} \, ds \, dt$$
$$- \int_{-\infty}^{T} (\psi g, \partial_t^{\alpha}\phi) \, dt - \alpha \int_{-\infty}^{T} \int_{-\infty}^{t} \frac{(\phi(t), g(s)[\psi(t) - \psi(s)])}{(t-s)^{1+\alpha}} \, ds \, dt$$
$$= \int_{-\infty}^{T} \frac{(g, \psi\phi)}{(T-t)^{\alpha}} \, dt + \alpha \int_{-\infty}^{T} \int_{-\infty}^{t} \frac{(g(t) - g(s), \psi(t)\phi(t) - \psi(s)\phi(s))}{(t-s)^{1+\alpha}} \, ds \, dt$$
$$- \int_{-\infty}^{T} (g, \partial_t^{\alpha}(\psi\phi)) \, dt.$$

We now use that $C^1(0, T; \mathscr{H})$ is dense in $H^{\alpha/2}(0, T; \mathscr{H})$ and since $u(t) = 0$ for $t \leq 0$, we may by approximation substitute u for g. The technical point is to show that

(2.3)
$$G(t) := \alpha \int_{-\infty}^{t} \frac{u(s)[\psi(t) - \psi(s)]}{(t-s)^{1+\alpha}} \, ds$$

is in $L^2(0,T;\mathscr{H})$. We have that

$$\|G(t)\|_{\mathscr{H}} \le \int_{-\infty}^{t} \frac{\|u(s)\|_{\mathscr{H}}|\psi(t)-\psi(s)|}{(t-s)^{1+\alpha}}\, ds \le C\int_{T-2\epsilon}^{t} \frac{\|u(s)\|_{\mathscr{H}}}{(t-s)^{\alpha}}\, ds.$$

By using Lemma 4.26 from [**9**] and the fact that $[T-2\epsilon, T]$ is a finite interval, we have that

$$\int_{T-2\epsilon}^{T}\left(\int_{T-2\epsilon}^{t} \frac{\|u(s)\|_{\mathscr{H}}}{(t-s)^{\alpha}}\, ds\right)^2 < \infty.$$

Then $G(t) \in L^2(0,T;\mathscr{H})$, and by approximation we obtain that (2.2) holds for u.

Finally, since $\psi(t)$ is constant for each fixed t, we have that $a(t,u,\psi\phi) = a(t,\psi u,\phi)$. By integrating in time and applying the conclusion from above, we obtain (2.2). □

We define the Steklov averages

$$\eta_{\bar{h}} = \frac{1}{h}\int_{t-h}^{t} \eta$$
$$\eta_h = \frac{1}{h}\int_{t}^{t+h} \eta.$$

Notice that if $\eta(t)$ and u_h both vanish in $[t-2h, T]$, then

$$\int_{-\infty}^{T} (u(t), \eta_{\bar{h}}(t))\, dt = \int_{-\infty}^{T-h} (u_h(t), \eta(t))\, dt = \int_{-\infty}^{T} (u_h(t), \eta(t))\, dt. \tag{2.4}$$

LEMMA 2.2. *If $\partial_t^\alpha \eta \in L^2(-\infty,T;\mathscr{H})$, and $\eta(t) = 0$ for $t < -M$ for some $M > 0$, then $\partial_t^{-1}\partial_t^\alpha \eta = \partial_t^\alpha \partial_t^{-1}\eta$ in $L^2(-\infty,T;\mathscr{H})$ where*

$$\partial_t^{-1}\eta := \int_{-\infty}^{t} \eta(\tau)\, d\tau.$$

PROOF. We first assume that $\eta \in C^1(-\infty,T;\mathscr{H})$ and use the notion of Caputo derivative given in (1.6). Then

$$\partial_t^\alpha \partial_t^{-1}\eta = c_\alpha \int_{-\infty}^{t} \frac{\eta(s)}{(t-s)^\alpha}\, ds.$$

Now

$$\begin{aligned}
\partial_t^{-1}\partial_t^\alpha \eta &= c_\alpha \int_{-\infty}^{t}\int_{-\infty}^{\tau} \frac{\eta'(s)}{(\tau-s)^\alpha}\, ds\, d\tau \\
&= c_\alpha \int_{-\infty}^{t}\int_{s}^{t} \frac{\eta'(s)}{(\tau-s)^\alpha}\, d\tau\, ds \\
&= \frac{-c_\alpha}{1-\alpha}\int_{-\infty}^{t}\int_{\tau}^{t} \eta'(s)(t-s)^{1-\alpha}\, ds \\
&= c_\alpha \int_{-\infty}^{t} \frac{\eta(s)}{(t-s)^\alpha}\, ds.
\end{aligned}$$

Alternatively, one may use the Laplace transform

$$\mathcal{L}[\partial_t^{-1}\partial_t^\alpha \eta] = s\mathcal{L}[\partial_t^\alpha \eta] = s^1 s^{-\alpha}\mathcal{L}[\eta] = s^{-\alpha}\mathcal{L}[\partial_t^{-1}\eta] = \mathcal{L}[\partial_t^\alpha \partial_t^{-1}\eta].$$

We then use that $C^1(-\infty,T;\mathscr{H})$ is dense in $L^2(-\infty,T;\mathscr{H})$. □

LEMMA 2.3. *If $\eta \in L^2(-\infty, T; \mathscr{H})$ and $\eta(t) = 0$ if $t \leq -M$ for some M, then*

$$\partial_t^\alpha \eta_{\bar{h}}(t) = (\partial_t^\alpha \eta)_{\bar{h}}(t)$$

PROOF. We notice that

$$\eta_{\bar{h}} = \frac{1}{h} \int_{-\infty}^{t} \eta \, d\tau - \frac{1}{h} \int_{-\infty}^{t-h} \eta \, d\tau.$$

From Lemma 2.2 the anti-derivative commutes with the Marchaud derivative. □

LEMMA 2.4. *Let $f \in L^2(0, T; \mathscr{V})$ (or $f \in L^2(0, T; \mathscr{H})$). Extend $f(x, t) = 0$ for $t \notin [0, T]$. Then $f_h \to f$ in $L^2(0, T; \mathscr{V})$ (or in $L^2(0, T; \mathscr{H})$).*

PROOF. We suppose that $f \in L^2(0, T; \mathscr{V})$, and remark that the proof when $f \in L^2(0, T; \mathscr{H})$ is the same.

$$\begin{aligned}
(f_h(t), f_h(t))_{\mathscr{V}} &= \left(\frac{1}{h} \int_t^{t+h} f(s) \, ds, \frac{1}{h} \int_t^{t+h} f(s) \, ds \right)_{\mathscr{V}} \\
&\leq \frac{1}{h} \int_t^{t+h} (f(s), f(s))_{\mathscr{V}} \, ds \\
&= ((f(t), f(t))_{\mathscr{V}})_h.
\end{aligned} \tag{2.5}$$

Now since $f(t) = 0$ for $t \notin [0, T]$ we have by changing the order of integration that

$$\begin{aligned}
\int_0^T \frac{1}{h} \int_t^{t+h} (f(s), f(s)) \, ds \, dt &= \int_0^{T+h} \frac{1}{h} \int_{\max\{s-h,0\}}^{\max\{s-h,T\}} (f(s), f(s)) \, dt \, ds \\
&= \int_0^T (f(s), f(s)) \, ds \\
&\quad + \frac{1}{h} \int_T^{T+h} \int_{s-h}^{\max\{s+h,T\}} (f(s), f(s)) \, ds.
\end{aligned} \tag{2.6}$$

Thus $\|((f, f))_h\|_{L^1(0,T;\mathscr{V})} \to \|(f, f)\|_{L^1(0,T;\mathscr{V})}$. Also we have that $((f, f))_h \to (f, f)$ pointwise for almost everywhere t. Now $(f_h, f_h) \to (f, f)$ pointwise for almost every t and $(f_h, f_h) \leq ((f, f))_h$ from (2.5). Then from the Lebesgue dominated convergence theorem for the Bochner intergral, we conclude that $\|f_h\|_{L^2(0,T;\mathscr{V})} \to \|f\|_{L^2(0,T;\mathscr{V})}$. It then follows that $f_h \to f$ in $L^2(0, T; \mathscr{V})$. □

LEMMA 2.5. *Let $a(t, \cdot, \cdot)$ be a bounded $\mathscr{V}$-coercive bilinear form. Let $u \in L^2(0, T; \mathscr{V})$ with $u(x, t) = 0$ for $t \notin [0, T]$. Then*

$$\lim_{h \to 0} \int_0^T \frac{1}{h} \int_s^{s+h} a(t, \psi u(t), \psi u_h(s)) \, dt \, ds = \int_0^T a(t, \psi u(t), \psi u(t)) \, dt. \tag{2.7}$$

PROOF. If $\|\cdot\|$ represents the norm in the space $\mathscr{V}$, then

$$\begin{aligned}
&\left|\int_0^T \frac{1}{h}\int_s^{s+h} a(t, \psi u(t), \psi u_h(s))\, dt\, ds\right| \\
&\quad \le \int_0^T \frac{1}{h}\int_s^{s+h} \Lambda \|\psi u(t)\| \|\psi u_h(s)\|\, dt\, ds \\
&\le \Lambda^2 \int_0^T \frac{1}{h}\int_s^{s+h} \|\psi u_h(s)\|^2\, dt\, ds + \int_0^T \frac{1}{h}\int_s^{s+h} \|\psi u(t)\|^2\, dt\, ds. \\
&= \Lambda^2 \int_0^T \|\psi u_h(s)\|^2\, ds + \int_0^T \frac{1}{h}\int_s^{s+h} \|\psi u(t)\|^2\, dt\, ds.
\end{aligned}$$

From Lemma 2.4 we have that

$$\lim_{h\to 0} \int_0^T \|u_h(s)\|^2\, ds = \int_0^T \|u\|^2\, ds,$$

so it will also be true that

$$\lim_{h\to 0} \int_0^T \|\psi u_h(s)\|^2\, ds = \int_0^T \|\psi u(s)\|^2\, ds.$$

Furthermore, it was shown in the proof of Lemma 2.4 that

$$\lim_{h\to 0} \int_0^T \frac{1}{h}\int_s^{s+h} \|\psi u(t)\|^2\, dt\, ds = \int_0^T \|\psi u(s)\|^2\, ds.$$

From the dominated convergence theorem for the Bochner integral, we conclude (2.7) is true. □

Before ending this section we establish one more identity. We note that if $u, \eta \in C^1(-\infty, T; \mathscr{H})$ and have compact support in $(-\infty, T-2h)$, then from (2.4) and (2.3) we obtain

$$\begin{aligned}
&\int_{-\infty}^T (u, \partial_t^\alpha \eta_{\bar{h}}) + ((\partial_t^\alpha u), \eta_{\bar{h}})\, dt \\
&\quad = \int_{-\infty}^T (u, (\partial_t^\alpha \eta)_{\bar{h}}) + ((\partial_t^\alpha u)_h, \eta)\, dt \\
&\quad = \int_{-\infty}^T (u_h, \partial_t^\alpha \eta) + ((\partial_t^\alpha u)_h, \eta)\, dt
\end{aligned}$$

As a consequence, we immediately have the equality

$$\begin{aligned}
&\int_{-\infty}^T \frac{(u, \eta_{\bar{h}})}{(T-t)^\alpha}\, dt + \alpha \int_{-\infty}^T \int_{-\infty}^t \frac{(u(t)-u(s), \eta_{\bar{h}}(t) - \eta_{\bar{h}}(s))}{(t-s)^{1+\alpha}}\, ds\, dt \\
&= \int_{-\infty}^T \frac{(u_h, \eta)}{(T-t)^\alpha}\, dt + \alpha \int_{-\infty}^T \int_{-\infty}^t \frac{(u_h(t)-u_h(s), \eta(t) - \eta(s))}{(t-s)^{1+\alpha}}\, ds\, dt.
\end{aligned} \tag{2.8}$$

If $u \in H^{\alpha/2}(0, T; \mathscr{H})$ with $u(t) = 0$ for $t \le 0$ and $T - 2h \le t \le T$, then by approximation (2.8) will also hold.

3. Uniqueness

In order to prove uniqueness of solutions, we will utilize the nonlocal structure of the Marchaud derivative. This is the content of the next Lemma. We will use a specific cut-off function ψ_δ which is piecewise linear.

$$\psi_\delta(t) := \begin{cases} 1 & \text{if } t \leq T-\epsilon-\delta \\ \delta^{-1}(T-\epsilon-t) & \text{if } T-\epsilon-\delta < t < T-\epsilon \\ 0 & \text{if } t \geq T-\epsilon. \end{cases} \tag{3.1}$$

LEMMA 3.1. *Let $u \in L^2(0,T;\mathscr{H})$, and let $u(t)=0$ for $t \leq 0$. Then for almost all $\epsilon \in (0,T)$, if ψ_δ is defined as in (3.1), then*

$$\lim_{\delta\to 0} \int_{-\infty}^{T}\int_{-\infty}^{t} \frac{(\psi_\delta u(t), u(s)[\psi_\delta(t)-\psi_\delta(s)])}{(t-s)^{1+\alpha}}\,ds\,dt = 0. \tag{3.2}$$

PROOF. Throughout this proof the norm $\|\cdot\|$ will refer to $\|\cdot\|_{\mathscr{H}}$. We have

$$\begin{aligned} &\left|\int_{-\infty}^{T}\int_{-\infty}^{t} \frac{(\psi_\delta u(t), u(s)[\psi_\delta(t)-\psi_\delta(s)])}{(t-s)^{1+\alpha}}\,ds\,dt\right| \\ &\leq \int_{0}^{T} \|\psi_\delta u(t)\| \int_{-\infty}^{t} \frac{\|u(s)\||\psi_\delta(t)-\psi_\delta(s)|}{(t-s)^{1+\alpha}}\,ds\,dt \\ &= \int_{T-\epsilon-\delta}^{T-\epsilon} \|\psi_\delta u(t)\| \int_{-\infty}^{t} \frac{\|u(s)\||\psi_\delta(t)-\psi_\delta(s)|}{(t-s)^{1+\alpha}}\,ds\,dt. \end{aligned} \tag{3.3}$$

The last inequality is due to the fact that $\|\psi_\delta u(t)\| = 0$ for $t > T-\epsilon-\delta$ and $\psi_\delta(t)-\psi_\delta(s) = 0$ for $t \leq T-\epsilon-\delta$. We define for $0 \leq t_0 < t \leq T$, the function

$$F(t,t_0) := \int_{t_0}^{t} \frac{\|u(s)\|}{(t-s)^{\alpha}}\,ds.$$

Since $\|u(s)\| \in L^2(0,T)$, then also $\|u(s)\| \in L^1(0,T)$. From Lemma 4.1 in [**9**], for almost every $t \in (0,T)$ we have $F(t,t_0) \leq F(t,0) < \infty$. Also, as explained earlier in the proof of Lemma 2.1, $F(t,t_0) \in L^2(0,T)$ for any $t_0 \in (T-2\epsilon, T)$. We choose ϵ so that $F(T-\epsilon,0) < \infty$ and $T-\epsilon$ is a Lebesgue point for $\|u\|$ and $F(T-\epsilon,0)$ and hence also for $F(T-\epsilon,t_0)$.

Now if we define

$$H(t) := \int_{-\infty}^{t} \frac{\|u(s)\||\psi_\delta(t)-\psi_\delta(s)|}{(t-s)^{1+\alpha}}\,ds,$$

then $H(t) = 0$ for $t \leq T-\epsilon-\delta$. Furthermore, if $t_0 < T-\epsilon-\delta$, then for $t \in [T-\epsilon-\delta, T-\epsilon]$ we have

$$\begin{aligned} H(t) &\leq +\frac{1}{\delta}\int_{t_0}^{t} \frac{\|u(s)\|}{(t-s)^{\alpha}}\,ds \\ &= \delta^{-1}F(t,t_0). \end{aligned} \tag{3.4}$$

We now multiply by $\|\psi_\delta u(t)\|$ and integrate over the variable t. Since $T-\epsilon$ is a Lebesgue point for $\|u\|$, we have

$$\lim_{\delta\to 0}\frac{1}{\delta}\int_{T-\epsilon-\delta}^{T-\epsilon} \|\psi_\delta u(t)\|F(t,t_0) \leq \|u(T-\epsilon)\|F(T-\epsilon,t_0).$$

Since $F(T-\epsilon,0)<\infty$ it follows that

$$\lim_{t_0\to T-\epsilon} F(T-\epsilon,t_0)\to 0. \tag{3.5}$$

Combining (3.3), (3.4), and (3.5) we conclude (3.2). □

We now give the proof of the Main Theorem.

PROOF OF THEOREM 1.1. Let u_1,u_2 be solutions to (1.14) with some fixed right hand side $f\in \mathscr{V}'$. Assume also that $u_1(t)=u_2(t)$ for $t\le 0$. Then $u=u_1(t)-u_2(t)$ is a solution to (1.14) with zero right hand side. Furthermore, $u(t)=0$ for $t\le 0$.

From Lemma 2.1 we have that ψu satisfies (2.2). We choose $h<\epsilon/2$ so that $\psi u(t)=0$ for $t\ge T-2h$. If we choose $\eta_{\bar h}$ as a test function with $\eta\in C_0^1(0,T-2h;\mathscr{V})$, then from (2.8), Lemma 2.3, and (2.4) we obtain

$$\begin{aligned}\int_{-\infty}^T \frac{((\psi u)_h,\eta)}{(T-t)^\alpha}\,dt+\alpha\int_{-\infty}^T\int_{-\infty}^t \frac{((\psi u)_h(t)-(\psi u)_h(s),\eta(t)-\eta(s))}{(t-s)^{1+\alpha}}\,ds\,dt\\ -\int_{-\infty}^T(\psi u)_h\partial_t^\alpha\eta\,dt+\int_0^T a(t,\psi u(t),\eta_{\bar h}(t))\,dt\\ =\alpha\int_{-\infty}^T\int_{-\infty}^t\frac{(\eta_{\bar h},u(s)[\psi(t)-\psi(s)])}{(t-s)^{1+\alpha}}\,ds\,dt.\end{aligned} \tag{3.6}$$

If $G(t)$ is as defined as in (2.3), and $\eta(t)=0$ for $t>T-2h$, then

$$\int_{-\infty}^T(\eta_{\bar h},G(t))\,ds=\int_{-\infty}^T(\eta,G_h(t))\,ds.$$

We also have that

$$\begin{aligned}\int_0^T a(t,\psi u(t),\eta_{\bar h}(t))\,dt&=\int_0^T\frac1h\int_{t-h}^t a(t,\psi u(t),\eta(s))\,ds\,dt\\ &=\int_0^T\frac1h\int_s^{\min\{s+h,T\}}a(t,\psi u(t),\eta(s))\,dt\,ds\\ &=\int_0^T\frac1h\int_s^{s+h}a(t,\psi u(t),\eta(s))\,dt\,ds.\end{aligned}$$

In the second equality above we have used that $\psi u=0$ for $t\ge T-\epsilon$ and extended $\psi u=0$ for $t\ge T$. Now since ψu_h is Lipschitz in time, $\partial_t^\alpha(\psi u_h)$ as given in (1.9) is well-defined and in $L^2(0,T;\mathscr{H})$. Then $(\psi u)_h$ is a valid test function, and if we let $\eta=(\psi u)_h$ in (3.6), then we obtain

$$\begin{aligned}\frac12\int_{-\infty}^T\frac{\|(\psi u)_h\|^2}{(T-t)^\alpha}\,dt+\frac\alpha2\int_{-\infty}^T\int_{-\infty}^t\frac{\|(\psi u)_h(t)-(\psi u)_h(s)\|^2}{(t-s)^{1+\alpha}}\,ds\,dt\\ +\int_0^T\frac1h\int_s^{\min\{s+h,T\}}a(t,\psi u(t),(\psi u)_h(s))\,dt\,ds\\ =\alpha\int_{-\infty}^T((\psi u)_h,G_h(t,s))\,dt.\end{aligned} \tag{3.7}$$

By omitting the first two positive terms we obtain the following inequality

$$\int_0^T\frac1h\int_s^{\min\{s+h,T\}}a(t,\psi u(t),(\psi u)_h(s))\,dt\,ds\le\alpha\int_{-\infty}^T((\psi u)_h,G_h(t,s))\,dt.$$

We showed earlier in the proof of Lemma 2.1 that $G \in L^2(0,T;\mathscr{H})$. Since also $\psi u \in L^2(0,T;\mathscr{H})$, we let $h \to 0$ and use Lemmas 2.4 and 2.5 to conclude

$$\int_0^T a(t,\psi u,\psi u)\, dt \leq \alpha \int_{-\infty}^T \int_{-\infty}^t \frac{(\psi u(t), u(s)[\psi(t)-\psi(s)])}{(t-s)^{1+\alpha}}\, ds\, dt.$$

If we let $\psi = \psi_\delta$ as defined in (3.1) and let $\delta \to 0$ we obtain from Lemma 3.1 that

$$\int_0^{T-\epsilon} \lambda \|u(t)\|_{\mathscr{V}}^2\, dt \leq \int_0^{T-\epsilon} a(t,u,u)\, dt \leq 0,$$

where λ is the coercivity constant for the bilinear form a. Then $u(t) = 0$ for $t \leq T-\epsilon$. Since ϵ can be chosen arbitrarily small, we conclude $u(t) \equiv 0$ on $[0,T]$. □

References

[1] Mark Allen, *Hölder regularity for nondivergence nonlocal parabolic equations*, Calc. Var. Partial Differential Equations **57** (2018), no. 4, 57:110, DOI 10.1007/s00526-018-1367-1. MR3826717

[2] Mark Allen, *A nondivergence parabolic problem with a fractional time derivative*, Differential Integral Equations **31** (2018), no. 3-4, 215–230. MR3738196

[3] Mark Allen, Luis Caffarelli, and Alexis Vasseur, *A parabolic problem with a fractional time derivative*, Arch. Ration. Mech. Anal. **221** (2016), no. 2, 603–630, DOI 10.1007/s00205-016-0969-z. MR3488533

[4] Mark Allen, Luis Caffarelli, and Alexis Vasseur, *Porous medium flow with both a fractional potential pressure and fractional time derivative*, Chin. Ann. Math. Ser. B **38** (2017), no. 1, 45–82, DOI 10.1007/s11401-016-1063-4. MR3592156

[5] Ana Bernardis, Francisco J. Martín-Reyes, Pablo Raúl Stinga, and José L. Torrea, *Maximum principles, extension problem and inversion for nonlocal one-sided equations*, J. Differential Equations **260** (2016), no. 7, 6333–6362, DOI 10.1016/j.jde.2015.12.042. MR3456835

[6] Eleonora Di Nezza, Giampiero Palatucci, and Enrico Valdinoci, *Hitchhiker's guide to the fractional Sobolev spaces*, Bull. Sci. Math. **136** (2012), no. 5, 521–573, DOI 10.1016/j.bulsci.2011.12.004. MR2944369

[7] Kai Diethelm, *The analysis of fractional differential equations*, Lecture Notes in Mathematics, vol. 2004, Springer-Verlag, Berlin, 2010. An application-oriented exposition using differential operators of Caputo type. MR2680847

[8] O. A. Ladyˇ zenskaja, V. A. Solonnikov, and N. N. Ural′ceva, *Linear and quasilinear equations of parabolic type*, Izdat. "Nauka", Mosc, 1968. MR0241821

[9] Boris Rubin, *Fractional integrals and potentials*, Pitman Monographs and Surveys in Pure and Applied Mathematics, vol. 82, Longman, Harlow, 1996. MR1428214

[10] Stefan G. Samko, Anatoly A. Kilbas, and Oleg I. Marichev, *Fractional integrals and derivatives*, Gordon and Breach Science Publishers, Yverdon, 1993. Theory and applications; Edited and with a foreword by S. M. Nikol′skiĭ; Translated from the 1987 Russian original; Revised by the authors. MR1347689

[11] Rico Zacher, *Weak solutions of abstract evolutionary integro-differential equations in Hilbert spaces*, Funkcial. Ekvac. **52** (2009), no. 1, 1–18, DOI 10.1619/fesi.52.1. MR2538276

[12] Rico Zacher, *A De Giorgi–Nash type theorem for time fractional diffusion equations*, Math. Ann. **356** (2013), no. 1, 99–146, DOI 10.1007/s00208-012-0834-9. MR3038123

Department of Mathematics, Brigham Young University, Provo, Utah 84602
Email address: allen@mathematics.byu.edu

Contemporary Mathematics
Volume **723**, 2019
https://doi.org/10.1090/conm/723/14549

Boundary regularity for the free boundary in the one-phase problem

Héctor Chang-Lara and Ovidiu Savin

ABSTRACT. We consider the Bernoulli one-phase free boundary problem in a domain Ω and show that the free boundary F is $C^{1,1/2}$ regular in a neighborhood of the fixed boundary $\partial\Omega$. We achieve this by relating the behavior of F near $\partial\Omega$ to a Signorini-type obstacle problem.

1. Introduction

The Bernoulli one-phase problem consists in finding a nonnegative function u which is fixed on the boundary $\partial\Omega$ of some given domain $\Omega \subseteq \mathbb{R}^n$, such that u is harmonic in its positive set $\Omega^+ = \{u > 0\} \cap \Omega$, and u has a prescribed gradient over the *free boundary* $F = \partial\Omega^+ \cap \Omega$. Precisely, given Ω and two functions $g \geq 0$ and $Q > 0$, we need to find u which satisfies

$$\begin{cases} \Delta u = 0 \text{ in } \Omega^+ = \{u > 0\} \cap \Omega, \\ |Du| = Q(x) \text{ on } F = \partial\Omega^+ \cap \Omega, \\ u = g \text{ on } \partial\Omega. \end{cases}$$

In hydrodynamics these equations can be found in models of jets and cavities where the solution u is the stream function for an incompressible and irrotational fluid [**11**].

Solutions can be constructed either variationally as critical points of the associated energy functional (see [**1**])

$$J(u) = \int_\Omega |Du|^2 + Q^2 \chi_{\{u>0\}} \, dx,$$

or by a viscosity solution approach using Perron's method [**6**].

The local regularity theory for the free boundary F at interior points of Ω is available for solutions u which satisfy an additional nondegeneracy condition which requires $u \sim dist(\cdot, F)$. This was developed by Caffarelli in a series of papers in the 80's. The nondegeneracy condition is satisfied for example if either a) u is a *minimizer* of the functional J or b) u is the *minimal supersolution*.

In this paper we address the regularity of the free boundary F near a portion of the fixed boundary $\partial\Omega$ where u vanishes.

The situation is the following. We assume that $g = 0$ over a portion of the boundary $Z \subseteq \partial\Omega$, with Z relatively open in the induced topology of $\partial\Omega$. Assume

2010 *Mathematics Subject Classification.* Primary 35R35.

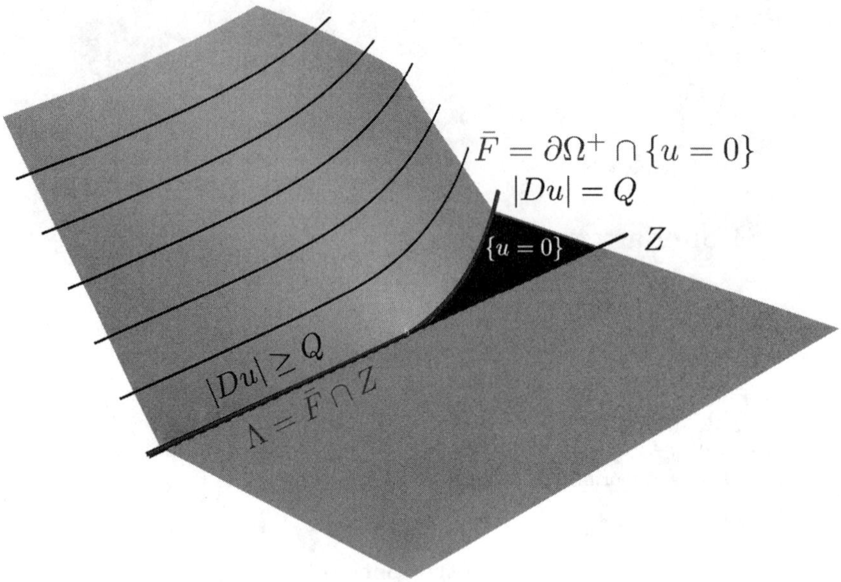

FIGURE 1. Graph of the solution u.

for simplicity that Z is locally a smooth hypersurface. We are interested in the behavior of the free boundary F near Z, or in other words how F separates from Z. Notice that Z acts as an obstacle for the "extension" $\bar{F}$ of the free boundary F to the whole $\bar{\Omega}$ which is defined as

$$\bar{F} = \partial\Omega^+ \cap \{u = 0\}.$$

Moreover, it is not difficult to check that if we are in either of the situations a) or b) above then

$$|Du| \geq Q(x) \text{ over } \Lambda = \bar{F} \cap Z.$$

This can be interpreted as a nondegeneracy condition for u on the coincidence set Λ, or equivalently as a stability condition for F. From the point of view of hydrodynamic models, the separation of $\bar{F}$ from Z describes how the fluid detaches from a fixed boundary with slip condition.

Our main result is the following.

THEOREM 1.1. *Let $\Omega \subseteq \mathbb{R}^n$ be a domain with a $C^{1,\alpha}$ boundary portion $Z \subseteq \partial\Omega$ for some $\alpha > 1/2$, and let $Q \in C^{0,1}(\bar{\Omega})$, $Q > 0$. Let $u : \bar{\Omega} \to \mathbb{R}^+$ be a viscosity solution of*

$$\begin{cases} \Delta u = 0 \text{ in } \Omega^+ = \{u > 0\} \cap \Omega, \\ u = 0 \text{ on } Z, \\ |Du| \geq Q(x) \text{ on } \bar{F} \cap Z, \\ |Du| = Q(x) \text{ on } F. \end{cases}$$

Then $\bar{F}$ is $C^{1,1/2}$ regular in a neighborhood of every $x_0 \in \Lambda = \bar{F} \cap Z$.

Next we illustrate the main idea of Theorem 1.1 by formally linearizing the one-phase problem near a point in Λ.

Assume for simplicity that $\Omega = B_1^+ = B_1 \cap \{x_n > 0\}$ and $Z = B_1' = B_1 \cap \{x_n = 0\}$, with $Q \equiv 1$ and say that $\bar{F}$ separates from Z at the origin. As a first order approximation, we expect

$$u = x_n + o(|x|).$$

Let $u = x_n - \varepsilon w$ in Ω^+. Then the perturbation w is harmonic, and moreover w is nonnegative over $\bar{F}$. The free boundary condition over F can be written as $|Du| = 1$ or, in terms of w, as

$$\partial_n w = \frac{\varepsilon}{2}|Dw|^2 \text{ on } F.$$

Additionally, $|Du| \geq 1$ on Λ means that

$$\partial_n w \leq 0 \text{ on } \Lambda.$$

As $\varepsilon \to 0$, we expect $\Omega^+ \to B_1^+$, $\bar{F} \to B_1'$, and w to solve

$$\begin{cases} \Delta w = 0 \text{ in } B_1^+, \\ w \geq 0 \text{ on } B_1', \\ \partial_n w = 0 \text{ on } \{w > 0\} \cap B_1', \\ \partial_n w \leq 0 \text{ on } B_1'. \end{cases} \tag{1.1}$$

These equations are known as the *Signorini problem* or the *thin obstacle problem.* Our main result states that $\bar{F} = \{x_n = \varepsilon w\}$ inherits the optimal regularity of the solution for the Signorini problem established by Athanasopoulos and Caffarelli in [**2**].

The regularity stated in our main result is optimal in terms of the regularity for Q. Consider in polar coordinates $u = (r \sin\theta - r^{3/2}\cos(3\theta/2))_+$ such that $F = \{\sin\theta = r^{1/2}\cos(3\theta/2)\}$ and over such set we get

$$|Du|^2 = 1 + 9r/4 - 3\frac{\sin\theta\sin(\theta/2)}{\cos(3\theta/2)},$$

From here we can extend Q to a global Lipschitz function such that u solves the corresponding one-phase problem on the upper half space.

1.1. Previous results and overview of the paper. For the one-phase problem, Alt and Caffarelli showed in [**1**] that F is smooth outside of a set of $\mathcal{H}^{n-1}$ measure zero. Their proof is inspired by the regularity theory of minimal surfaces. The key estimate in [**1**] states that the free boundary is $C^{1,\alpha}$ regular provided a flatness hypothesis. A more general theory for two-phase problems was later developed by Caffarelli in [**5–7**] based on a viscosity solution approach.

Following the methods in [**5–7**], several authors extended the results in different directions, for instance the case of variable coefficients with several types of regularity. At this point we would like to highlight one of the recent results due to De Silva, Ferrari, and Salsa [**10**] as it will be relevant for our work. The main theorem in [**10**] establishes that flat free boundaries are $C^{1,\alpha}$ regular in the case of divergence operators with Hölder continuous coefficients. The strategy is based in a compactness approach started by De Silva in [**9**].

The optimal regularity for the solution of the Signorini problem was first established by Athanasopoulos and Caffarelli in [**2**]. The regularity of the free boundary around regular points was established by Athanasopoulos, Caffarelli and Salsa in [**3**].

In this last reference the authors follow a blow-up procedure based on the monotonicity of the Almgren frequency formula which also plays an important role in our theorem.

We combine some of the recent strategies for the one-phase and the Signorini problem to prove our regularity result. In Section 3 we obtain the $C^{1,\beta}$ regularity of $\partial\Omega^+ \cap B_1$ for every $\beta \in (0, 1/2)$ by following the compactness approach from [**9**]. In Section 4 we establish the monotonicity of an Almgren's type frequency formula in order to achieve the optimal $C^{1,1/2}$ regularity for $\partial\Omega^+ \cap B_1$. This section bears some similarities with work by Guillén [**13**] and Garofalo, Smith Vega Garcia [**12**] for the Signorini problem with variable coefficients.

2. Preliminaries

In this section we state the notion of solutions for the one-phase problem in the viscosity sense. A change of variables allows us to reformulate the problem over a convenient geometry. As a trade off we need to consider operators with variable coefficients.

Let a^{ij} be symmetric and uniformly elliptic with respect to some fixed $\lambda > 0$

$$\lambda|\xi|^2 \geq a^{ij}(x)\xi_i\xi_j \geq \lambda^{-1}|\xi|^2.$$

We denote

$$Lu = \partial_i(a^{ij}(x)\partial_j u) \qquad \text{and} \qquad |D_a u| = \sqrt{a^{ij}(x)\partial_i u \partial_j u}.$$

We say that u is *L-superharmonic (L-subharmonic or L-harmonic)* if $Lu \leq (\geq$ or $=)$ 0 holds in the weak sense.

From now on we fix Q continuous such that

$$Q_{min} \leq Q(x) \leq Q_{max} \text{ for some fixed } 0 < Q_{min} \leq Q_{max}.$$

Let $\varphi \in C(B_r(x_0))$ be nonnegative and L-superharmonic over $\Omega^+ = \{\varphi > 0\} \cap B_r(x_0)$. We call φ a *strict comparison supersolution (subsolution) of the one-phase problem* if $\varphi \in C^1(\overline{\Omega^+})$ and

$$|D_a\varphi| < (>)\ Q(x) \text{ on } \partial\Omega^+ \cap B_r(x_0).$$

Given $u, \varphi \in C(S)$, we say that φ *touches u from above (below) at $x_0 \in S$* if

$$u(x_0) = \varphi(x_0) \text{ and } u \leq (\geq)\ \varphi \text{ in } S.$$

Let $u \in C(\Omega)$ be nonnegative and L-subharmonic in $\Omega^+ = \{u > 0\} \cap \Omega$. We call u a *viscosity subsolution (supersolution) of the one-phase problem*

$$\begin{cases} Lu = 0 \text{ in } \Omega^+ = \{u > 0\} \cap \Omega, \\ |D_a u| = Q(x) \text{ on } F = \partial\Omega^+ \cap \Omega, \end{cases} \tag{2.2}$$

if there is no strict comparison supersolution (subsolution) that touches u from above (below). We call u is a *viscosity solution* of (2.2) if u is simultaneously a subsolution and a supersolution.

Next we define the notion of viscosity solution up to the boundary. We assume for simplicity that Ω is a Lipschitz domain and $Z \subseteq \partial\Omega$ is a relatively open set in $\partial\Omega$ which is locally a $C^{1,\alpha}$ hypersurface. As in the introduction we denote by $\bar{F}$ the extension of F to $\bar{\Omega}$,

$$\bar{F} = \partial\Omega^+ \cap \{u = 0\}.$$

Following the terminology of the obstacle problem, we define the *contact set* as

$$\Lambda = \bar{F} \cap Z.$$

Finally we denote by $\partial' \Lambda$ the (thin) boundary of Λ relative to Z.

DEFINITION 2.1. Let $u \in C(\bar{\Omega})$ be nonnegative which vanishes on Z. We say that u is a viscosity solution of the one-phase problem up to Z

$$\begin{cases} Lu = 0 \text{ in } \Omega^+, \\ u = 0 \text{ on } Z, \\ |D_a u| = Q(x) \text{ on } F, \\ |D_a u| \geq Q(x) \text{ on } \Lambda, \end{cases} \tag{2.3}$$

if it is a viscosity solution of the one-phase problem on Ω and the last inequality is satisfied in the viscosity sense, i.e. u cannot be touched from above at $x_0 \in \Lambda$ by a C^1 function φ with $|D_a\varphi(x_0)| < Q(x_0)$.

REMARK 2.2. An equivalent definition can be given by extending Ω to some domain $U \supset \Omega$ such that $Z = \partial\Omega \cap U$. Then $u \in C(\bar{U})$ nonnegative which vanishes on $U \setminus \Omega$ is a solution of (2.3) if is a solution of the one-phase problem on Ω and a subsolution on U.

2.1. Existence. Variational solutions for the one-phase problem can be constructed as minimizers of the following functional with $g \in H^1(\Omega)$ nonnegative

$$J(u) = \int_\Omega a^{ij}\partial_i u \partial_j u + Q^2 \chi_{\{u>0\}}\, dx \quad \text{over} \quad K = \{u \in H^1(\Omega) : u - g \in H^1_0(\Omega)\}.$$

In [**1**] it was shown that such minimizers of J are viscosity solutions of the one-phase problem. We remarked in the introduction that they also satisfy Definition 2.1.

LEMMA 2.3. *Let $g = 0$ on Z. A minimizer of J is a viscosity solution of* (2.3).

Caffarelli developed in [**6**] the Perron's method for viscosity solutions of a family of free boundary problems, including the one-phase problem. The idea is to construct a *minimal viscosity solution* as the infimum over a family of *admissible supersolutions* above a given *subsolution minorant.* The main Theorem in [**6**] states that the *minimal viscosity solution* is a viscosity solution. We refer to [**6**] for the precise definitions.

LEMMA 2.4. *Let $g = 0$ on Z. The minimal viscosity solution above a subsolution minorant that vanishes over Z is a viscosity solution of* (2.3).

The proof of both lemmas can be achieved in similar ways by a contradiction argument. For instance, if there is a test function φ touching u from above at some $x_0 \in \Lambda$ such that $|D_a\varphi(x_0)| < Q(x_0)$, then by applying a inward deformation of Ω^+ as in [**6**, Lemma 9] we obtain an admissible supersolution smaller than u. The same deformation decreases the energy J.

From now on we assume that

$$a^{ij}, Q \in C^\alpha, \text{ and } Z \text{ is } C^{1,\alpha} \text{ regular},$$

and we focus our attention in a neighborhood of a point $x_0 \in \Lambda$. After a domain deformation, we may reduce our analysis to the case

$$\Omega = B_1^+, \quad Z = B_1', \quad 0 \in \Lambda.$$

In this case we will frequently consider u to be defined over B_1 such that u vanishes over $B_1 \setminus B_1^+$ and it is a subsolution of the one-phase problem over B_1, see Remark 2.2.

2.2. Lipschitz regularity and flatness of u. Let $u \in C(B_1^+)$, nonnegative, harmonic in $\Omega^+ = \{u > 0\} \cap B_1$, with $0 \in \partial\Omega^+$. If Ω^+ satisfies either the interior or exterior ball condition at 0, then a barrier argument shows that u has a linear asymptotic behavior at 0. From this observation we get to define the non-tangential gradient $Du(0)$ such that

$$u(x) = (Du(0) \cdot x)_+ + o(|x|) \text{ as } x \to 0 \text{ non-tangentially in } \Omega^+,$$

See [**4**, Chapter 11]. The same result can be reproduced for L with $a^{ij} \in C^\alpha$ thanks to the Schauder estimates.

In the case of u being a solution of (2.3) with $\Omega = B_1^+$, $Z = B_1'$, we get that all points in Λ are regular from outside and it is then possible to construct barriers to bound $|Du|$ in terms of $\|u\|_{L^\infty(B_1^+)}$. This ultimately implies the Lipschitz regularity of the solution up to Λ.

LEMMA 2.5. *Let u be a viscosity solution of* (2.3) *with $\Omega = B_1^+$, $Z = B_1'$. Then*

$$\|u\|_{C^{0,1}\left(B_{1/2}^+\right)} \leq C\left(1 + \|u\|_{L^\infty(B_1^+)}\right).$$

On the other hand, the slope $|Du(0)|$ is bounded from below by $Q_{min} > 0$. The asymptotic expansion and the Lipschitz regularity allows us to deduce the following flatness result.

LEMMA 2.6. *Let u be a viscosity solution of* (2.3) *with $\Omega = B_1$, $Z = B_1'$ such that $0 \in \Lambda$. Then, given $\varepsilon > 0$ there exists $\delta > 0$ such that*

$$(|Du(0)| + \varepsilon)x_n \geq u \geq |Du(0)|(x_n - \varepsilon\delta)_+ \text{ in } B_\delta^+.$$

2.3. Interior regularity of flat free boundaries. Finally we would like to recall one of the main results proved by Alt and Caffarelli in [**1**]: sufficiently flat free boundaries of the one-phase problem are $C^{1,\beta}$. See also the recent results by De Silva, Ferrari, and Salsa for (two-phase) problems with divergence operators [**10**].

In the following we suppose $Q \in C^{0,1}$, and $a^{ij} \in C^\alpha$ such that for some $\varepsilon > 0$,

$$a^{ij}(0) = \delta^{ij}, \qquad Q(0) = 1, \qquad \|a^{ij} - \delta^{ij}\|_{C^\alpha(B_1)} + \|Q - 1\|_{C^{0,1}(B_1)} \leq \varepsilon^2.$$

We assume $u \in C(B_1)$ to be a viscosity solution of

$$\begin{cases} Lu = 0 \text{ in } \Omega^+ = \{u > 0\} \cap B_1, \\ |D_a u| = Q(x) \text{ on } F = \partial\Omega^+ \cap B_1, \end{cases}$$

such that

$$B_1 \cap \{x_n > -\varepsilon\} \supseteq \Omega^+ \supseteq B_1 \cap \{x_n > \varepsilon\}.$$

THEOREM 2.7 (DFS). *For any $\beta \in (0,1)$, there exists $\varepsilon_0 \in (0,1)$ such that if $\varepsilon \in (0, \varepsilon_0)$ then F is parametrized by a $C^{1,\beta}$ function in $B_{1/2}'$*

$$F \cap B_{1/2} = \{x_n = \varepsilon\bar{u}(x') : x' \in B_{1/2}'\},$$

with the following estimate for some universal $C > 0$,

$$\|\bar{u}\|_{C^{1,\beta}(B_{1/2}')} \leq C.$$

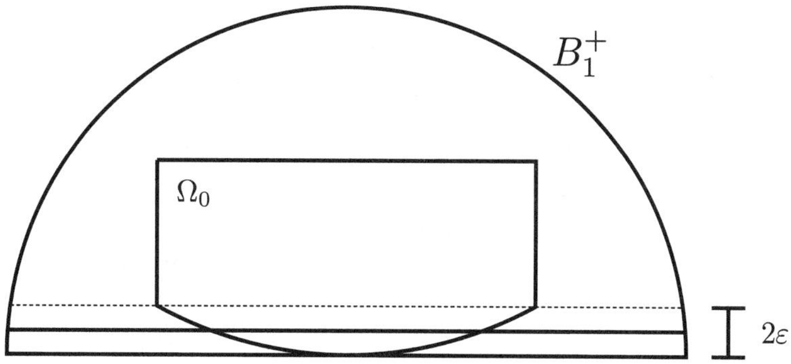

FIGURE 2. Configuration for the proof of Lemma 3.1

3. Almost optimal regularity

In this section we show that $\bar{F}$ has almost optimal regularity in a neighborhood of Z. Precisely, we will show that if $x_0 \in \Lambda$, then F is a $C^{1,\beta}$ regular surface in a neighborhood of x_0 for any $\beta \in (0, \min(1/2, \alpha))$, see Proposition 3.8.

After a domain deformation and a dilation we assume as before that u is a viscosity solution of (2.3) with

$$\Omega = B_1^+, \quad Z = B_1', \quad 0 \in \Lambda,$$

and that for $\alpha \in (0,1)$ the following smallness hypothesis for the coefficients hold

$$(C_{\varepsilon,\delta}) \quad a^{ij}(0) = \delta^{ij}, \qquad Q(0) = 1, \qquad \|a^{ij} - \delta^{ij}\|_{C^\alpha(B_1)} + \|Q-1\|_{C^\alpha(B_1)} \le \delta\varepsilon.$$

for some $\varepsilon \in (0, \varepsilon_0)$ and with δ and ε_0 small, universal, to be made precise later.

The next Lemma will be used to show that $\{|Du| > Q\} \cap \Lambda$ is open relative to $\{x_n = 0\}$.

LEMMA 3.1. *Given $\eta > 0$ there exists $\varepsilon_0 > 0$ such that if $\varepsilon \in (0, \varepsilon_0)$ then*

$$u > (Q(0) + \eta)(x_n - \varepsilon)_+ \text{ in } B_1^+ \qquad \Rightarrow \qquad |D_a u(0)| > Q(0).$$

PROOF. Let Ω_0 be the domain above the parabola $P := \{x_n = 8\varepsilon|x'|^2\}$ that lies inside the cylinder $B'_{1/2} \times [0, 1/2]$, i.e.

$$\Omega_0 := \{1/2 > x_n > 8\varepsilon|x'|^2\} \cap \{|x'| < 1/2\}.$$

Define φ_0 in Ω_0 as the solution to

$$\begin{cases} L\varphi_0 = 0 \text{ in } \Omega_0, \\ \varphi_0 = x_n - 8\varepsilon|x'|^2 \text{ on } \partial\Omega_0. \end{cases}$$

By $(C_{\varepsilon,\delta})$ we easily get that φ_0 is an ε-perturbation of x_n and by the Schauder estimates up to the boundary we obtain that

$$||D_a\varphi_0| - 1| \le C\varepsilon \text{ over } P \cap \{x_n \le \varepsilon\}.$$

The hypothesis implies that $\psi_0 := (1 + \eta/2)\varphi_0$ is below u on $\partial\Omega_0 \cap \{x_n > \varepsilon\}$, and the inequality above says that $|D_a\psi_0| > 1 + \varepsilon > Q$ on the remaining part of the boundary, provided that $\varepsilon < c\eta$.

Let $\psi_t(x) := \psi_0(x - te_n)$ be the translation of ψ_0 by te_n. Notice that the graph of ψ_ε is below the graph of u. Then we slide the graph of ψ_ε in the $-e_n$ direction till it coincides with ψ_0 (i.e. decrease t from ε to 0). The graph of ψ_t cannot touch the graph of u neither on the free boundary, nor on the remaining part of the boundary of $\partial\Omega_t \cap \{x_n > \varepsilon\}$. In conclusion $u \geq \psi_0$ which gives the desired claim. □

Thanks to Lemma 2.6 any point in $\{|Du| > Q\} \cap \Lambda$ satisfies the hypothesis of the previous lemma after a sufficiently large dilation (with η depending on $|Du| - Q$). By applying the previous result centered at points in a sufficiently small neighborhood we conclude that $\{|Du| > Q\} \cap \Lambda$ is open relative to $\{x_n = 0\}$. To prove Theorem 1.1 we can now focus on the case where $0 \in \Lambda$ with $Du(0) = e_n$, i.e. when 0 belongs to the thin boundary $\partial'\Lambda$. By invoking once again Lemma 2.6, we get that after a sufficiently large dilation we can start with a flatness hypothesis of the form

$$(F_\varepsilon) \qquad x_n + \varepsilon \geq u \geq (x_n - \varepsilon)_+ \text{ in } B_1^+,$$

for some small ε. Let us recall the Harnack inequality from [**10**, Theorem 4.1].

LEMMA 3.2. *Let v be a viscosity* ***solution*** *of* (2.2) *in B_1. There exist $\varepsilon_0, \theta \in (0,1)$ such that if for $a, b \in (0, \varepsilon_0)$,*

$$(x_n + a)_+ \geq v \geq (x_n - b)_+ \text{ in } B_1$$

then in $B_{1/2}$ either

$$(x_n + a - \theta c)_+ \geq v \qquad or \qquad v \geq (x_n - b + \theta c)_+ \qquad (c = (a+b)/2)$$

Let us briefly recall the ideas from [**10**] to prove Lemma 3.2. Let $P_+(x) = (x_n + a)_+$, $P_-(x) = (x_n - b)_+$, and $P = (x_n + d)_+$ where $d = (a - b)/2$. One has two consider two possible cases, either $u(e_n/2) \geq P(e_n/2)$ or the opposite inequality holds. In the former case one gets to improve the lower bound, i.e. $v \geq (x_n - b + \theta c)_+$, and in the latter one gets to improve the upper bound by a similar argument. Assuming that $u(e_n/2) \geq P(e_n/2)$, the idea is to apply the classical Harnack inequality to $v - P_-$ around $e_n/2$ and construct a barrier that propagates the improvement beyond $\{x_n = b\}$ thanks to the comparison principle.

In the case that u is only a subsolution of (2.2) in B_1, the barrier argument to improve the upper bound of u still applies. If u is a supersolution of (2.2) restricted to B_1^+, the barrier argument to improve the lower bound can be performed if we assume that the free boundary of the barrier does not reach $\{x_n = 0\}$, where u is no longer a supersolution. In this case we can get an improvement proportional to $c = (a+b)/2$ if we assume $b \geq a$.

COROLLARY 3.3. *There exist $\varepsilon_0, \theta \in (0,1)$ such that if for $0 < a \leq b < \varepsilon_0$,*

$$(x_n + a)_+ \geq u \geq (x_n - b)_+ \text{ in } B_1^+$$

then in $B_{1/2}^+$ either

$$(x_n + a - \theta c)_+ \geq u \qquad or \qquad u \geq (x_n - b + \theta c)_+ \qquad (c = (a+b)/2)$$

By iterating the previous corollary we get the following diminish of oscillation.

LEMMA 3.4. *There exist $\varepsilon_0, \mu, \theta \in (0,1)$ such that if for $\varepsilon \in (0, \varepsilon_0)$ and (F_ε) holds, then in B_μ^+ either*

$$x_n \geq u \qquad or \qquad u \geq (x_n - (1-\theta)\varepsilon)_+.$$

We remark that in the Corollary 3.3 and Lemma 3.4 above we do not assume that $0 \in \partial'\Lambda$ but only that (F_ε) holds. If the first alternative of Lemma 3.4 holds then $\bar{F}$ is unconstrained in B_μ^+ and we fall in situation of the interior case as in [**10**]. If the second alternative holds then u satisfies a version of (F_ε) in which we replace B_1^+ by B_μ^+ and ε by $(1-\theta)\varepsilon$.

PROOF. Let $\bar{\varepsilon}_0, \bar{\mu} = 1/2, \theta \in (0,1)$ the constants corresponding to Corollary 3.3 and let $\varepsilon_0, \mu \in (0,1)$ to be fixed later in the proof. As we will be iterating Corollary 3.3 a finite number of times let us actually say that for some $k \in \mathbb{N}$ to be determined $\mu = \bar{\mu}^k$.

Let $u_i(x) = \bar{\mu}^{-i}u(\bar{\mu}^i x)$. We have that for any $i = 0, 1, 2, \dots$

$$(x_n + \varepsilon)_+ \geq (1+\varepsilon)x_n \geq u_i \text{ in } B_1^+$$

Let $C_0 = (1-\theta/2)/\bar{\mu} > 1$ and $b_i = (2\varepsilon/\theta)C_0^i$ for $i \in \{0, 1, 2, \dots, (k-1)\}$. Assume by induction that

$$u_i \geq (x_n - b_i)_+ \text{ in } B_1^+$$

By Corollary 3.3, we get that in $B_{\bar{\mu}}^+$ either

$$(x_n + \varepsilon - \theta(\varepsilon + b_i)/2)_+ \geq u_i \qquad \text{or} \qquad u_i \geq (x_n - b_i + \theta(\varepsilon + b_i)/2)$$

The first alternative implies $(x_n)_+ \geq u$ in B_μ^+ and would settle the proof. On the other hand, the second option implies the subsequent step in the induction, $u_{i+1} \geq (x_n - b_{i+1})_+$ in B_1^+.

In order to iterate Corollary 3.3 up to $i = k$ we need $b_{k-1} \leq \bar{\varepsilon}_0$ which follows by taking $\varepsilon_0 = \theta\bar{\varepsilon}_0/(2C_0^{k-1})$.

We finally fix k sufficiently large such that $1 - \theta \geq (2/\theta)(1-\theta/2)^k$. Hence $u_k \geq (x_n - b_k)_+$ in B_1^+ implies $u \geq (x_n - (1-\theta)\varepsilon)_+$ in B_μ^+. □

Next we define the function w in $\overline{\Omega^+}$ as

$$w := \frac{x_n - u}{\varepsilon},$$

and clearly $w \geq 0$ on $\bar{F}$ and $\|w\|_{L^\infty} \leq 1$ if (F_ε) holds. Lemmas 3.2 and 3.3 provide a diminish of oscillation for w as we restrict to a smaller ball. By iterating these lemmas (and the standard Harnack inequality at points away from $\{x_n = 0\}$) several times we obtain an almost uniform Holder modulus of continuity for w (except for points at smaller and smaller scales). A version of Arzela-Ascoli theorem gives the compactness of a family of w's as $\varepsilon, \delta \to 0$.

Precisely, let us consider a sequence a sequence of solutions $\{u_k\}$ satisfying $(F_\varepsilon F_{\varepsilon_k})$ and $(C_{\varepsilon,\delta} C_{\varepsilon_k,\delta_k})$ with $\varepsilon_k, \delta_k \to 0$, and the graphs of the corresponding w_k restricted to the cylinder $\bar{B}_{1/2} \times \mathbb{R}$,

$$G_k := \{(x, w_k(x)) | \, x \in \overline{\Omega_k^+} \cap \bar{B}_{1/2}\}.$$

COROLLARY 3.5. *There exists a subsequence of G_k's which converges (in the Hausdorff distance) to the graph of a Holder continuous function $\bar{w} \in C(\bar{B}_{1/2}^+)$.*

Notice that in the previous corollary the domains of definition of w_k vary with k, however they converge to $\bar{B}_1^+$.

LEMMA 3.6. *The function $\bar{w}$ solves the Signorini Problem* (1.1) *(in the viscosity sense).*

PROOF. Since $u_k = x_n - \varepsilon_k w_k$ and $L_k u_k = 0$ we find that

$$L_k w_k = \frac{1}{\varepsilon_k} L_k x_n = \frac{1}{\varepsilon_k} \partial_i (a_k^{in} - \delta^{in}).$$

From $(C_{\varepsilon,\delta} C_{\varepsilon_k,\delta_k})$ we see that as $w_k \to \bar{w}$, $\delta_k, \varepsilon_k \to 0$ we obtain $\Delta \bar{w} = 0$ in B_1^+.

Since $w_k \geq 0$ on $\bar{F}_k$ we obtain that $\bar{w} \geq 0$ on B_1'.

It remains to check that on B_1' we satisfy the Signorini condition $\partial_n \bar{w} \leq 0$ in the viscosity sense and we have equality over the positivity set of $\bar{w}$. Assume that $a + p \cdot x - C|x|^2$ touches $\bar{w}$ from below at a point $x_0 \in B_{1/2}'$, and assume for simplicity of notation that $x_0 = 0$. We need to show that $p_n \leq 0$.

Given $\eta > 0$ we may assume that the polynomial $P(x) = a + p \cdot x - \eta x_n - C(|x'|^2 - n x_n^2)$ touches w strictly from below at 0 in B_r^+ for some small r.

Let $\varphi_k = x_n - \varepsilon_k P$ and $\widetilde{\varphi}_k$ such that $L_k \widetilde{\varphi}_k = \Delta \varphi_k$ in B_r^+, and $\widetilde{\varphi}_k = \varphi_k$ on ∂B_r^+. By $(C_{\varepsilon,\delta} C_{\varepsilon_k,\delta_k})$ and Schauder estimates we have $\|\widetilde{\varphi}_k - \varphi_k\|_{C^{1,\alpha}(B_{r/2}^+)} \leq C \delta_k \varepsilon_k$.

By the convergence of G_k to the graph of w, we get that for k sufficiently large and some $d_k \in (-\xi, \xi)$, $P + (\varphi_k - \widetilde{\varphi}_k)/\varepsilon_k + d_k$ touches w_k from below at some $x_k \in (\Omega_k^+ \cup F_k) \cap B_{r/2}$ with $x_k \to 0$. In other words, $\widetilde{\varphi}_k - \varepsilon_k d_k$ touches u_k from above. Given that $L_k \widetilde{\varphi}_k = -\varepsilon_k \Delta P < 0$ we have that $x_k \in F_k \cap B_{r/2}$. By the free boundary condition

$$1 - 2\delta_k \varepsilon_k \leq |D_{a_k} \widetilde{\varphi}_k|^2 \leq |D\varphi_k|^2 + C\delta_k \varepsilon_k \leq 1 - 2\varepsilon_k (p_n - \eta) + C(\delta_k \varepsilon_k + \varepsilon_k^2),$$

which implies the desired bound for p_n after we let $k \to \infty$ and then $\eta \to 0$.

A similar argument shows that $\partial_n w \geq 0$ over $\{w > 0\} \cap B_1'$. □

If we assume that $0 \in \partial' \Lambda_k$ then $w_k(0) = 0$ and $\bar{w}(0) = 0$. Since $\|\bar{w}\|_{L^\infty} \leq 1$, the optimal $C^{1,1/2}$ regularity for the Signorini problem implies that

$$|\bar{w}(x)| \leq C|x|^{3/2},$$

for some C universal. This implies that given $\beta \in (0, 1/2)$, there exists μ small depending on β and the other universal constants such that for all k large

$$|w_k| \leq \mu^{1+\beta} \quad \text{in} \quad \overline{\Omega_k^+} \cap B_\mu. \tag{3.4}$$

We have established the following improvement of flatness result.

LEMMA 3.7. *Given $\beta \in (0, 1/2)$, there exist $\varepsilon_0, \delta, \mu$ depending on β, and the other universal constants such that if $0 \in \partial' \Lambda$ and $(F_\varepsilon F_\varepsilon)$ and $(C_{\varepsilon,\delta} C_{\varepsilon,\delta})$ for some $\varepsilon \in (0, \varepsilon_0)$ then*

$$x_n + \varepsilon \mu^{1+\beta} \geq u \geq (x_n - \varepsilon \mu^{1+\beta})_+ \text{ in } B_\mu^+,$$

i.e., the rescaling $\tilde{u}(x) := \mu^{-1} u(\mu x)$ satisfies $(F_\varepsilon F_{\tilde{\varepsilon}})$ with $\tilde{\varepsilon} = \varepsilon \mu^\beta$.

The proof of the lemma follows by contradiction and compactness. If the statement fails for a sequence of u_k's, and with corresponding ε_k, $\delta_k \to 0$, then we argue as above and find from (3.4) that the u_k's do satisfy the conclusion of the lemma for all large k.

We can iterate the lemma above provided that $\beta \leq \alpha$ so that hypothesis $(C_{\varepsilon,\delta})$ scales accordingly. We obtain that u is pointwise $C^{1,\beta}$ at $0 \in \partial' \Lambda$ in the domain of definition, i.e.

$$|u - x_n| \leq C\varepsilon |x|^{1+\beta} \text{ in } \overline{\Omega^+}.$$

Now it is standard to extend the $C^{1,\beta}$ regularity from $\partial' \Lambda$ to the whole domain of definition.

PROPOSITION 3.8. *Let $\beta \in (0, \min(1/2, \alpha))$ and assume that u satisfies $(C_{\varepsilon,\delta})$, (F_ε) for some $\varepsilon \in (0, \varepsilon_0)$. Then*

$$\|u\|_{C^{1,\beta}(\overline{\Omega^+} \cap B_{1/2})} \leq C.$$

Notice that the estimate above implies that the free boundary $\bar{F} \in C^{1,\beta}$ as well.

PROOF. It suffices to show that u is pointwise $C^{1,\beta}$ at all points $y \in \overline{\Omega^+} \cap B_{1/2}$. We look at the distance r from y to $\partial'\Lambda$, and assume for simplicity that the distance is realized at $0 \in \partial'\Lambda$. We assume without loss of generality that $F \cap B_{1/2} \subseteq \{x_n < |x'|\}$, which follows from Lemma 3.7 after a suitable dilation.

By Lemma 3.7 u is approximated in a $C^{1,\beta}$ fashion by x_n in balls of radius greater than r centered at y. To check that u is approximated at scales smaller than r we distinguish three cases.

If $y_n > |y'|/2$ then the desired conclusion follows by interior Schauder estimates.

If $y_n \leq |y'|/2$ and $B'_{|y'|}(y') \subset \Lambda$ then the conclusion follows by Schauder estimates up to the boundary.

If $y_n \leq |y'|/2$ and $B'_{|y'|}(y') \cap \Lambda = \emptyset$ then F is unconstrained in $B^+_{r/2}(y', 0)$. Now the estimates in [**10**] apply, or alternatively we could repeat the arguments of Lemma 3.7 in the unconstrained setting. □

REMARK 3.9. In terms of the function w the estimate we obtained is

$$\varepsilon\|w\|_{L^\infty(\Omega^+ \cap B_1)} \leq \varepsilon_0 \qquad \Rightarrow \qquad \|w\|_{C^{1,\beta}(\overline{\Omega^+} \cap B_{1/2})} \leq C\|w\|_{L^\infty(\Omega^+ \cap B_1)}.$$

4. Optimal regularity

In this section we will establish Theorem 1.1. We assume that u is a solution of (2.3) for

$$a^{ij} = \delta^{ij}, \quad \Omega = B_1 \cap \{x_n > g(x')\}, \quad \text{and} \quad Z = \{x_n = g(x')\} \cap B_1,$$

where $g \in C^{1,1/2+\sigma}(B'_1)$ for some small $\sigma > 0$ and

$$g(0) = 0, \qquad D'g(0) = 0, \qquad \|g\|_{C^{1,1/2+\sigma}(B'_1)} \leq 1.$$

We consider $Q \in C^{0,1}(B_1)$ satisfying

$$Q(0) = 1, \qquad \|Q - 1\|_{C^{0,1}(B_1)} \leq 1,$$

and assume $0 \in \partial'\Lambda$, hence $Du(0) = e_n$.

In view of the previous section $u \in C^{1,\beta}(\Omega^+ \cap \bar{F})$ for some $\beta \in (0, 1/2)$ that we choose sufficiently close to $1/2$ so that $\beta \in (1/2 - \sigma/10, 1/2)$.

To establish the $C^{1,1/2}$ regularity of $\bar{F}$ we follow the strategy from [**8,13**] applied to the function w defined in the previous section as

$$w := x_n - u,$$

and we suppose without loss of generality that

$$\|w\|_{C^{1,\beta}(\bar{\Omega})} \leq 1.$$

Since $w(0) = 0$, $Dw(0) = 0$ we have

$$w = O(r^{1+\beta}), \quad |Dw| = O(r^\beta) \quad \text{in} \quad \bar{\Omega} \cap B_r. \tag{4.5}$$

Moreover, the free boundary condition $|Du| = Q$ on F implies $\partial_n w = O(r^{2\beta})$ on $F \cap B_r$, or

$$\partial_\nu w = O(r^{2\beta}) \quad \text{and} \quad w \geq 0 \quad \text{on} \quad F \cap B_r, \tag{4.6}$$

where ν is the outward normal to Ω^+. On the remaining part Λ of $\bar{F} \cap B_r$ (where $\bar{F}$ coincides with Z) we have $w = g(x')$ and $|Du| \geq Q$. We easily deduce

$$w = O(r^{\frac{3}{2}+\sigma}), \quad Dw \cdot x = O(r^{\frac{3}{2}+\sigma}), \quad \partial_\nu w \geq -Cr^{2\beta} \quad \text{on} \quad \Lambda \cap B_r. \tag{4.7}$$

Combining the inequalities above we find

$$w\partial_\nu w = O(r^{1+3\beta} + r^{\frac{3}{2}+\beta+\sigma}) = O(r^{2+\sigma/2}) \quad \text{on} \quad \bar{F} \cap B_r. \tag{4.8}$$

The main goal is to use Almgren's monotonicity formula and show that for r sufficiently small

$$H(r) := \frac{1}{r^{n-1}} \int_{\partial B_r(x_0) \cap \Omega^+} w^2 \leq Cr^3, \tag{4.9}$$

from which we can easily deduce that $w = O(r^{\frac{3}{2}})$.

Below we use the following convention for various average integrals over sets $E \subset \bar{B}_r$,

$$⨍_E f = \frac{1}{r^d} \int_E f, \qquad \text{where } d = \dim(E),$$

hence

$$H(r) = ⨍_{\partial B_r \cap \Omega^+} w^2.$$

4.1. Almgren's frequency formula. If w is a homogeneous function we get that the homogeneity of w can be computed from the *frequency functional*

$$N(r) = r\frac{d}{dr} \ln \left(⨍_{\partial B_r} w^2 \right)^{1/2}.$$

Almgren's monotonicity formula says that if w is harmonic near the origin, then N is nondecreasing. Moreover, if N remains constant, then w is homogeneous of degree N.

Let us compute straightaway the derivative of H. In the following ∂_r denotes the radial derivative.

$$\begin{aligned} H'(r) &= 2⨍_{\partial B_r \cap \Omega^+} w\partial_r w - \frac{1}{r^2} ⨍_{\partial B_r \cap \bar{F}} w^2 (x \cdot \nu) \\ &= 2r ⨍_{B_r \cap \Omega^+} |Dw|^2 - 2⨍_{B_r \cap \bar{F}} w\, \partial_\nu w - \frac{1}{r^2} ⨍_{\partial B_r \cap \bar{F}} w^2 (x \cdot \nu). \end{aligned}$$

In order to get an exact formula for the second derivatives we consider the following perturbation of H,

$$\widetilde{H}(r) := H(r) + \int_0^r (E_1(\rho) + E_2(\rho)) \frac{d\rho}{\rho},$$

with (see (4.5)-(4.8))

$$E_1(r) := \frac{1}{r} ⨍_{\partial B_r \cap \bar{F}} w^2 (x \cdot \nu) = O(r^{2+3\beta}),$$

$$E_2(r) := 2r ⨍_{B_r \cap \bar{F}} w\partial_\nu w = O(r^{3+\sigma/2}).$$

Thus,

$$\widetilde{H}(r) = H(r) + O(r^{3+\sigma/2}), \qquad \widetilde{H}'(r) = 2r \fint_{B_r\cap\Omega^+} |Dw|^2, \tag{4.10}$$

and we also have

$$\frac{n-1}{r}\widetilde{H}'(r) + \widetilde{H}''(r) = 2\fint_{\partial B_r\cap\Omega^+} |Dw|^2.$$

By the Rellich's identity

$$(n-2)|Dw|^2 - 2(Dw\cdot x)\Delta w = \operatorname{div}\left(|Dw|^2 x - 2(Dw\cdot x)Dw\right),$$

we obtain

$$(n-2)\int_{B_r\cap\Omega^+} |Dw|^2 = r\int_{\partial B_r\cap\Omega^+} \left(|Dw|^2 - 2(\partial_r w)^2\right) + \int_{B_r\cap\bar{F}} \left(|Dw|^2(x\cdot\nu) - 2(Dw\cdot x)\,\partial_\nu w\right).$$

Using (4.5)-(4.7) we find that on $\bar{F}\cap B_r$

$$|Dw|^2(x\cdot\nu) = O(r^{1+3\beta}), \quad (Dw\cdot x)\,\partial_\nu w = O(r^{\frac{3}{2}+\sigma+\beta}),$$

hence

$$(n-2)\fint_{B_r\cap\Omega^+} |Dw|^2 = \fint_{\partial B_r\cap\Omega^+} \left(|Dw|^2 - 2(\partial_r w)^2\right) + O(r^{1+\sigma/2}),$$

which gives

$$\widetilde{H}''(r) + \frac{1}{r}\widetilde{H}'(r) = 4\fint_{\partial B_r\cap\Omega^+} (\partial_r w)^2 + O(r^{1+\sigma/2}). \tag{4.11}$$

As in [**8, 13**] we consider now a truncated type of frequency

$$\widetilde{N}(r) = \frac{r}{2}\frac{d}{dr}\ln\max(\widetilde{H}(r), r^{3+\sigma/10}),$$

and show that it is almost monotone.

LEMMA 4.1.

$$\widetilde{N}'(r) \geq -Cr^{-1+\sigma/10}\widetilde{N}(r).$$

First we establish an auxiliary result needed in the proof of Lemma 4.1.

LEMMA 4.2. *If* $\widetilde{H}(r) \geq r^{3+\sigma/10}$ *then*

$$\widetilde{H}'(r) = 2r\fint_{B_r\cap\Omega^+} |Dw|^2 \geq c_0 r^{2+\sigma/10}. \tag{4.12}$$

PROOF. We obtain the lower bound in two steps. Using that $w \geq -r^{3/2+\sigma}$ over $\bar{F}\cap B_r$, we get that thanks to the Sobolev and trace inequality

(4.13)

$$r^2\fint_{B_r\cap\Omega^+} |Dw|^2 \geq c\fint_{\partial B_r\cap\Omega^+} [(w + r^{3/2+\sigma})^-]^2 \geq c\fint_{\partial B_r\cap\Omega^+} (w^-)^2 - Cr^{3+2\sigma}.$$

Next we consider the harmonic function h in $\Omega^+\cap B_r$ such that

$$\Delta h = 0 \text{ in } B_r\cap\Omega^+, \qquad \partial_\nu h = 0 \text{ on } B_r\cap\bar{F}, \qquad h = w \text{ on } \partial B_r\cap\Omega^+,$$

and notice that

$$\int_{B_r\cap\Omega^+} |Dw|^2 \geq \int_{B_r\cap\Omega^+} |Dh|^2.$$

By the maximum principle and using that $\partial_\nu w \geq -Cr^{2\beta}$ on $\bar{F}$, we get that for some $C > 0$,

$$h + Cr^{2\beta}(x_n - r) \leq w.$$

Since $w(0) = 0$, we find

$$h(0) \leq Cr^{1+2\beta} \leq Cr^{3/2+\sigma}. \tag{4.14}$$

Let us assume by contradiction that the conclusion does not hold. Then, by the standard L^2 estimates,

$$c_0 r^{3+\sigma/10} \geq r^2 \fint_{B_r\cap\Omega^+} |Dh|^2 \geq c\|h - \bar{h}\|^2_{L^\infty(B_{r/2}\cap\bar{\Omega}^+)},$$

where $\bar{h}$ denotes the average of h over $B_r\cap\Omega^+$. From (4.14) we find $\bar{h} \leq Cc_0 r^{3/2+\sigma/10}$, and by the Poincaré and trace inequality we obtain that

$$r^2 \fint_{B_r\cap\Omega^+} |Dh|^2 \geq c \fint_{\partial B_r\cap\Omega^+} (h-\bar{h})^2 \geq c \fint_{\partial B_r\cap\Omega^+} (w^+)^2 - Cc_0 r^{3+\sigma/10}. \tag{4.15}$$

Now we reach a contradiction by combining (4.13) and (4.15), provided that c_0 is chosen sufficiently small. □

COROLLARY 4.3. *If $\widetilde{H}(r) \geq r^{3+\sigma/10}$ then*

$$\fint_{\partial B_r\cap\Omega^+} w\partial_r w \geq c\, r^{2+\sigma/10}$$

The corollary follows from Lemma 4.2 by noticing that the difference between $\frac{1}{2}\widetilde{H}'(r)$ and the left-hand side above is $\fint_{\bar{F}\cap B_r} w\,\partial_\nu w = O(r^{2+\sigma/2})$.

PROOF OF LEMMA 4.1. We focus on the case $\widetilde{H}(r) > r^{3+\sigma/2}$ such that Lemma 4.2 and its corollary apply. We compute the logarithmic derivative by using (4.11),

$$\begin{aligned}
\frac{\widetilde{N}'(r)}{\widetilde{N}(r)} &= \frac{1}{r} + \frac{\widetilde{H}''}{\widetilde{H}'} - \frac{\widetilde{H}'}{\widetilde{H}} \\
&\geq \frac{4\fint_{\partial B_r\cap\Omega^+}(\partial_r w)^2 - Cr^{1+\sigma/2}}{\widetilde{H}'(r)} - \frac{2\fint_{\partial B_r\cap\Omega^+} w\partial_r w + 2\fint_{B_r\cap F} w\,\partial_\nu w}{\widetilde{H}(r)} \\
&\geq \frac{4\fint_{\partial B_r\cap\Omega^+}(\partial_r w)^2}{\widetilde{H}'(r)} - \frac{2\fint_{\partial B_r\cap\Omega^+} w\partial_r w}{\widetilde{H}(r)} - Cr^{-1+\sigma/10}.
\end{aligned} \tag{4.16}$$

We use that

$$\widetilde{H}' = 2\fint_{\partial B_r\cap\Omega^+} w\partial_r w + O(r^{2+\sigma/2}), \qquad \widetilde{H} = \fint_{\partial B_r\cap\Omega^+} w^2 + O(r^{3+\sigma/2}),$$

together with Lemma 4.2, Corollary 4.3, and obtain by Cauchy-Schwartz inequality

$$\frac{1}{2}\frac{\widetilde{N}'(r)}{\widetilde{N}(r)} \geq \frac{\fint_{\partial B_r\cap\Omega^+}(\partial_r w)^2}{\fint_{\partial B_r\cap\Omega^+} w\partial_r w} - \frac{\fint_{\partial B_r\cap\Omega^+} w\partial_r w}{\fint_{\partial B_r\cap\Omega^+} w^2} - Cr^{-1+\sigma/10} \geq -Cr^{-1+\sigma/10}.$$

□

We have the following consequence of Lemma 4.2.

COROLLARY 4.4. *$(1+Cr^{\sigma/10})\widetilde{N}(r)$ is nondecreasing.*

4.2. Blowup. Our next goal is to show the lower bound

$$\widetilde{N}(0^+) \geq \frac{3}{2}. \tag{4.17}$$

We achieve this by blowing up w at the origin so that the blowup limit is a nontrivial homogeneous global solution of the Signorini problem. Then we will obtain the desired bound for H from (4.17) by integrating $(1+Cr^{\sigma/10})\widetilde{N}(r) \geq 3/2$.

Let

$$u_r(x) = \frac{u(rx)}{r} \qquad \varepsilon_r = \frac{H(r)^{1/2}}{r} \qquad w_r(x) = \frac{w(rx)}{H(r)^{1/2}} = \frac{x_n - u_r(x)}{\varepsilon_r}$$

with the corresponding domains

$$\Omega_r^+ = r^{-1}\Omega^+, \qquad \bar{F}_r = r^{-1}\bar{F}.$$

By construction the L^2 norm of w_r over $\partial B_1 \cap \Omega_r^+$ is one. Also from Remark 3.9 we have that given $K \subset\subset B_1$, there exist $\varepsilon_0 \in (0,1)$ and $C>0$ (depending also on K) such that

$$\varepsilon_r\|w_r\|_{L^\infty(B_1\cap\Omega^+)} \in (0,\varepsilon_0) \qquad \Rightarrow \qquad \|w_r\|_{C^{1,\beta}(K\cap\bar{\Omega}^+)} \leq C\|w_r\|_{L^\infty(B_1\cap\Omega^+)}. \tag{4.18}$$

The following lemma establishes the existence of a homogeneous blowup limit for w_r's.

LEMMA 4.5. *If*

$$\liminf_{r\to 0^+} \frac{\varepsilon_r}{r^{1/2+\sigma/20}} > 1,$$

then there exists a blowup limit $w_0 \in C^{1,\beta}(\bar{B}_1^+)$ such that for some sequence $r_k \to 0^+$, the graphs of $w_k = w_{r_k}$ converge on compact sets of $B_1 \times \mathbb{R}$ (in the $C^{1,\beta}$ topology) to the graph of w_0. Moreover, w_0 is a nontrivial solution of the Signorini problem, homogeneous of degree $\widetilde{N}(0^+)$.

First we show that the L^∞ norm of w_r can be controlled by the H^1 norm in $B_1 \cap \Omega_r^+$.

LEMMA 4.6. *Assume $\widetilde{H}(r) \geq r^{3+\sigma/10}$. Given $K \subset\subset B_1$ there exist C (depending on K) such that*

$$\|w_r\|_{L^\infty(K\cap\Omega_r^+)} \leq C(\|w_r\|_{H^1(B_1\cap\Omega_r^+)} + 1).$$

PROOF. Consider $h \geq 0$ such that

$$\Delta h = 0 \text{ in } B_1 \cap \Omega_r^+, \qquad \partial_\nu h = 0 \text{ on } B_1 \cap F_r, \qquad h = w_r^+ \text{ on } \partial B_1 \cap \Omega_r^+.$$

Notice that, $\partial_\nu w_r = O(r^{1/2-\sigma/4})$ over F_r, meanwhile $w_r \leq r^{19\sigma/20}$ over Λ_r. By the comparison principle we get that $w_r \leq h + 1 + C(1-x_n)$. Given that h is bounded on K in terms of the H^1 norm of h in $B_1 \cap \Omega_r$, which in turn is bounded by the H^1 norm of w_r^+ in the same domain, we get the desired bound from above.

To obtain the bound from below we consider instead $v \leq 0$ such that

$$\Delta v = 0 \text{ in } B_1 \cap \Omega_r^+ \qquad v = 0 \text{ on } B_1 \cap F_r \qquad v = (w_r+1)^- \text{ on } \partial B_1 \cap \Omega_r^+$$

Using that $w_r \geq -r^{19\sigma/20}$ over F we get that $v - 1 \leq w_r$. Since on K, v is bounded by $\|v\|_{H^1}$ which in turn is bounded by $\|(w_r+1)^-\|_{H^1}$ we deduce the desired lower bound. □

REMARK 4.7. The same proof applies to $\widetilde{w}_r(x) = r^{-3/2}w(rx)$ (without the assumption $\widetilde{H}(r) \geq r^{3+\sigma/10}$).

PROOF OF LEMMA 4.5. Let $r \in (0, r_0)$ with r_0 sufficiently small such that $\varepsilon_r > r^{1/2+\sigma/20}$. Given that $\widetilde{H}(r) = H(r) + O(r^{3+\sigma/2})$, this implies that, for r_0 possibly smaller, we have $\widetilde{H}(r) \geq r^{3+\sigma/10}$, and

$$\fint_{B_1 \cap \Omega_r^+} |Dw_r|^2 = \widetilde{N}(r) \leq C\widetilde{N}(1).$$

Because $\|w_r\|_{L^2(\partial B_1 \cap \Omega_r^+)} = 1$ we recover that the H^1 norm of w_r on $B_1 \cap \Omega_r^+$ is uniformly bounded. Indeed, one can use that for $(y', y_n) \in (\partial B_1)^+$

$$\int_{B_1 \cap \Omega_r^+ \cap \{x'=y'\}} w_r^2 \, dx_n \leq C \left(w_r^2(y', y_n) + \int_{B_1 \cap \Omega_r^+ \cap \{x'=y'\}} |Dw_r|^2 dx_n \right).$$

Then the bound follows after integrating over $y' \in B_1'$.

Consider now an extension to B_1 still denoted by w_r and uniformly bounded in $H^1(B_1)$. This means that some sequence $w_k = w_{r_k}$ converges to w_0 weakly in $H^1(B_1)$ and strongly in $L^2(\partial B_1)$. Moreover w_0 is nontrivial because $\|w_0\|_{L^2((\partial B_1)^+)} = 1$.

The almost optimal regularity Proposition 3.8 gives that $\varepsilon_r \to 0$. From the L^∞ bound in Lemma 4.6 and (4.18) we deduce that w_k's are uniformly bounded in $C^{1,\beta}$ in the interior and the convergence to w_0 holds in the $C^{1,\beta}$ norm on compact sets. Clearly w_0 solves the Signorini problem in B_1^+, and from the convergence in $C^1_{loc}(B_1^+)$ we get that for any $r \in (0,1)$

$$r^2 \frac{\fint_{B_r^+} |Dw_0|^2}{\fint_{(\partial B_r)^+} w_0^2} = \lim_{k\to\infty} r^2 \frac{\fint_{B_r \cap \Omega_{r_k}^+} |Dw_{r_k}|^2}{\fint_{\partial B_r \cap \Omega_{r_k}^+} w_{r_k}^2} = \lim_{k\to\infty} \widetilde{N}(rr_k) = \widetilde{N}(0^+).$$

Given that the standard frequency of w_0 is constant we get that it is necessarily a homogeneous function. □

The minimum homogeneity of a nontrivial solution of the Signorini problem is 3/2. On the other hand if the hypothesis about $\liminf_{r\to 0^+} \varepsilon_r / r^{1/2+\sigma/20}$ is not satisfied, then $\widetilde{N}(0+) = 3/2 + \sigma/20$. In conclusion we have established the claim (4.17).

At this point we are ready to settle our main result.

PROOF OF THEOREM 1.1. After a sufficiently large dilation we can arrange u to satisfy all the hypotheses of this section. Thus, by (4.17), $(1+Cr^{\sigma/10})\widetilde{N}(r) \geq 3/2$ and by integrating,

$$\widetilde{H}(r) = H(r) + O(r^{3+\sigma/2}) \leq Cr^3 \qquad \Rightarrow \qquad H(r) \leq Cr^3.$$

Now we consider $\widetilde{w}_r(x) = r^{-3/2}w(rx)$, it is not difficult to check (see (4.10)) that its H^1 norm is uniformly bounded. Remark 4.7 gives the desired modulus of continuity at 0,

$$\sup_{r\in(0,r_0)} r^{-3}H(r) \leq C \qquad \Rightarrow \qquad \sup_{r\in(0,r_0)} r^{-3/2} \operatorname*{osc}_{\Omega^+ \cap B_r} w \leq C,$$

and we established the pointwise $C^{1,1/2}$ of u at $0 \in \partial'\Lambda$. As in the proof of Proposition 3.8, this can be easily extended to $\bar{B}_{1/2} \cap \bar{\Omega}^+$ by standard arguments. Finally the regularity of the free boundary follows from the regularity of u by the implicit function theorem. □

References

[1] H. W. Alt and L. A. Caffarelli, *Existence and regularity for a minimum problem with free boundary*, J. Reine Angew. Math. **325** (1981), 105–144. MR618549

[2] I. Athanasopoulos and L. A. Caffarelli, *Optimal regularity of lower dimensional obstacle problems* (English, with English and Russian summaries), Zap. Nauchn. Sem. S.-Peterburg. Otdel. Mat. Inst. Steklov. (POMI) **310** (2004), no. Kraev. Zadachi Mat. Fiz. i Smezh. Vopr. Teor. Funkts. 35 [34], 49–66, 226, DOI 10.1007/s10958-005-0496-1; English transl., J. Math. Sci. (N.Y.) **132** (2006), no. 3, 274–284. MR2120184

[3] I. Athanasopoulos, L. A. Caffarelli, and S. Salsa, *The structure of the free boundary for lower dimensional obstacle problems*, Amer. J. Math. **130** (2008), no. 2, 485–498, DOI 10.1353/ajm.2008.0016. MR2405165

[4] Luis Caffarelli and Sandro Salsa, *A geometric approach to free boundary problems*, Graduate Studies in Mathematics, vol. 68, American Mathematical Society, Providence, RI, 2005. MR2145284

[5] Luis A. Caffarelli, *A Harnack inequality approach to the regularity of free boundaries. I. Lipschitz free boundaries are* $C^{1,\alpha}$, Rev. Mat. Iberoamericana **3** (1987), no. 2, 139–162, DOI 10.4171/RMI/47. MR990856

[6] Luis A. Caffarelli, *A Harnack inequality approach to the regularity of free boundaries. III. Existence theory, compactness, and dependence on* X, Ann. Scuola Norm. Sup. Pisa Cl. Sci. (4) **15** (1988), no. 4, 583–602 (1989). MR1029856

[7] Luis A. Caffarelli, *A Harnack inequality approach to the regularity of free boundaries. II. Flat free boundaries are Lipschitz*, Comm. Pure Appl. Math. **42** (1989), no. 1, 55–78, DOI 10.1002/cpa.3160420105. MR973745

[8] Luis A. Caffarelli, Sandro Salsa, and Luis Silvestre, *Regularity estimates for the solution and the free boundary of the obstacle problem for the fractional Laplacian*, Invent. Math. **171** (2008), no. 2, 425–461, DOI 10.1007/s00222-007-0086-6. MR2367025

[9] D. De Silva, *Free boundary regularity for a problem with right hand side*, Interfaces Free Bound. **13** (2011), no. 2, 223–238, DOI 10.4171/IFB/255. MR2813524

[10] Daniela De Silva, Fausto Ferrari, and Sandro Salsa, *Regularity of the free boundary for two-phase problems governed by divergence form equations and applications*, Nonlinear Anal. **138** (2016), 3–30, DOI 10.1016/j.na.2015.11.013. MR3485136

[11] Avner Friedman, *Variational principles and free-boundary problems*, Pure and Applied Mathematics, John Wiley & Sons, Inc., New York, 1982. A Wiley-Interscience Publication. MR679313

[12] Nicola Garofalo and Mariana Smit Vega Garcia, *New monotonicity formulas and the optimal regularity in the Signorini problem with variable coefficients*, Adv. Math. **262** (2014), 682–750, DOI 10.1016/j.aim.2014.05.021. MR3228440

[13] Nestor Guillen, *Optimal regularity for the Signorini problem*, Calc. Var. Partial Differential Equations **36** (2009), no. 4, 533–546, DOI 10.1007/s00526-009-0242-5. MR2558329

Department of Mathematics, Columbia University, New York, New York 10027
Email address: changlara@math.columbia.edu

Department of Mathematics, Columbia University, New York, New York 10027
Email address: osavin@math.columbia.edu

Contemporary Mathematics
Volume **723**, 2019
https://doi.org/10.1090/conm/723/14545

Fractional Laplacians on the sphere, the Minakshisundaram zeta function and semigroups

Pablo Luis De Nápoli and Pablo Raúl Stinga

ABSTRACT. In this paper we show novel underlying connections between fractional powers of the Laplacian on the unit sphere and functions from analytic number theory and differential geometry, like the Hurwitz zeta function and the Minakshisundaram zeta function. Inspired by Minakshisundaram's ideas, we find a precise pointwise description of $(-\Delta_{\mathbb{S}^{n-1}})^s u(x)$ in terms of fractional powers of the Dirichlet-to-Neumann map on the sphere. The Poisson kernel for the unit ball will be essential for this part of the analysis. On the other hand, by using the heat semigroup on the sphere, additional pointwise integro-differential formulas are obtained. Finally, we prove a characterization with a local extension problem and the interior Harnack inequality.

1. Introduction

Nonlinear problems with fractional Laplacians have been receiving a lot of attention for the past 12 years. Fractional nonlocal equations appear in several areas of pure and applied mathematics, see for instance [**1, 4, 5, 19, 22**].

For a function $u : \mathbb{R}^n \to \mathbb{R}$, $n \geq 1$, the fractional Laplacian $(-\Delta)^s u$ with $0 < s < 1$ is naturally defined in a spectral way by using the Fourier transform as

$$\widehat{(-\Delta)^s u}(\xi) = |\xi|^{2s}\, \widehat{u}(\xi), \quad \xi \in \mathbb{R}^n.$$

The following equivalent semigroup formula holds:

$$(-\Delta)^s u(X) = \frac{1}{\Gamma(-s)} \int_0^\infty \left(e^{t\Delta}u(X) - u(X)\right) \frac{dt}{t^{1+s}}, \quad X \in \mathbb{R}^n.$$

Here Γ is the Gamma function and $\{e^{t\Delta}\}_{t\geq 0}$ is the classical heat semigroup on $\mathbb{R}^n$. This implies the pointwise integro-differential formula

$$(-\Delta)^s u(X) = \frac{4^s\Gamma(n/2+s)}{\pi^{n/2}|\Gamma(-s)|}\,\mathrm{P.V.}\int_{\mathbb{R}^n} \frac{u(X)-u(Y)}{|X-Y|^{n+2s}}\,dY, \quad X \in \mathbb{R}^n,$$

2010 *Mathematics Subject Classification.* Primary 11M41, 26A33, 35R11; Secondary 11M35, 35K08, 47D06.

Key words and phrases. Fractional Laplacian on the sphere, zeta function, method of semigroups, spherical harmonics, Harnack inequality.

The first author was supported by ANPCyT under grant PICT 2014-1771 and by Universidad de Buenos Aires under grant 2002016010002BA. He is a member of Conicet, Argentina. The second author was supported by Simons Foundation grant 580911 and by grant MTM2015-66157-C2-1-P, MINECO/FEDER, EU, from Government of Spain.

see [20]. Clearly, $(-\Delta)^s$ is a nonlocal operator. The Caffarelli–Silvestre extension theorem [3] establishes that if $U = U(X, y)$ is the solution to

$$\begin{cases} \Delta U + \frac{1-2s}{y}\partial_y U + \partial_{yy}U = 0, & \text{for } X \in \mathbb{R}^n,\ y > 0, \\ U(X,0) = u(X), & \text{for } X \in \mathbb{R}^n, \end{cases}$$

then

$$-y^{1-2s}\partial_y U(X,y)\big|_{y=0^+} = \frac{\Gamma(1-s)}{4^{s-1/2}\Gamma(s)}(-\Delta)^s u(X).$$

Moreover, the solution U is given explicitly as (see [20])

$$\begin{aligned} U(X,y) &= \frac{y^{2s}}{4^s\Gamma(s)}\int_0^\infty e^{-y^2/(4t)}e^{t\Delta}u(X)\,\frac{dt}{t^{1+s}} \\ &= \frac{\Gamma(n/2+s)}{\pi^{n/2}\Gamma(s)}\int_{\mathbb{R}^n}\frac{y^{2s}}{(|X-Y|^2+y^2)^{\frac{n+2s}{2}}}u(Y)\,dY. \end{aligned}$$

See [3, 10, 20] for more details about the extension problem and its applications.

In this paper we present several descriptions of the fractional powers of the Laplacian $\Delta_{\mathbb{S}^{n-1}}$ on the unit sphere

$$\mathbb{S}^{n-1} = \{X \in \mathbb{R}^n : |X| = 1\}, \quad n \geq 2.$$

The Laplacian on the sphere is defined in the following simple way. If $u = u(x)$ is a real function on $\mathbb{S}^{n-1}$ then we denote by $\tilde{u}$ the extension of u to $\mathbb{R}^n \setminus \{0\}$ that coincides with u on $\mathbb{S}^{n-1}$ and is constant along the lines normal to $\mathbb{S}^{n-1}$, namely, $\tilde{u}(X) = u(X/|X|)$, $X \neq 0$. Then $\Delta_{\mathbb{S}^{n-1}}u$ is the restriction of the function $\Delta\tilde{u}$ to $\mathbb{S}^{n-1}$. As a Riemannian manifold, the sphere has a natural Laplace–Beltrami operator, that coincides with $\Delta_{\mathbb{S}^{n-1}}$ as obtained above.

Similarly to the fractional Laplacian on $\mathbb{R}^n$, the fractional powers of $-\Delta_{\mathbb{S}^{n-1}}$ are defined in a spectral way. Indeed, the spectral decomposition of the Laplacian on the sphere is given by the spherical harmonics, that we briefly describe next (see [9, 18]). We have $L^2(\mathbb{S}^{n-1}) = \bigoplus_{k=0}^\infty SH^k$, where SH^k denotes the space of spherical harmonics of degree $k \geq 0$ of dimension $d_k = \frac{(2k+n-2)(k+n-3)!}{k!(n-2)!}$. From now on and for the rest of the paper we fix an orthonormal basis $\{Y_{k,l} : 1 \leq l \leq d_k\}$ of SH^k consisting of real spherical harmonics $Y_{k,l}$ of degree $k \geq 0$. The spherical harmonics are the eigenfunctions of the Laplacian on the sphere, namely,

$$-\Delta_{\mathbb{S}^{n-1}}Y_{k,l}(x) = \lambda_k Y_{k,l}(x), \quad x \in \mathbb{S}^{n-1},$$

with eigenvalues

$$\lambda_k = k(k+n-2),$$

of multiplicity d_k, for $k \geq 0$. If u has an expansion into spherical harmonics as

$$u(x) = \sum_{k=0}^\infty\sum_{l=1}^{d_k} c_{k,l}(u)Y_{k,l}(x), \quad c_{k,l}(u) = \int_{\mathbb{S}^{n-1}} u(y)Y_{k,l}(y)\,d\mathcal{H}^{n-1}(y), \tag{1.1}$$

where $d\mathcal{H}^{n-1}$ is the $(n-1)$-dimensional Hausdorff measure restricted to $\mathbb{S}^{n-1}$, then for any $\sigma \in \mathbb{C}$ such that $s = \mathrm{Re}(\sigma) > 0$, we can define

$$(-\Delta_{\mathbb{S}^{n-1}})^{\pm\sigma}u(x) = \sum_{k=0}^\infty \lambda_k^{\pm\sigma}\sum_{l=1}^{d_k} c_{k,l}(u)Y_{k,l}(x). \tag{1.2}$$

The orthogonal projector onto the subspace SH^k of $L^2(\mathbb{S}^{n-1})$ is given by

$$P_k u(x) = \sum_{l=1}^{d_k} c_{k,l}(u) Y_{k,l}(x). \tag{1.3}$$

Hence

$$(-\Delta_{\mathbb{S}^{n-1}})^{\pm\sigma} u(x) = \sum_{k=0}^{\infty} \lambda_k^{\pm\sigma} P_k u(x), \tag{1.4}$$

so the definition of $(-\Delta_{\mathbb{S}^{n-1}})^{\pm\sigma} u$ is independent of the basis.

Some care is needed for the formulas above to be well defined. For $(-\Delta_{\mathbb{S}^{n-1}})^{-\sigma} u$ to make sense we need $\int_{\mathbb{S}^{n-1}} u = 0$. This condition says that the term $P_0 u(x)$, which corresponds to the eigenvalue $\lambda_0 = 0$ in (1.2) and (1.4), vanishes, so we may consider the sums as starting from $k = 1$. On the other hand, $(-\Delta_{\mathbb{S}^{n-1}})^{\sigma} u \in L^2(\mathbb{S}^{n-1})$ if and only if $u \in \operatorname{Dom}((-\Delta_{\mathbb{S}^{n-1}})^{\sigma}) \equiv \left\{u \in L^2(\mathbb{S}^{n-1}) : \sum_{k=0}^{\infty} \lambda_k^{2s} \sum_{l=1}^{d_k} |c_{k,l}(u)|^2 < \infty\right\}$. In particular, (1.2) and (1.4) make sense for functions $u \in C^\infty(\mathbb{S}^{n-1})$, see (4.2). Here $u \in C^m(\mathbb{S}^{n-1})$, $m \in \mathbb{N}_0 \cup \{\infty\}$, means that the extension $\tilde{u}$ of u along normal lines to the sphere is a C^m function in a neighborhood of the sphere. In general, $(-\Delta_{\mathbb{S}^{n-1}})^{\sigma} u$ is defined as a distribution in $H^{-\sigma}(\mathbb{S}^{n-1})$, for every $u \in H^{\sigma}(\mathbb{S}^{n-1})$. In addition, if $u \in H^{-\sigma}(\mathbb{S}^{n-1})$ satisfies $\langle u, 1\rangle = 0$ then $(-\Delta_{\mathbb{S}^{n-1}})^{-\sigma} u \in H^{\sigma}(\mathbb{S}^{n-1})$ with $\int_{\mathbb{S}^{n-1}} (-\Delta_{\mathbb{S}^{n-1}})^{-\sigma} u = 0$. In this paper we are mainly interested in describing pointwise formulas and the extension problem for these fractional operators in connection with functions from number theory and differential geometry, and semigroups. Therefore we will always assume that all our functions are smooth.

The fractional Laplacian on the sphere is a natural object to consider since it is the simplest example of a fractional nonlocal operator on a compact Riemannian manifold. Moreover, the sphere has a rich structure related to the fact that it is a homogenous space under the action of the Lie group $SO(n)$. Such a rich theory and some clever formulas connected with the Minakshisundaram zeta function in combination with the method of semigroups will allow us to obtain rather explicit expressions for the fractional operators in terms of the Dirichlet-to-Neumann map and the heat semigroup on the sphere. For applications, see for example [**1**]. For hypersingular integrals and potential operators on the sphere (which are not the same as the fractional powers of the Laplacian on the sphere) the reader can consult [**16**, **17**]. Certainly one can define fractional powers of Laplace–Beltrami operators in Riemannian manifolds by using the spectral theorem. In those cases in which estimates for the corresponding heat semigroup kernel are available, one could try to derive pointwise expressions and kernel estimates for fractional operators by using the method of semigroups from [**10**, **20**], in an analogous way as explained here for the case of the sphere (see section 7). Observe that for this general semigroup technique to apply we do not need the manifold to be necessarily compact.

A description of the contents of the paper follows.

(a) In sections 2 and 3 we first consider the cases of the negative and positive powers of the Laplacian on the unit circle, respectively, in connection with the Hurwitz zeta function the functions defined by Fine in [**8**].

(b) In section 4 we define the Dirichlet-to-Neumann operator L for the Laplacian in the unit ball. The semigroup generated by L is obtained from the

Poisson kernel for the ball. We show how this kernel can be deduced by using the Funk–Hecke identity and the generating formula for Gegenbauer polynomials.

(c) In section 5 we relate the negative powers of $-\Delta_{\mathbb{S}^{n-1}}$ in dimension $n \geq 3$ with the Minakshisundaram Riemann zeta function on the sphere. We use this connection and the Poisson kernel to find precise estimates for the kernel of $(-\Delta_{\mathbb{S}^{n-1}})^{-s}$.

(d) In section 6, inspired by Minakshisundaram's ideas, we prove a numerical formula (Lemma 6.1) that permits us to express the fractional Laplacian on the sphere $(-\Delta_{\mathbb{S}^{n-1}})^s$ in terms of the fractional Dirichlet-to-Neumann operator $(L + \frac{n-2}{2})^{2s}$. Precise kernel estimates are also shown.

(e) In section 7 we use the heat semigroup on the sphere to give equivalent formulas for $(-\Delta_{\mathbb{S}^{n-1}})^{\pm s}u(x)$ with kernel estimates. We also present extension problem characterizations and, as an application, the interior Harnack inequality.

Notation. Throughout the paper $\sigma \in \mathbb{C}_+$ denotes a complex number with positive real part $\mathrm{Re}(\sigma) = s > 0$. For two positive quantities A and B we write $A \sim B$ to mean that there exist constants $c_{n,\sigma}$ and $C_{n,\sigma}$ such that $c_{n,\sigma} \leq A/B \leq C_{n,\sigma}$. If $x, y \in \mathbb{S}^{n-1}$ then $d(x,y) = \cos^{-1}(x \cdot y)$ is the geodesic distance between x and y. Notice that $|x-y|^2 = 2(1 - x \cdot y) \sim d(x,y)^2$, for all $x, y \in \mathbb{S}^{n-1}$. We will always denote by u a smooth real function on the sphere.

2. Negative powers in the unit circle and the Hurwitz zeta function

In this section and the next one we consider the one-dimensional case of the unit circle $\mathbb{T} = \mathbb{S}^1 \subset \mathbb{R}^2$. As usual, we identify functions on $\mathbb{T}$ with periodic functions on the interval $[0,1]$, so that $\Delta_{\mathbb{T}} = \frac{d^2}{dx^2}$. The spherical harmonics become the complex exponentials $\{e^{2\pi ikx}\}_{k\in\mathbb{Z}}$, $x \in [0,1]$. We have $-\Delta_{\mathbb{T}} e^{2\pi ikx} = (2\pi k)^2 e^{2\pi ikx}$.

When $\int_{\mathbb{T}} u = 0$ and $\sigma \in \mathbb{C}_+$ with $s = \mathrm{Re}(\sigma) > 0$, we can write

$$
\begin{aligned}
(-\Delta_{\mathbb{T}})^{-\sigma/2} u(x) &= \sum_{k\in\mathbb{Z}\setminus\{0\}} (2\pi|k|)^{-\sigma} c_k(u) e^{2\pi ikx} \\
&= \sum_{k\in\mathbb{Z}\setminus\{0\}} (2\pi|k|)^{-\sigma} \int_0^1 u(y) e^{2\pi ik(x-y)}\, dy \\
&= \int_0^1 K_{-\sigma}(x-y) u(y)\, dy,
\end{aligned}
$$

where the kernel is given by

$$
K_{-\sigma}(x) = \sum_{k\in\mathbb{Z}\setminus\{0\}} \frac{e^{2\pi ikx}}{(2\pi|k|)^{\sigma}}. \tag{2.1}
$$

Our main interest will be to understand the behavior of $K_{-\sigma}(x)$ as $x \to 0, 1$. Estimates for this kernel can be obtained, for example, by using the semigroup method of [**20**] or the transference principle from [**15**]. Nevertheless, here we take a different approach by connecting formula (2.1) with some useful functions from analytic number theory.

The Hurwitz zeta function (see [**2**, Chapter 12], also [**8, 13**]) is initially defined for $0 < x \leq 1$ and $\sigma \in \mathbb{C}_+$ such that $s = \operatorname{Re}(\sigma) > 1$ by

$$\zeta(\sigma, x) = \sum_{k=0}^{\infty} \frac{1}{(k+x)^\sigma}. \tag{2.2}$$

When $x = 1$ this reduces to the Riemann zeta function $\zeta(\sigma) = \zeta(\sigma, 1)$. Moreover,

$$\zeta(\sigma, x) = x^{-\sigma} + \zeta(\sigma, x+1). \tag{2.3}$$

Using the Gamma function it is possible to extend $\zeta(\sigma, x)$ as an analytic function of $\sigma \in \mathbb{C}$ except for a simple pole at $\sigma = 1$, see [**2**, Theorem 12.4].

A related function is the periodic zeta function defined for $x \in \mathbb{R}$ and $s > 1$ by

$$F(x, \sigma) = \sum_{k=1}^{\infty} \frac{e^{2\pi i k x}}{k^\sigma}.$$

This series converges absolutely if $s > 1$. If x is not an integer the series also converges conditionally whenever $s > 0$. Notice that $F(x, \sigma)$ is a periodic function with period 1 that coincides with the Riemann zeta function $\zeta(\sigma)$ at $x = 1$. It is clear then that

$$K_{-\sigma}(x) = \frac{1}{(2\pi)^\sigma} \big[F(x, \sigma) + F(-x, \sigma)\big].$$

The two functions $\zeta(\sigma, x)$ and $F(x, \sigma)$ are related by the following formula due to Hurwitz, see [**2**, Theorem 12.6], also [**8**]: if $0 < x \leq 1$ and $s > 1$ then

$$\zeta(1-\sigma, x) = \frac{\Gamma(\sigma)}{(2\pi)^\sigma} \big[e^{-\pi i \sigma/2} F(x, \sigma) + e^{\pi i \sigma/2} F(-x, \sigma)\big].$$

If $x \neq 1$ this representation is also valid for $s > 0$. If we substitute x by $1 - x$, then

$$\zeta(1-\sigma, 1-x) = \frac{\Gamma(\sigma)}{(2\pi)^\sigma} \big[e^{-\pi i \sigma/2} F(-x, \sigma) + e^{\pi i \sigma/2} F(x, \sigma)\big].$$

Therefore

$$\zeta(1-\sigma, x) + \zeta(1-\sigma, 1-x) = \frac{\Gamma(\sigma)}{(2\pi)^\sigma} 2\cos\left(\tfrac{\pi\sigma}{2}\right) \big[F(x, \sigma) + F(-x, \sigma)\big],$$

or

$$K_{-\sigma}(x) = \frac{1}{2\Gamma(\sigma)\cos\left(\frac{\pi\sigma}{2}\right)} \big[\zeta(1-\sigma, x) + \zeta(1-\sigma, 1-x)\big]. \tag{2.4}$$

To estimate $K_{-\sigma}(x)$ we use (2.4) and well known asymptotic expansions for the Hurwitz zeta function. Recall that $\zeta(1-\sigma, 1) = \zeta(1-\sigma)$. For $\sigma \neq 1$ fixed,

$$\zeta(\sigma, x+1) = \zeta(\sigma) - \sigma\zeta(\sigma+1)x + O(x^2), \quad \text{as } x \to 0,$$

see [**13**]. By replacing σ by $1-\sigma$ in (2.3) and using the asymptotic expansion above,

$$\begin{aligned} \zeta(1-\sigma, x) &= x^{\sigma-1} + \zeta(1-\sigma, x+1) \\ &= x^{\sigma-1} + \zeta(1-\sigma) - (1-\sigma)\zeta(2-\sigma)x + O(1), \end{aligned}$$

as $x \to 0$. If we substitute x by $1 - x$ in the latter expansion, we get

$$\zeta(1-\sigma, 1-x) = (1-x)^{\sigma-1} + \zeta(1-\sigma) - (1-\sigma)\zeta(2-\sigma)(1-x) + O(1),$$

as $x \to 1$. We plug these estimates into (2.4) to deduce the asymptotic formulas

$$K_{-\sigma}(x) = C_\sigma x^{\sigma-1} + O(1), \quad \text{as } x \to 0,$$

and

$$K_{-\sigma}(x) = C_\sigma(1-x)^{\sigma-1} + O(1), \quad \text{as } x \to 1.$$

We conclude that

$$K_{-\sigma}(x) \sim \frac{1}{x^{1-\sigma}} + \frac{1}{(1-x)^{1-\sigma}}, \quad \text{as } x \to 0, 1.$$

3. Positive powers in the unit circle and the Hurwitz zeta and Fine functions

In this section we continue our analysis of the one-dimensional case of the unit circle $\mathbb{T} = \mathbb{S}^1 \subset \mathbb{R}^2$, $\mathbb{T} \equiv [0,1]$. We study the kernel of the fractional power operator $(-\Delta_{\mathbb{T}})^{\sigma/2}$, when $\sigma \in \mathbb{C}_+$ is such that $0 < \operatorname{Re}(\sigma) < 2$. As in the previous section we denote by $\zeta = \zeta(\sigma, x)$ the Hurwitz zeta function (2.2).

The key idea is to use the heat semigroup on the circle. This approach for the case of the sphere will be developed in detail in section 7. It is easy to check that the proof of Theorem 7.1 is valid when $n = 2$ and s is replaced by $\sigma/2$, see also [**10, 14, 15**]. Then, for any $x \in \mathbb{T}$,

$$(-\Delta_{\mathbb{T}})^{\sigma/2} u(x) = \text{P. V.} \int_{\mathbb{T}} (u(y) - u(x)) K_\sigma(x-y)\, dy,$$

where the kernel K_σ is given by

$$K_\sigma(x) = \frac{1}{\Gamma(-\sigma/2)} \int_0^\infty W_t(x) \frac{dt}{t^{1+\sigma/2}}.$$

Here $W_t(x)$ is the heat kernel in the unit circle:

$$\begin{aligned} W_t(x) &= \sum_{k \in \mathbb{Z}} e^{-4\pi^2 k^2 t} e^{2\pi i k x} \\ &= 1 + 2\sum_{k=1}^\infty e^{-4\pi^2 k^2 t} \cos(2\pi k x). \end{aligned} \tag{3.1}$$

Notice that this kernel is essentially the so-called Jacobi theta function.

By following Fine [**8**], we define, for $0 < x < 1$ and $t > 0$, the function

$$f(x,t) = 1 + 2\sum_{k=1}^\infty e^{-\pi k^2 t} \cos(2\pi k x).$$

From (3.1) it follows that

$$W_t(x) = f(x, 4\pi t). \tag{3.2}$$

Next, Fine observes in [**8**] that the functions

$$F(x,\omega) = \int_0^1 f(x,t) t^{\omega/2-1}\, dt$$

and

$$G(x,\omega) = \int_1^\infty (f(x,t) - 1) t^{\omega/2-1}\, dt$$

are entire functions in ω. Additionally, Fine introduces the function

$$H(x,\omega) = F(x,\omega) + G(x,\omega) - \frac{2}{\omega}$$

and proves the following relation with the Hurwitz zeta function: if $\operatorname{Re}(\omega) < 0$ then

$$H(x,\omega) = \frac{\Gamma(\frac{1-\omega}{2})}{\pi^{\frac{1-\omega}{2}}}\big[\zeta(1-\omega,x)+\zeta(1-\omega,1-x)\big], \tag{3.3}$$

see [**8**, eq. (11)]. Thus, by using (3.2), the change of variables $4\pi t \to t$ and the functions F, G and H, together with the functional identity (3.3), we find that

$$\begin{aligned}
K_\sigma(x) &= \frac{(4\pi)^{\sigma/2}}{\Gamma(-\sigma/2)}\int_0^\infty f(x,t)\,\frac{dt}{t^{1+\sigma/2}}\\
&= \frac{(4\pi)^{\sigma/2}}{\Gamma(-\sigma/2)}\left[\int_0^1 f(x,t)\,\frac{dt}{t^{1+\sigma/2}} + \int_1^\infty \big(f(x,t)-1\big)\,\frac{dt}{t^{1+\sigma/2}} + \frac{2}{\sigma}\right]\\
&= \frac{(4\pi)^{\sigma/2}}{\Gamma(-\sigma/2)}\left[F(x,-\sigma)+G(x,-\sigma)+\frac{2}{\sigma}\right]\\
&= \frac{(4\pi)^{\sigma/2}}{\Gamma(-\sigma/2)}H(x,-\sigma)\\
&= \frac{4^{\sigma/2}\Gamma(\frac{1+\sigma}{2})}{\Gamma(-\sigma/2)\sqrt{\pi}}\big[\zeta(1+\sigma,x)+\zeta(1+\sigma,1-x)\big].
\end{aligned}$$

We can use the following properties of the Gamma function

$$\Gamma(\tfrac{1}{2}+z)\Gamma(\tfrac{1}{2}-z) = \frac{\pi}{\cos(\pi z)} \qquad \Gamma(2z) = \frac{2^{2z-1}}{\sqrt{\pi}}\Gamma(z)\Gamma(z+\tfrac{1}{2}) \tag{3.4}$$

with $z = -\sigma/2$ to further simplify the expression above to

$$K_\sigma(x) = \frac{1}{2\Gamma(-\sigma)\cos(-\frac{\pi\sigma}{2})}\big[\zeta(1+\sigma,x)+\zeta(1+\sigma,1-x)\big].$$

This is perfectly consistent with (2.4). By performing an asymptotic analysis with of the Hurwitz zeta function analogous to the one we did in section 2, we conclude that

$$K_\sigma(x) \sim \frac{1}{x^{1+\sigma}} + \frac{1}{(1-x)^{1+\sigma}}, \qquad \text{as } x \to 0, 1.$$

4. The semigroup generated by the Dirichlet-to-Neumann map

In this section we introduce the Dirichlet-to-Neumann map L for the Laplacian in the unit ball $\mathbb{B} = \{X \in \mathbb{R}^n : |X| < 1\}$, $n \geq 2$. We show that the kernel of the semigroup generated by L is obtained from the Poisson kernel for the unit ball. These objects will be very useful in the description of the fractional operators in sections 5 and 6.

Let u be a smooth function on the sphere with series expansion (1.1). We define the Dirichlet-to-Neumann operator L on the sphere $\mathbb{S}^{n-1}$ by

$$Lu(x) = \sum_{k=0}^\infty k \sum_{l=1}^{d_k} c_{k,l}(u)Y_{k,l}(x) = \sum_{k=0}^\infty kP_k u(x),$$

for $x \in \mathbb{S}^{n-1}$, where P_k is the orthogonal projector (1.3). The series is absolutely convergent and can be differentiated term by term. Indeed, for each multi-index $\alpha = (\alpha_1, \ldots, \alpha_n) \in \mathbb{N}_0^n$, there exists $C_{\alpha,n} > 0$ such that

$$|D^\alpha Y_{k,l}(X/|X|)|^2 \leq (C_{\alpha,n})^2 k^{2|\alpha|+n-2}, \quad \text{for all } 1 \leq l \leq d_k, \tag{4.1}$$

where $D^\alpha = \frac{\partial^{|\alpha|}}{\partial X_1^{\alpha_1}\cdots\partial X_n^{\alpha_n}}$, $|\alpha| = \alpha_1 + \cdots + \alpha_n$, see [**18**]. Then, for any $m \in \mathbb{N}$, by the symmetry of $\Delta_{\mathbb{S}^{n-1}}$ and the Cauchy-Schwartz inequality,

$$
\begin{aligned}
|c_{k,l}(u)| &= \frac{1}{\lambda_k^m}\left|\int_{\mathbb{S}^{n-1}} (-\Delta_{\mathbb{S}^{n-1}})^m u(x) Y_{k,l}(x)\, d\mathcal{H}^{n-1}(x)\right| \\
&\leq \frac{C_{u,m}}{(k(k+n-2))^m}.
\end{aligned} \tag{4.2}
$$

Therefore the estimates in (4.1) and (4.2) give the conclusion.

The semigroup $\{e^{-tL}\}_{t\geq 0}$ generated by the Dirichlet-to-Neumann operator is related to the solution $w = w(X)$ of the Dirichlet problem in the unit ball

$$
\begin{cases} \Delta w = 0, & \text{in } \mathbb{B}, \\ w = u, & \text{on } \mathbb{S}^{n-1}, \end{cases}
$$

in the following way. It is known that w can be recovered from u by using the Poisson integral formula for the unit ball:

$$
w(X) = \int_{\mathbb{S}^{n-1}} \frac{1 - |X|^2}{\omega_{n-1}(1 - 2X\cdot y + |X|^2)^{n/2}} u(y)\, d\mathcal{H}^{n-1}(y), \tag{4.3}
$$

where $\omega_{n-1} = \frac{2\pi^{n/2}}{\Gamma(n/2)}$ is the surface area of $\mathbb{S}^{n-1}$. On the other hand, if we introduce polar coordinates $X = rx$, $0 < r \leq 1$, $x \in \mathbb{S}^{n-1}$, then,

$$
w(X) = \sum_{k=0}^\infty r^k \sum_{l=1}^{d_k} c_{k,l}(u) Y_{k,l}(x) = \sum_{k=0}^\infty r^k P_k u(x).
$$

See [**9**] for details. Now, by making the change of parameters $r = e^{-t}$, $t \geq 0$, we can regard $w(X)$ as a function of $t \geq 0$ and $x \in \mathbb{S}^{n-1}$:

$$
e^{-tL}u(x) \equiv w(X) = \sum_{k=0}^\infty e^{-tk} \sum_{l=1}^{d_k} c_{k,l}(u) Y_{k,l}(x) = \sum_{k=0}^\infty e^{-tk} P_k u(x). \tag{4.4}
$$

This is in fact the heat-diffusion semigroup generated by L. Indeed, we can differentiate the series in (4.4) to get

$$
-\partial_t e^{-tL}u(x)\big|_{t=0} = r\partial_r w(X)\big|_{r=1} = \partial_\nu w(X) = Lu(x),
$$

which also shows that the name "Dirichlet-to-Neumann operator" for L is fully justified. Also, $e^{-tL}u \to u$, as $t \to 0$, uniformly and in $L^p(\mathbb{S}^{n-1})$, for $1 \leq p < \infty$.

By letting $r = e^{-t}$ in (4.3), we find that the semigroup e^{-tL} admits an expression as a convolution on the sphere with the Poisson kernel:

$$
\begin{aligned}
e^{-tL}u(x) &= \int_{\mathbb{S}^{n-1}} \frac{1 - e^{-2t}}{\omega_{n-1}(1 - 2e^{-t}x\cdot y + e^{-2t})^{n/2}} u(y)\, d\mathcal{H}^{n-1}(y) \\
&= \int_{\mathbb{S}^{n-1}} \frac{1 - e^{-2t}}{\omega_{n-1}((1 - e^{-t})^2 + 2e^{-t}(1 - x\cdot y))^{n/2}} u(y)\, d\mathcal{H}^{n-1}(y) \\
&\equiv \int_{\mathbb{S}^{n-1}} P_{e^{-t}}(x\cdot y) u(y)\, d\mathcal{H}^{n-1}(y).
\end{aligned} \tag{4.5}
$$

LEMMA 4.1. *There exists a constant $C > 0$ depending only on u and n such that $|e^{-tL}u(x) - u(x)| \leq Ct$, for any $0 < t < 1$, for all $x \in \mathbb{S}^{n-1}$.*

PROOF. By the mean value theorem,

$$|e^{-tL}u(x) - u(x)| = |e^{-tL}u(x) - e^{-0L}u(x)| = |\partial_t e^{-tL}u(x)|\big|_{t=\xi} t,$$

where ξ is an intermediate point between 0 and t. But then, using (4.4) together with (4.1) and (4.2), it readily follows that $\sup_{t,x} |\partial_t e^{-tL}u(x)| \leq C$. □

To relate L with the Laplacian on the sphere we recall that the eigenvalues of $-\Delta_{\mathbb{S}^{n-1}}$ are $\lambda_k = k(k+n-2)$. Observe that, unlike the case of the classical Poisson integral in the upper half-plane, the Poisson integral in the ball (4.3) is *not* the Poisson semigroup $\{e^{-t(-\Delta_{\mathbb{S}^{n-1}})^{1/2}}\}_{t\geq 0}$ generated by $-\Delta_{\mathbb{S}^{n-1}}$. In fact, $e^{-t(-\Delta_{\mathbb{S}^{n-1}})^{1/2}} \neq e^{-tL}$, except for the unit circle, that is, if $n=2$. Moreover,

$$-\Delta_{\mathbb{S}^{n-1}} = L(L + (n-2)\,\mathrm{I}),$$

where I is the identity operator. It then follows that

$$(-\Delta_{\mathbb{S}^{n-1}})^{\pm\sigma} = L^{\pm\sigma}(L + (n-2)\,\mathrm{I})^{\pm\sigma}, \qquad \sigma \in \mathbb{C}_+.$$

The explicit formula for the Poisson kernel in (4.5) comes from the generating formula for the Gegenbauer (or ultraspherical) polynomials. The Funk–Hecke theorem (see [**18**]) states that if $F(\tau)(1-\tau^2)^{(n-3)/2}$ is an integrable function on the interval $(-1,1)$ then, for each spherical harmonic $Y_{k,l}$ and $x \in \mathbb{S}^{n-1}$,

$$\begin{aligned}(4.6)\qquad &\int_{\mathbb{S}^{n-1}} F(x\cdot y) Y_{k,l}(y)\, d\mathcal{H}^{n-1}(y) \\ &\qquad = Y_{k,l}(x) \frac{\omega_{n-2}}{C_k(1)} \int_{-1}^{1} F(\tau) C_k(\tau)(1-\tau^2)^{(n-3)/2}\, d\tau,\end{aligned}$$

where $C_k(\tau)$ is the Gegenbauer polynomial $C_k^\nu(\tau)$ with parameter $\nu = (n-2)/2$. For $n=2$ they are the Chebyshev polynomials and for $n=3$ they are the Legendre polynomials (see [**9, 11, 13**]). Using the generating formula

$$(1 - 2\tau r + r^2)^{-\nu} = \sum_{k=0}^{\infty} C_k^\nu(\tau) r^k,$$

see [**11**, eq. (5.12.7)], it easily follows that

$$(4.7)\qquad \sum_{k=0}^{\infty} \frac{k+\nu}{\nu} C_k^\nu(\tau) r^k = \frac{1-r^2}{(1-2r\tau + r^2)^{\nu+1}}.$$

For each fixed $x \in \mathbb{S}^{n-1}$, $C_k(x\cdot y)$ (as a function of $y \in \mathbb{S}^{n-1}$) is in SH^k, see [**9**]. By expressing $C_k(x\cdot y)$ in terms of the orthonormal basis $\{Y_{k,l} : l = 1,\dots,d_k\}$ of SH^k, and applying the Funk–Hecke formula (4.6) in combination with properties of Gegenbauer polynomials and the Gamma function, we can see that

$$(4.8)\qquad \frac{1}{\omega_{n-1}} \cdot \frac{k + \frac{n-2}{2}}{\frac{n-2}{2}} C_k(x\cdot y) = \sum_{l=1}^{d_k} Y_{k,l}(x) Y_{k,l}(y).$$

Plugging this into the first sum in (4.4) we get

$$e^{-tL}u(x) = \int_{\mathbb{S}^{n-1}} \frac{1}{\omega_{n-1}} \bigg(\sum_{k=0}^{\infty} \frac{k+\frac{n-2}{2}}{\frac{n-2}{2}} C_k(x\cdot y) r^k \bigg) u(y)\, d\mathcal{H}^{n-1}(y).$$

Thus (4.5) with $e^{-t} = r$ follows by using the generating formula (4.7) with $\nu = \frac{n-2}{2}$.

5. Negative powers and the Minakshisundaram zeta function

Throughout this section we always assume that $\int_{\mathbb{S}^{n-1}} u = 0$. Recall that we have set $s = \mathrm{Re}(\sigma) > 0$ and that the eigenvalues of $-\Delta_{\mathbb{S}^{n-1}}$ are $\lambda_k = k(k+n-2)$, $k \geq 0$. By using (4.8) in (1.2) we find that the negative powers of the Laplacian on the sphere have an integral representation as an spherical convolution

$$(5.1)\qquad (-\Delta_{\mathbb{S}^{n-1}})^{-\sigma}u(x) = \int_{\mathbb{S}^{n-1}} K_{-\sigma}(x\cdot y)u(y)\,d\mathcal{H}^{n-1}(y),$$

where the kernel is given in terms of the Gegenbauer polynomials as

$$(5.2)\qquad K_{-\sigma}(x\cdot y) = \frac{1}{\omega_{n-1}}\sum_{k=1}^{\infty}\lambda_k^{-\sigma}\cdot\frac{k+\frac{n-2}{2}}{\frac{n-2}{2}}C_k(x\cdot y).$$

The series in (5.2) is the zeta function on the sphere of S. Minakshisundaram, see [**12**][1]. In [**12**] Minakshisundaram analyzed this function with methods similar to those used in analytic number theory for studying the Riemann zeta function. Using that $|C_k(x\cdot y)| \leq |C_k(1)| = \binom{k+n-3}{k}$ we get

$$C_k(x\cdot y) = O(k^{n-3}).$$

From here it is easy to see that the Dirichlet series (5.2) converges absolutely and uniformly in compact sets when $s = \mathrm{Re}(\sigma) > \frac{n-1}{2}$. For this range of s, $K_{-\sigma}(x\cdot y)$ represents an analytic function of σ. However, Minakshisundaram showed that (as it happens with Riemann zeta function) the function $K_{-\sigma}(x\cdot y)$ can be continued as a meromorphic function of σ to the whole complex plane. In order to do so, he proved the following integral representation of his zeta function

$$(5.3)\qquad \begin{aligned} K_{-\sigma}(x\cdot y) &= \frac{1}{\omega_{n-1}}\left(\frac{2}{\nu}\right)^{\sigma-1/2}\frac{\Gamma(\sigma+\frac12)}{\Gamma(2\sigma)} \\ &\quad\times\int_0^\infty\left(\frac{1-e^{-2t}}{(1-2e^{-t}x\cdot y+e^{-2t})^{\nu+1}}-1\right)\mathcal{I}_{\sigma-1/2}(\nu t)e^{-\nu t}\frac{dt}{t^{1/2-\sigma}},\end{aligned}$$

where $\nu = \frac{n-2}{2}$ and $\mathcal{I}_{\sigma-1/2}$ is the modified Bessel function of the first kind

$$(5.4)\qquad \mathcal{I}_\rho(z) = \sum_{m=0}^{\infty}\frac{(z/2)^{2m+\rho}}{\Gamma(m+1)\Gamma(m+\rho+1)}.$$

Recall that $\mathcal{I}_\rho(z)$ is an analytic function of $z \in \mathbb{C}\setminus(-\infty,0)$ and an entire function of ρ. Moreover, the series in (5.4) is uniformly convergent in any disk $|z| < R$, $|\rho| < N$. See [**11**, Chapter 5] and [**13**] for details. It is easy to see that the integral (5.3) defines an analytic function of σ in the half plane $\mathrm{Re}(\sigma) > 0$. Then, Minakshisundaram obtained the meromorphic continuation of his zeta function by performing the usual trick of replacing the domain of integration in (5.3) by a loop integral around the real axis, with a small circle around the origin.

Thanks to Minakshisundaram's integral formula (5.3) we are able to estimate the integral kernel in the pointwise formula (5.1). For the sake of simplicity and because of our interest on fractional nonlocal equations, we will only consider the case when σ is real, namely, $\sigma = s > 0$. Recall the semigroup kernel $P_{e^{-t}}$ in (4.5).

[1]We have a slightly different normalization of Gegenbauer polynomials with respect to Minakshisundaram. Compare (4.7) with [**12**, Lemma 1]. Nevertheless, the zeta functions coincide up to the normalizing factor ω_{n-1}.

THEOREM 5.1 (Fractional integrals on the sphere). *Let $s > 0$. Then*

$$(-\Delta_{\mathbb{S}^{n-1}})^{-s}u(x) = \int_{\mathbb{S}^{n-1}} K_{-s}(x\cdot y)u(y)\,d\mathcal{H}^{n-1}(y),$$

for $x \in \mathbb{S}^{n-1}$. The kernel $K_{-s}(x\cdot y)$ is given by

$$\begin{aligned} K_{-s}(x\cdot y) &= \left(\frac{4}{n-2}\right)^{s-1/2}\frac{\Gamma(s+\frac{1}{2})}{\Gamma(2s)} \\ &\quad\times \int_0^\infty e^{-t(\frac{n-2}{2})}\big(P_{e^{-t}}(x\cdot y) - \tfrac{1}{\omega_{n-1}}\big)\mathcal{I}_{s-1/2}(\tfrac{n-2}{2}t)\,\frac{dt}{t^{1/2-s}}, \end{aligned} \tag{5.5}$$

and satisfies the estimates

$$|K_{-s}(x\cdot y)| \le c_{n,s}\begin{cases} 1/d(x,y)^{(n-1)-2s}, & \text{if } s < \frac{n-1}{2}, \\ \ln(1+(1-x\cdot y)^{-n/2}), & \text{if } s = \frac{n-1}{2}, \\ 1, & \text{if } s > \frac{n-1}{2}, \end{cases}$$

for all $x, y \in \mathbb{S}^{n-1}$, $x \neq y$.

PROOF. We split the integral in (5.5) as the sum of two integrals $I + II$, where

$$I = \int_0^1 e^{-t(\frac{n-2}{2})}\big(P_{e^{-t}}(x\cdot y) - \tfrac{1}{\omega_{n-1}}\big)\mathcal{I}_{s-1/2}(\tfrac{n-2}{2}t)\,\frac{dt}{t^{1/2-s}}.$$

Let us estimate I. By using the asymptotic behavior $\mathcal{I}_\rho(z) \approx \frac{1}{\Gamma(\rho+1)}(\frac{1}{2}z)^\rho$, valid for $z \to 0$ and any $\rho \neq -1, -2, \dots$ (see [**11, 13**]) we see that

$$\int_0^1 e^{-t(\frac{n-2}{2})}\frac{1}{\omega_{n-1}}\mathcal{I}_{s-1/2}(\tfrac{n-2}{2}t)\frac{dt}{t^{1/2-s}} \sim \int_0^1 t^{2s-1}\,dt \sim 1.$$

On the other hand,

$$\begin{aligned} &\int_0^1 e^{-t(\frac{n-2}{2})}P_{e^{-t}}(x\cdot y)\mathcal{I}_{s-1/2}(\tfrac{n-2}{2}t)\,\frac{dt}{t^{1/2-s}} \\ &\quad\sim \int_0^1 \frac{1-e^{-2t}}{((1-e^{-t})^2 + 2e^{-t}(1-x\cdot y))^{n/2}}\,t^{2s-1}\,dt \\ &\quad\sim \int_0^1 \frac{t^{2s}}{(t^2+(1-x\cdot y))^{n/2}}\,dt =: I_1. \end{aligned}$$

If $s < \frac{n-1}{2}$ then

$$\begin{aligned} I_1 &= \frac{1}{(1-x\cdot y)^{n/2}}\int_0^1 \frac{t^{2s}}{(\frac{t^2}{1-t}+1)^{n/2}}\,dt \\ &= \frac{1}{(1-x\cdot y)^{\frac{(n-1)-2s}{2}}}\int_0^{\frac{1}{(1-x\cdot y)^{1/2}}} \frac{\omega^{2s}}{(\omega^2+1)^{n/2}}\,d\omega \\ &\sim \frac{1}{d(x,y)^{(n-1)-2s}}. \end{aligned}$$

In the case $s = \frac{n-1}{2}$ we notice that

$$\begin{aligned} I_1 &= \int_0^1 \frac{t^{n-1}}{(t^2 + (1 - x \cdot y))^{n/2}}\,dt \\ &\sim \int_0^1 \frac{t^{n-1}}{t^n + (1 - x \cdot y)^{n/2}}\,dt \\ &= \frac{1}{n}\ln(1 + (1 - x \cdot y)^{-n/2}). \end{aligned}$$

Finally, if $s > \frac{n-1}{2}$ then $2s = n - 1 + \varepsilon$ for some $\varepsilon > 0$ and

$$I_1 = \int_0^1 \frac{t^{n-1+\varepsilon}}{(t^2 + (1 - x \cdot y))^{n/2}}\,dt \sim \int_0^1 \frac{t^{n-1+\varepsilon}}{t^n + (1 - x \cdot y)^{n/2}}\,dt \sim 1,$$

where the last estimate can be checked by considering the cases $1 - x \cdot y < 1$ and $1 \leq 1 - x \cdot y < 2$. We conclude that I satisfies estimates as in the statement.

Consider now II. For $1 < t < \infty$ we have $0 < r = e^{-t} < e^{-1} < 1$. We then need to estimate $|P_r(x \cdot y) - \frac{1}{\omega_{n-1}}|$ uniformly in $x \cdot y \in [-1, 1)$, for every $0 < r < e^{-1}$. Fix any such r. For each $x \cdot y \in [-1, 1)$, by the mean value theorem, $\left|P_r(x \cdot y) - \frac{1}{\omega_{n-1}}\right| = |P_r(x \cdot y) - P_0(x \cdot y)| = |\partial_r P_r(x \cdot y)|\,\big|_{r=\xi} r$, for some ξ between 0 and r. We have

$$\partial_r P_r(x \cdot y)\big|_{r=\xi} = \frac{-2r + (x \cdot y - r)[4r^2 + n(1 - r^2)] + 2r^3}{\omega_{n-1}((1 - r)^2 + 2r(1 - x \cdot y))^{n/2+1}}\bigg|_{r=\xi}.$$

It is clear that $|\partial_r P_r(x \cdot y)|\big|_{r=\xi} \leq c_n$, uniformly in $x \cdot y$. Hence, by going back to $r = e^{-t}$, we get $|P_{e^{-t}}(x \cdot y) - \frac{1}{\omega_{n-1}}| \leq c_n e^{-t}$, uniformly in $x \cdot y$. This and the asymptotic expansion $\mathcal{I}_\rho(z) \approx \frac{1}{(2\pi z)^{1/2}} e^z (1 + O(z^{-1}))$ as $|z| \to \infty$ (see [**11, 13**]) imply

$$II \leq c_{n,s} \int_1^\infty e^{-t(\frac{n-2}{2})} e^{-t} \frac{e^{t(\frac{n-2}{2})}}{t^{1/2}} \frac{dt}{t^{1/2-s}} = c_{n,s}.$$

By pasting together the estimates for I and II the conclusions follow. □

6. Positive powers and the Dirichlet-to-Neumann map

In this section we let $\sigma = s$ to be real, with $0 < s < 1$. Recall that

$$\begin{aligned} (-\Delta_{\mathbb{S}^{n-1}})^s u(x) &= \sum_{k=0}^\infty \lambda_k^s P_k u(x) \\ &= \sum_{k=0}^\infty (k(k + n - 2))^s \sum_{l=1}^{d_k} c_{k,l}(u) Y_{k,l}(x). \end{aligned} \tag{6.1}$$

If $u \in C^\infty(\mathbb{S}^{n-1})$ then (4.1) and (4.2) imply that the sums in (6.1) are absolutely convergent and can be differentiated term by term, so $(-\Delta_{\mathbb{S}^{n-1}})^s u \in C^\infty(\mathbb{S}^{n-1})$.

Our aim is to prove a numerical identity (Lemma 6.1) that will permit us to express the operator $(-\Delta_{\mathbb{S}^{n-1}})^s$ in terms of its principal part, namely, the fractional power of the Dirichlet-to-Neumann map $(L + \frac{n-2}{2})^{2s}$, plus a remainder, smoothing operator S^s. Since L is an operator of order one, $(L + \frac{n-2}{2})^{2s}$ is an operator of order $2s$, which is also the order of $(-\Delta_{\mathbb{S}^{n-1}})^s$. The remainder operator S^s will act as a fractional integral operator of order $2 - 2s$. The semigroup e^{-tL} generated by L (see section 4) will play a key role throughout the analysis.

Let us begin with our new numerical identity. Here we were inspired by the work of Minakshisundaram. Indeed, he proved a similar formula but for negative values of s in [**12**, Lemma 2], which allowed him in turn to find (5.3).

LEMMA 6.1. *Let $k \geq 0$, $\nu \geq 0$, with $k+\nu > 0$, and $0 < s < 1$. Then*

$$(k(k+2\nu))^s = (k+\nu)^{2s} + \frac{(2\nu)^{s+1/2}}{\pi^{-1/2}\Gamma(-s)} \int_0^\infty e^{-t(k+\nu)} \mathcal{B}_{-s-1/2}(\nu t) \, \frac{dt}{t^{1/2+s}}.$$

Here $\mathcal{B}_\rho(z)$ is the modified Bessel function of the first kind $\mathcal{I}_\rho(z)$ minus the first term of its Taylor series expansion, namely,

$$\mathcal{B}_\rho(z) = \mathcal{I}_\rho(z) - \frac{(z/2)^\rho}{\Gamma(\rho+1)} = \sum_{m=1}^\infty \frac{(z/2)^{2m+\rho}}{\Gamma(m+1)\Gamma(m+\rho+1)}.$$

PROOF. The following estimates, direct consequences of the series above and the asymptotic formula for $\mathcal{I}_\rho(z)$ for large $|z|$, hold:

$$|\mathcal{B}_{-s-1/2}(\nu t)| \leq c_s \begin{cases} (\nu t)^{2-s-1/2}, & \text{as } t \to 0, \\ \dfrac{e^{\nu t}}{(\nu t)^{1/2}}, & \text{as } t \to \infty. \end{cases} \tag{6.2}$$

These show that the integral in the statement is absolutely convergent. Let $|\tau| < 1$. From the binomial theorem and the symmetry formula for the quotient of Gamma functions we get

$$\begin{aligned}(1-\tau)^s &= \sum_{m=0}^\infty (-1)^m \frac{\Gamma(s+1)}{\Gamma(m+1)\Gamma(s-m+1)} \tau^m \\ &= 1 + \sum_{m=1}^\infty \frac{\Gamma(m-s)}{\Gamma(m+1)\Gamma(-s)} \tau^m\end{aligned}$$

For any $\lambda, \alpha > 0$ it is easy to check that the following identity holds:

$$\lambda^{-\alpha} = \frac{1}{\Gamma(\alpha)} \int_0^\infty e^{-t\lambda} \frac{dt}{t^{1-\alpha}}.$$

Let us take $\tau = \frac{\nu^2}{(k+\nu)^2}$, $\lambda = k+\nu$, $\alpha = 2(m-s)$, and apply the duplication formula for the Gamma function (the second identity in (3.4)) with $z = m-s$ to obtain

$$\begin{aligned}(k(k+2\nu))^s &= (k+\nu)^{2s}\left(1 - \frac{\nu^2}{(k+\nu)^2}\right)^s \\ &= (k+\nu)^{2s} + \sum_{m=1}^\infty \frac{\Gamma(m-s)}{\Gamma(m+1)\Gamma(-s)} \nu^{2m}(k+\nu)^{-2(m-s)} \\ &= (k+\nu)^{2s} \\ &\quad + \sum_{m=1}^\infty \frac{\Gamma(m-s)\nu^{2m}}{\Gamma(m+1)\Gamma(-s)\Gamma(2(m-s))} \int_0^\infty e^{-t(k+\nu)} \frac{dt}{t^{1-2(m-s)}} \\ &= (k+\nu)^{2s} \\ &\quad + \frac{(2\nu)^{s+1/2}}{\pi^{-1/2}\Gamma(-s)} \int_0^\infty e^{-t(k+\nu)} \left[\sum_{m=1}^\infty \frac{(\nu t/2)^{2m+(-s-1/2)}}{\Gamma(m+1)\Gamma(m-s+1/2)}\right] \frac{dt}{t^{1/2+s}}.\end{aligned}$$

The sum inside the brackets is exactly $\mathcal{B}_{-s-1/2}(\nu t)$. ∎

Let u be as in (1.1), see also (1.3). The fractional power of order $2s$ of the Dirichlet-to-Neumann operator $L+\frac{n-2}{2}$ is given by

$$(6.3)\qquad \begin{aligned}(L+\tfrac{n-2}{2})^{2s}u(x) &= \sum_{k=0}^{\infty}(k+\tfrac{n-2}{2})^{2s}P_k u(x)\\ &= \sum_{k=0}^{\infty}(k+\tfrac{n-2}{2})^{2s}\sum_{l=1}^{d_k}c_{k,l}(u)Y_{k,l}(x).\end{aligned}$$

By applying the numerical identity from Lemma 6.1 with $\nu=\frac{n-2}{2}$ to the spectral definition of $(-\Delta_{\mathbb{S}^{n-1}})^s u(x)$ in (6.1) and recalling the definition of $e^{-tL}u(x)$ in (4.4), we finally obtain the desired relation between the fractional Laplacian on the sphere and the fractional Dirichlet-to-Neumann map (6.3).

THEOREM 6.2 (Fractional Laplacian on the sphere and fractional Dirichlet–to-Neumann map). *Let $0<s<1$. Then*

$$(-\Delta_{\mathbb{S}^{n-1}})^s u(x) = (L+\tfrac{n-2}{2})^{2s}u(x)+S^s u(x),$$

for $x\in\mathbb{S}^{n-1}$. The operator S^s is given by

$$S^s u(x) = \frac{(n-2)^{s+1/2}}{\pi^{-1/2}\Gamma(-s)}\int_0^\infty e^{-t(\frac{n-2}{2})}e^{-tL}u(x)\mathcal{B}_{-s-1/2}(\tfrac{n-2}{2}t)\,\frac{dt}{t^{1/2+s}}.$$

We pointed out in section 4 that $-\Delta_{\mathbb{S}^{n-1}}=L$ only when $n=2$. Notice that if we let $n=2$ in Theorem 6.2 then $S^s u=0$ and $(-\Delta_{\mathbb{T}})^s u = L^{2s}u$. The fractional Laplacian on the torus, which includes this case $n=2$, has been extensively studied in [**14, 15**]. Hence for the rest of this section we will focus on the case $n\geq 3$. We see from Theorem 6.2 that in order to understand the fractional Laplacian on the sphere in terms of the fractional powers of the Dirichlet-to-Neumann map, we need to study separately $(L+\frac{n-2}{2})^{2s}u$ and the fractional integral operator $S^s u$.

6.1. The fractional Dirichlet-to-Neumann operator $(L+\frac{n-2}{2})^{2s}$. As it was expected, the regularization effects of $(-\Delta_{\mathbb{S}^{n-1}})^s$ are given by the fractional operator $(L+\frac{n-2}{2})^{2s}$. Since L is an operator of order one, it is enough to restrict our analysis to the case of powers $2s\in(0,1)$, in which $(L+\frac{n-2}{2})^{2s}$ becomes an operator of order $2s$. When $2s\in[1,2)$ we can just simply write

$$(L+\tfrac{n-2}{2})^{2s}u = (L+\tfrac{n-2}{2})^{2s-1}(Lu+\tfrac{n-2}{2}u),$$

and $2s-1\in[0,1)$.

THEOREM 6.3 (Fractional Dirichlet-to-Neumann map). *Let $0<2s<1$. Then*

$$(L+\tfrac{n-2}{2})^{2s}u(x) = \int_{\mathbb{S}^{n-1}}(u(x)-u(y))L_{2s}(x\cdot y)\,d\mathcal{H}^{n-1}(y)+\left(\tfrac{n-2}{2}\right)^{2s}u(x),$$

for $x\in\mathbb{S}^{n-1}$, where the integral is absolutely convergent. The kernel $L_{2s}(x\cdot y)$ is given by

$$L_{2s}(x\cdot y) = \frac{1}{|\Gamma(-2s)|}\int_0^\infty e^{-t(\frac{n-2}{2})}P_{e^{-t}}(x\cdot y)\,\frac{dt}{t^{1+2s}},$$

and satisfies the estimate

$$L_{2s}(x\cdot y)\sim\frac{1}{d(x,y)^{(n-1)+2s}},$$

for all $x,y\in\mathbb{S}^{n-1}$, $x\neq y$.

PROOF. By applying the numerical formula

$$\lambda^{2s} = \frac{1}{\Gamma(-2s)} \int_0^\infty \left(e^{-t\lambda} - 1\right) \frac{dt}{t^{1+2s}}, \tag{6.4}$$

with $\lambda = (k + \frac{n-2}{2}) > 0$ to (6.3) and recalling (4.4) we obtain the semigroup formula

$$(L + \tfrac{n-2}{2})^{2s} u(x) = \frac{1}{\Gamma(-2s)} \int_0^\infty \left(e^{-t(\frac{n-2}{2})} e^{-tL} u(x) - u(x)\right) \frac{dt}{t^{1+2s}}.$$

Lemma 4.1 shows that this integral is absolutely convergent. From (4.5) and the fact that

$$e^{-tL} 1(x) = \int_{\mathbb{S}^{n-1}} P_{e^{-t}}(x \cdot y)\, d\mathcal{H}^{n-1}(y) \equiv 1,$$

for all $t > 0$ and $x \in \mathbb{S}^{n-1}$, we arrive to

$$\begin{aligned} &(L + \tfrac{n-2}{2})^{2s} u(x) \\ &= \frac{1}{|\Gamma(-2s)|} \int_0^\infty \int_{\mathbb{S}^{n-1}} (u(x) - u(y)) e^{-t(\frac{n-2}{2})} P_{e^{-t}}(x \cdot y)\, d\mathcal{H}^{n-1}(y) \frac{dt}{t^{1+2s}} \\ &\quad + u(x) \frac{1}{\Gamma(-2s)} \int_0^\infty \left(e^{-t(\frac{n-2}{2})} - 1\right) \frac{dt}{t^{1+2s}}. \end{aligned} \tag{6.5}$$

The identity in (6.4) with $\lambda = \frac{n-2}{2}$ implies that the second term in (6.5) is equal to $(\frac{n-2}{2})^{2s} u(x)$. We want to interchange the order of the integrals in the first term of (6.5). On one hand, since $2s < 1$,

$$\begin{aligned} \int_0^1 e^{-t(\frac{n-2}{2})} P_{e^{-t}}(x \cdot y) \frac{dt}{t^{1+2s}} &\sim \int_0^1 \frac{t}{(t^2 + (1 - x \cdot y))^{n/2}} \frac{dt}{t^{1+2s}} \\ &\sim \frac{1}{(1 - x \cdot y)^{\frac{(n-1)+2s}{s}}} \sim \frac{1}{d(x,y)^{(n-1)+2s}}. \end{aligned}$$

Then, by the regularity of u and the Funk–Hecke formula (4.6),

$$\begin{aligned} \int_0^1 \int_{\mathbb{S}^{n-1}} |u(x) - u(y)| e^{-t(\frac{n-2}{2})} P_{e^{-t}}(x \cdot y)\, d\mathcal{H}^{n-1}(y) \frac{dt}{t^{1+2s}} \\ \leq c_{n,s} \int_{\mathbb{S}^{n-1}} \frac{(1 - x \cdot y)^{1/2}}{(1 - x \cdot y)^{\frac{(n-1)+2s}{2}}}\, d\mathcal{H}^{n-1}(y) \\ = c_{n,s} \int_{-1}^1 (1 - \tau)^{\frac{(1-2s)-(n-1)}{2}} (1 - \tau^2)^{\frac{n-3}{2}}\, d\tau < \infty, \end{aligned}$$

because $2s < 1$. On the other hand,

$$\int_1^\infty e^{-t(\frac{n-2}{2})} P_{e^{-t}}(x \cdot y) \frac{dt}{t^{1+2s}} \sim \int_1^\infty e^{-t(\frac{n-2}{2})} \frac{dt}{t^{1+2s}} \sim 1,$$

which gives

$$\int_1^\infty \int_{\mathbb{S}^{n-1}} |u(x) - u(y)| e^{-t(\frac{n-2}{2})} P_{e^{-t}}(x \cdot y)\, d\mathcal{H}^{n-1}(y) \frac{dt}{t^{1+2s}} \leq c_{n,s} \|u\|_{L^\infty(\mathbb{S}^{n-1})}.$$

Thus the double integral in (6.5) is absolutely convergent. The integral representation in the statement then follows from Fubini's theorem. In addition, the computations above prove the estimate for the kernel $L_{2s}(x \cdot y)$. □

6.2. The fractional integral operator S^s. The remaining operator $S^s u$ in Theorem 6.2 is a fractional integral operator. This is in some sense consistent with the numerical identities involved in the proof of Lemma 6.1.

THEOREM 6.4 (Fractional integral operator S^s). *Let $0 < s < 1$. Then*

$$S^s u(x) = \int_{\mathbb{S}^{n-1}} S_s(x \cdot y) u(y) \, d\mathcal{H}^{n-1}(y),$$

for $x \in \mathbb{S}^{n-1}$. The kernel $S_s(x \cdot y)$ is given by

$$S_s(x \cdot y) = \frac{(n-2)^{s+1/2}}{\pi^{-1/2}\Gamma(-s)} \int_0^\infty e^{-t(\frac{n-2}{2})} P_{e^{-t}}(x \cdot y) \mathcal{B}_{-s-1/2}(\tfrac{n-2}{2}t) \frac{dt}{t^{1/2+s}},$$

and satisfies the estimate

$$|S_s(x \cdot y)| \leq \frac{c_{n,s}}{d(x,y)^{(n-1)-(2-2s)}},$$

for all $x, y \in \mathbb{S}^{n-1}$, $x \neq y$.

PROOF. The spherical convolution formula for $S_s u(x)$ in the statement can be derived by using the kernel representation of $e^{-tL}u(x)$ and Fubini's theorem. To prove the estimate for the kernel we proceed as we did for the kernel of the negative powers $(-\Delta_{\mathbb{S}^{n-1}})^{-s}$ in the proof of Theorem 5.1. That is, we split the integral that defines $S_s(x \cdot y)$ as the sum of two integrals $I + II$ and we follow analogous computations. Recall that $n \geq 3$. By the estimates for $\mathcal{B}_{-s-1/2}$ in (6.2),

$$\begin{aligned} |I| &\leq c_{n,s} \int_0^1 \frac{t^{2-2s}}{t^n + d(x,y)^n} \, dt \\ &\leq \frac{c_{n,s}}{d(x,y)^n} \int_0^1 \frac{t^{2-2s}}{(\frac{t}{d(x,y)})^n + 1} \, dt \\ &= \frac{c_{n,s}}{d(x,y)^{(n-1)-(2-2s)}} \int_0^{\frac{1}{d(x,y)}} \frac{\omega^{2-2s}}{\omega^n + 1} \, d\omega \\ &\leq \frac{c_{n,s}}{d(x,y)^{(n-1)-(2-2s)}}, \end{aligned}$$

because $2 - 2s - n + 1 < 0$. For the integral II, notice that $|P_{e^{-t}}(x \cdot y)| \leq c_n$, uniformly in $x \cdot y$, for all $t > 1$. With this and (6.2) we get

$$|II| \leq c_{n,s} \int_1^\infty e^{-t(\frac{n-2}{2})} \frac{e^{t(\frac{n-2}{2})}}{t^{1/2}} \frac{dt}{t^{1/2+s}} = c_{n,s}.$$

□

7. Fractional Laplacians and the heat semigroup on the sphere. Extension problem and Harnack inequality

We have shown how the Minakshisudaram zeta function and our Lemma 6.1, in combination with a careful manipulation of the kernel $P_{e^{-t}}(x \cdot y)$ of the semigroup $e^{-tL}u$ generated by the Dirichlet-to-Neumann map L, permit us to obtain integro-differential formulas for fractional powers of the Laplacian on the sphere, with precise kernel estimates. In this section we present another technique to finding pointwise formulas, namely, by means of the method of heat semigroups. This method was first introduced in [**20**] and later on extended to the most general case in [**10**]. With this technique we can also prove an extension problem characterization, which implies the interior Harnack inequality. The reader should recall that in section 3 we used the heat semigroup (which in that case is the Jacobi theta function) to analyze the kernel of the fractional Laplacian on the unit circle in connection with the Hurwitz zeta function. Further applications of these ideas can be found, for example, in [**3–5, 14, 21**].

7.1. Fractional Laplacians on the sphere and the heat semigroup. The solution $v = v(t,x)$ to the heat equation on the sphere

$$\begin{cases} \partial_t v = \Delta_{\mathbb{S}^{n-1}} v, & \text{for } t>0,\ x\in\mathbb{S}^{n-1},\\ v(0,x) = u(x), & \text{for } x\in\mathbb{S}^{n-1}, \end{cases}$$

is given by the heat semigroup generated by $-\Delta_{\mathbb{S}^{n-1}}$:

$$\begin{aligned} e^{t\Delta_{\mathbb{S}^{n-1}}}u(x) \equiv v(t,x) &= \sum_{k=0}^{\infty} e^{-t\lambda_k} P_k u(x)\\ &= \sum_{k=0}^{\infty} e^{-t(k(k+n-2))} \sum_{l=0}^{d_k} c_{k,l}(u) Y_{k,l}(x), \end{aligned} \tag{7.1}$$

where P_k is the orthogonal projector (1.3). By writing down the definition of $c_{k,l}(u)$ and interchanging sums and integral, we see that the heat semigroup on the sphere can be written as a spherical convolution

$$e^{t\Delta_{\mathbb{S}^{n-1}}}u(x) = \int_{\mathbb{S}^{n-1}} W_t(x\cdot y) u(y)\, d\mathcal{H}^{n-1}(y).$$

The heat kernel $W_t(x\cdot y)$ is given by

$$\begin{aligned} W_t(x\cdot y) &= \sum_{k=0}^{\infty} e^{-t(k(k+n-2))} \sum_{l=0}^{d_k} Y_{k,l}(x) Y_{k,l}(y)\\ &= \frac{1}{\omega_{n-1}} \sum_{k=0}^{\infty} e^{-t(k(k+n-2))} \frac{k+\frac{n-2}{2}}{\frac{n-2}{2}} C_k(x\cdot y), \end{aligned}$$

where in the second identity we applied (4.8). This kernel turns out to be a smooth, positive function of $t>0$, $x,y\in\mathbb{S}^{n-1}$, see [**6**, Chapter 5]. Moreover, for any $t>0$ and $x\in\mathbb{S}^{n-1}$,

$$e^{t\Delta_{\mathbb{S}^{n-1}}}1(x) = \int_{\mathbb{S}^{n-1}} W_t(x\cdot y)\, d\mathcal{H}^{n-1}(y) \equiv 1. \tag{7.2}$$

In addition, $W_t(x\cdot y)$ satisfies two-sided Gaussian estimates. In fact, there is a constant $C>0$ that depends only on n such that

$$W_t(x\cdot y) \le \frac{C}{t^{(n-1)/2}} e^{-d(x,y)^2/(8t)}, \tag{7.3}$$

and

$$W_t(x\cdot y) \ge \frac{C^{-1}}{t^{(n-1)/2}} e^{-d(x,y)^2/(4t)}, \tag{7.4}$$

for all $t>0$, for any $x,y\in\mathbb{S}^{n-1}$, see [**6**, Theorems 5.5.6 and 5.6.1].

The heat semigroup formulas for the fractional operators are obtained as follows. For any $\lambda \geq 0$ and $0 < s < 1$ we have

$$\lambda^s = \frac{1}{\Gamma(-s)} \int_0^\infty \left(e^{-t\lambda} - 1\right) \frac{dt}{t^{1+s}}.$$

By taking $\lambda = \lambda_k = k(k+n-2)$ and recalling the spectral definition of the fractional Laplacian on the sphere (1.2) we immediately see that

$$(-\Delta_{\mathbb{S}^{n-1}})^s u(x) = \frac{1}{\Gamma(-s)} \int_0^\infty \left(e^{t\Delta_{\mathbb{S}^{n-1}}} u(x) - u(x)\right) \frac{dt}{t^{1+s}}. \tag{7.5}$$

For the negative fractional powers we use that, for any $\lambda, s > 0$,

$$\lambda^{-s} = \frac{1}{\Gamma(s)} \int_0^\infty e^{-t\lambda} \frac{dt}{t^{1-s}}.$$

If $\int_{\mathbb{S}^{n-1}} u = 0$ then we can take $\lambda = \lambda_k$ in the formula above and use the spectral definition (1.2) to infer that

$$(-\Delta_{\mathbb{S}^{n-1}})^{-s} u(x) = \frac{1}{\Gamma(s)} \int_0^\infty e^{t\Delta_{\mathbb{S}^{n-1}}} u(x) \frac{dt}{t^{1-s}}. \tag{7.6}$$

THEOREM 7.1 (Fractional Laplacians and heat semigroup). *Let $s > 0$.*

(1) *If $0 < s < 1/2$ then*

$$(-\Delta_{\mathbb{S}^{n-1}})^s u(x) = \int_{\mathbb{S}^{n-1}} (u(x) - u(y)) K_s(x \cdot y) \, d\mathcal{H}^{n-1}(y),$$

where the integral is absolutely convergent. If $1/2 \leq s < 1$ then

$$\begin{aligned}(-\Delta_{\mathbb{S}^{n-1}})^s u(x) &= \mathrm{P.V.} \int_{\mathbb{S}^{n-1}} (u(x) - u(y)) K_s(x \cdot y) \, d\mathcal{H}^{n-1}(y) \\ &= \int_{\mathbb{S}^{n-1}} (u(x) - u(y) - \nabla_{\mathbb{S}^{n-1}} u(x) \cdot (x - y)) K_s(x \cdot y) \, d\mathcal{H}^{n-1}(y),\end{aligned}$$

where the second integral is absolutely convergent. In both cases the kernel $K_s(x \cdot y)$ is given by

$$K_s(x \cdot y) = \frac{1}{|\Gamma(-s)|} \int_0^\infty W_t(x \cdot y) \frac{dt}{t^{1+s}} > 0,$$

and satisfies the estimate

$$K_s(x \cdot y) \sim \frac{1}{d(x,y)^{(n-1)+2s}}, \tag{7.7}$$

for all $x, y \in \mathbb{S}^{n-1}$, $x \neq y$.

(2) *If $0 < s < \frac{n-1}{2}$ and $\int_{\mathbb{S}^{n-1}} u = 0$ then*

$$(-\Delta_{\mathbb{S}^{n-1}})^{-s} u(x) = \int_{\mathbb{S}^{n-1}} K_{-s}(x \cdot y) u(y) \, d\mathcal{H}^{n-1}(y),$$

as in Theorem 5.1. An equivalent formula for the kernel $K_{-s}(x \cdot y)$ in (5.5) is

$$K_{-s}(x \cdot y) = \frac{1}{\Gamma(s)} \int_0^\infty W_t(x \cdot y) \frac{dt}{t^{1-s}} > 0,$$

from which we can deduce the estimate

$$K_{-s}(x\cdot y)\sim\frac{1}{d(x,y)^{(n-1)-2s}},\tag{7.8}$$

for all $x,y\in\mathbb{S}^{n-1}$, $x\neq y$.

PROOF. Let us begin by proving (1). We first estimate the kernel $K_s(x\cdot y)$ by applying (7.3)–(7.4). Indeed, by using the change of variables $r=d(x,y)^2/(8t)$,

$$\begin{aligned}\int_0^\infty W_t(x\cdot y)\,\frac{dt}{t^{1+s}}&\le C\int_0^\infty\frac{1}{t^{(n-1)/2}}e^{-d(x,y)^2/(8t)}\,\frac{dt}{t^{1+s}}\\&=\frac{c_{n,s}}{d(x,y)^{(n-1)+2s}}\int_0^\infty e^{-r}r^{(n-1)/2+s}\,\frac{dr}{r}\\&=\frac{c_{n,s}}{d(x,y)^{(n-1)+2s}}.\end{aligned}$$

The lower bound follows analogously via the change of variables $r=d(x,y)^2/(4t)$. Therefore (7.7) follows.

From the semigroup formula (7.5) and by using (7.2) we get

$$(-\Delta_{\mathbb{S}^{n-1}})^su(x)=\frac{1}{|\Gamma(-s)|}\int_0^\infty\int_{\mathbb{S}^{n-1}}(u(x)-u(y))W_t(x\cdot y)d\mathcal{H}^{n-1}(y)\frac{dt}{t^{1+s}}.\tag{7.9}$$

The pointwise formulas in (1) will follow after we apply Fubini's theorem.

Suppose that $0<s<1/2$. Then the double integral in (7.9) is absolutely convergent because of (7.7) and the estimate $|u(x)-u(y)|\le C|x-y|\sim Cd(x,y)$. Thus Fubini's theorem can be applied to conclude.

Suppose next that $1/2\le s<1$. Observe that $W_t(x\cdot y)=F(|x-y|)$ for some function $F:\mathbb{R}\to\mathbb{R}$. This implies that

$$\int_{\mathbb{S}^{n-1}\setminus B_\varepsilon(x)}(x_i-y_i)W_t(x\cdot y)\,d\mathcal{H}^{n-1}(y)=0,\tag{7.10}$$

for any $i=1,\dots,n$, for every $\varepsilon\ge0$. Hence the double integral in (7.9) can be written as

$$\int_0^\infty\int_{\mathbb{S}^{n-1}}(u(x)-u(y)-\nabla_{\mathbb{S}^{n-1}}u(x)\cdot(x-y))W_t(x\cdot y)d\mathcal{H}^{n-1}(y)\frac{dt}{t^{1+\sigma}}.$$

By (7.7) and the estimate $|u(x)-u(y)-\nabla_{\mathbb{S}^{n-1}}u(x)\cdot(x-y)|\le C|x-y|^2\sim Cd(x,y)^2$, this double integral is absolutely convergent, so that Fubini's theorem can be used. The principal value formula follows by applying (7.10) for each $\varepsilon>0$.

Let us continue with the proof of (2). The pointwise formula for $(-\Delta_{\mathbb{S}^{n-1}})^{-s}u(x)$ follows by writing down the heat kernel in (7.6) and using Fubini's theorem. This shows also the equivalent heat semigroup formula for the kernel $K_{-s}(x\cdot y)$. The estimate in (7.8) is obtained by using the heat kernel bounds as above. We just do the upper bound, the lower bound is completely analogous. We have

$$\begin{aligned}K_{-s}(x\cdot y)&\le C\int_0^\infty\frac{1}{t^{(n-1)/2}}e^{-d(x,y)^2/(8t)}\,\frac{dt}{t^{1-s}}\\&=\frac{c_{n,s}}{d(x,y)^{(n-1)-2s}}\int_0^\infty e^{-r}r^{(n-1)/2-s}\,\frac{dr}{r},\end{aligned}$$

and the last integral is finite because $(n-1)/2-s>0$. □

7.2. Extension problem and Harnack inequality. The extension problem for the fractional Laplacian on the sphere is a particular case of the general extension problem proved in [**20**], see also [**10**]. We present the proof here for the convenience of the reader. As an application, the interior Harnack inequality is proved.

THEOREM 7.2 (Extension problem). *Let $0 < s < 1$. Define*

$$U(x,y) = \frac{y^{2s}}{4^s\Gamma(s)} \int_0^\infty e^{-y^2/(4t)} e^{t\Delta_{\mathbb{S}^{n-1}}} u(x)\, \frac{dt}{t^{1+s}},$$

for $x \in \mathbb{S}^{n-1}$, $y > 0$. Then U solves

$$\begin{cases} \Delta_{\mathbb{S}^{n-1}} U + \frac{1-2s}{y}\partial_y U + \partial_{yy} U = 0, & \text{for } x \in \mathbb{S}^{n-1},\ y > 0, \\ U(x,0) = u(x), & \text{for } x \in \mathbb{S}^{n-1}, \\ -y^{1-2s}\partial_y U(x,y)\big|_{y=0^+} = \frac{\Gamma(1-s)}{4^{s-1/2}\Gamma(s)}(-\Delta_{\mathbb{S}^{n-1}})^s u(x), & \text{for } x \in \mathbb{S}^{n-1}. \end{cases}$$

Moreover, if $\displaystyle\int_{\mathbb{S}^{n-1}} u = 0$ *then*

$$U(x,y) = \frac{1}{\Gamma(s)} \int_0^\infty e^{-y^2/(4t)} e^{t\Delta_{\mathbb{S}^{n-1}}} \big((-\Delta_{\mathbb{S}^{n-1}})^s u\big)(x)\, \frac{dt}{t^{1-s}}. \tag{7.11}$$

REMARK 7.3 (Extension problem for negative powers). Let f be a function on the sphere such that $\displaystyle\int_{\mathbb{S}^{n-1}} f = 0$. Consider the solution u to

$$(-\Delta_{\mathbb{S}^{n-1}})^s u = f, \quad 0 < s < 1,$$

such that $\displaystyle\int_{\mathbb{S}^{n-1}} u = 0$. Then (7.11) reads

$$U(x,y) = \frac{1}{\Gamma(s)} \int_0^\infty e^{-y^2/(4t)} e^{t\Delta_{\mathbb{S}^{n-1}}} f(x)\, \frac{dt}{t^{1-s}},$$

which solves the Neumann problem

$$\begin{cases} \Delta_{\mathbb{S}^{n-1}} U + \frac{1-2s}{y}\partial_y U + \partial_{yy} U = 0, & \text{for } x \in \mathbb{S}^{n-1},\ y > 0, \\ -y^{1-2s}\partial_y U(x,y)\big|_{y=0^+} = \frac{\Gamma(1-s)}{4^{s-1/2}\Gamma(s)} f(x), & \text{for } x \in \mathbb{S}^{n-1}. \end{cases}$$

This is the extension problem for $(-\Delta_{\mathbb{S}^{n-1}})^{-s}$. Indeed,

$$U(x,0) = u(x) = (-\Delta_{\mathbb{S}^{n-1}})^{-s} f(x).$$

PROOF OF THEOREM 7.2. By using the change of variables $y^2/(4t) = r$ in the definition of U, we have the equivalent formula

$$U(x,y) = \frac{1}{\Gamma(s)} \int_0^\infty e^{-r} e^{\frac{y^2}{4r}\Delta_{\mathbb{S}^{n-1}}} u(x)\, \frac{dr}{r^{1-s}},$$

from which immediately follows that $U(x,0) = u(x)$. Let $\mathcal{K}_\rho(z)$ denote the modified Bessel function of second kind of order ρ, see [**11, 13**]. By using its integral representation in [**11**, eq. (5.10.25)] and the spectral definition of $e^{t\Delta_{\mathbb{S}^{n-1}}} u(x)$ in (7.1) we can write

$$\begin{aligned} U(x,y) &= \frac{y^{2s}}{4^s\Gamma(s)} \sum_{k=0}^{\infty} \left[\int_0^\infty e^{-y^2/(4t)} e^{-t\lambda_k} \frac{dt}{t^{1+s}} \right] P_k u(x) \\ &= \frac{2^{1-s}}{\Gamma(s)} \sum_{k=0}^{\infty} (y\lambda_k^{1/2})^s \mathcal{K}_s(y\lambda_k^{1/2}) P_k u(x) \\ &\equiv \frac{2^{1-s}}{\Gamma(s)} (y(-\Delta_{\mathbb{S}^{n-1}})^{1/2})^s \mathcal{K}_s(y(-\Delta_{\mathbb{S}^{n-1}})^{1/2}) u(x). \end{aligned}$$

With any of these formulas and the identities for the derivatives of $\mathcal{K}_s(z)$ it is easy to check that U satisfies the extension equation. By noticing that

$$\frac{y^{2s}}{4^s\Gamma(s)} \int_0^\infty e^{-y^2/(4t)} \frac{dt}{t^{1+s}} = 1,$$

we get

$$\begin{aligned} -y^{1-2s}\partial_y U(x,y) &= \frac{1}{4^s\Gamma(s)} \int_0^\infty \left(\frac{y^2}{2t} - 2s \right) e^{-y^2/(4t)} e^{t\Delta_{\mathbb{S}^{n-1}}} u(x) \frac{dt}{t^{1+s}} \\ &= \frac{1}{4^s\Gamma(s)} \int_0^\infty \left(\frac{y^2}{2t} - 2s \right) e^{-y^2/(4t)} \left(e^{t\Delta_{\mathbb{S}^{n-1}}} u(x) - u(x) \right) \frac{dt}{t^{1+s}} \\ &\longrightarrow -\frac{2s}{4^s\Gamma(s)} \int_0^\infty \left(e^{t\Delta_{\mathbb{S}^{n-1}}} u(x) - u(x) \right) \frac{dt}{t^{1+s}} \\ &= \frac{\Gamma(1-s)}{4^{s-1/2}\Gamma(s)} (-\Delta_{\mathbb{S}^{n-1}})^s u(x), \quad \text{as } y \to 0^+, \end{aligned}$$

where in the last identity we used the semigroup formula (7.5). If $\displaystyle\int_{\mathbb{S}^{n-1}} u = 0$ then $P_0 u(x) = 0$ and, by the change of variables $r = y^2/(4t\lambda_k)$, $k \geq 1$, we obtain

$$\begin{aligned} U(x,y) &= \frac{y^{2s}}{4^s\Gamma(s)} \sum_{k=1}^{\infty} \left[\int_0^\infty e^{-y^2/(4t)} e^{-t\lambda_k} \frac{dt}{t^{1+s}} \right] P_k u(x) \\ &= \frac{1}{\Gamma(s)} \sum_{k=1}^{\infty} \left[\int_0^\infty e^{-y^2/(4r)} e^{-r\lambda_k} \lambda_k^s P_k u(x) \frac{dt}{t^{1+s}} \right], \end{aligned}$$

which gives (7.11). □

THEOREM 7.4 (Harnack inequality). *Let $\Omega' \subset\subset \Omega \subset \mathbb{S}^{n-1}$ be open sets. There is a constant $C > 0$ depending on Ω', Ω, n and s such that*

$$\sup_{\Omega'} u \leq C \inf_{\Omega'} u,$$

for any solution u to

$$\begin{cases} (-\Delta_{\mathbb{S}^{n-1}})^s u = 0, & \text{in } \Omega, \\ u \geq 0, & \text{in } \mathbb{S}^{n-1}. \end{cases}$$

PROOF. The idea is to use the extension problem as done in other contexts, see for example [**3, 14, 20, 21**]. We sketch the steps next. Details are left to the interested reader. For any u as in the statement, let U be the solution to the extension problem given by Theorem 7.2. Since $u \geq 0$ in $\mathbb{S}^{n-1}$ and the heat kernel

$W_t(x\cdot y)$ is positive, we have $e^{t\Delta_{\mathbb{S}^{n-1}}}u\geq 0$ and thus $U\geq 0$ in $\mathbb{S}^{n-1}\times[0,\infty)$. The extension equation can be written as

$$\operatorname{div}_{(\mathbb{S}^{n-1},y)}(y^{1-2s}\nabla_{(\mathbb{S}^{n-1},y)}U)=0,$$

where $\nabla_{(\mathbb{S}^{n-1},y)}=(\nabla_{\mathbb{S}^{n-1}},\partial_y)$ and $\operatorname{div}_{(\mathbb{S}^{n-1},y)}$ is the corresponding divergence operator. Let $\bar{U}(x,y)=U(x,|y|)\geq 0$, for $x\in\mathbb{S}^{n-1}$ and $y\in\mathbb{R}$. By using that

$$-y^{1-2s}\partial_yU(x,y)\big|_{y=0^+}=\frac{\Gamma(1-s)}{4^{s-1/2}\Gamma(s)}(-\Delta_{\mathbb{S}^{n-1}})^su=0,\quad \text{in }\Omega,$$

it can be checked that $\bar{U}$ is a weak solution the the degenerate elliptic equation

$$\operatorname{div}_{(\mathbb{S}^{n-1},y)}(|y|^{1-2s}\nabla_{(\mathbb{S}^{n-1},y)}\bar{U})=0,$$

with Muckenhoupt weight $\omega(x,y)=|y|^{1-2s}\in A_2$, in $\{(x,y):x\in\Omega,-1/2<y<1/2\}$. These degenerate elliptic equations admit interior Harnack inequality (see [**7**]), so there exists a constant $C>0$ depending only on Ω, Ω', n and s such that

$$\sup_{\Omega'\times(-1/4,1/4)}\bar{U}\leq C\inf_{\Omega'\times(-1/4,1/4)}\bar{U}.$$

By restricting $\bar{U}$ back to $y=0$ the Harnack inequality for u follows. □

Acknowledgments. The authors would like to thank Sundaram Thangavelu for providing a copy of Minakshisundaram's paper [**12**]. They are also grateful to the referee for useful comments and questions.

References

[1] Ricardo Alonso and Weiran Sun, *The radiative transfer equation in the forward-peaked regime*, Comm. Math. Phys. **338** (2015), no. 3, 1233–1286, DOI 10.1007/s00220-015-2395-8. MR3355814

[2] Tom M. Apostol, *Introduction to analytic number theory*, Springer-Verlag, New York-Heidelberg, 1976. Undergraduate Texts in Mathematics. MR0434929

[3] Luis Caffarelli and Luis Silvestre, *An extension problem related to the fractional Laplacian*, Comm. Partial Differential Equations **32** (2007), no. 7-9, 1245–1260, DOI 10.1080/03605300600987306. MR2354493

[4] Luis A. Caffarelli and Pablo Raúl Stinga, *Fractional elliptic equations, Caccioppoli estimates and regularity*, Ann. Inst. H. Poincaré Anal. Non Linéaire **33** (2016), no. 3, 767–807, DOI 10.1016/j.anihpc.2015.01.004. MR3489634

[5] Luis A. Caffarelli and Alexis Vasseur, *Drift diffusion equations with fractional diffusion and the quasi-geostrophic equation*, Ann. of Math. (2) **171** (2010), no. 3, 1903–1930, DOI 10.4007/annals.2010.171.1903. MR2680400

[6] E. B. Davies, *Heat kernels and spectral theory*, Cambridge Tracts in Mathematics, vol. 92, Cambridge University Press, Cambridge, 1989. MR990239

[7] Eugene B. Fabes, Carlos E. Kenig, and Raul P. Serapioni, *The local regularity of solutions of degenerate elliptic equations*, Comm. Partial Differential Equations **7** (1982), no. 1, 77–116, DOI 10.1080/03605308208820218. MR643158

[8] N. J. Fine, *Note on the Hurwitz zeta-function*, Proc. Amer. Math. Soc. **2** (1951), 361–364, DOI 10.2307/2031757. MR0043194

[9] Gerald B. Folland, *Introduction to partial differential equations*, 2nd ed., Princeton University Press, Princeton, NJ, 1995. MR1357411

[10] José E. Galé, Pedro J. Miana, and Pablo Raúl Stinga, *Extension problem and fractional operators: semigroups and wave equations*, J. Evol. Equ. **13** (2013), no. 2, 343–368, DOI 10.1007/s00028-013-0182-6. MR3056307

[11] N. N. Lebedev, *Special functions and their applications*, Revised English edition. Translated and edited by Richard A. Silverman, Prentice-Hall, Inc., Englewood Cliffs, N.J., 1965. MR0174795

[12] S. Minakshisundaram, *Zeta functions on the sphere*, J. Indian Math. Soc. (N.S.) **13** (1949), 41–48. MR0031146

[13] NIST Digital Library of Mathematical Functions, version 1.0.15, last visited 08/30/2017. `http://dlmf.nist.gov/`.

[14] Luz Roncal and Pablo Raúl Stinga, *Fractional Laplacian on the torus*, Commun. Contemp. Math. **18** (2016), no. 3, 1550033, 26, DOI 10.1142/S0219199715500339. MR3477397

[15] Luz Roncal and Pablo Raúl Stinga, *Transference of fractional Laplacian regularity*, Special functions, partial differential equations, and harmonic analysis, Springer Proc. Math. Stat., vol. 108, Springer, Cham, 2014, pp. 203–212, DOI 10.1007/978-3-319-10545-1_14. MR3297661

[16] Boris Rubin, *Fractional integrals and potentials*, Pitman Monographs and Surveys in Pure and Applied Mathematics, vol. 82, Longman, Harlow, 1996. MR1428214

[17] Stefan G. Samko, *Hypersingular integrals and their applications*, Analytical Methods and Special Functions, vol. 5, Taylor & Francis, Ltd., London, 2002. MR1918790

[18] R. T. Seeley, *Spherical harmonics*, Amer. Math. Monthly **73** (1966), no. 4, 115–121, DOI 10.2307/2313760. MR0201695

[19] Luis Silvestre, *Regularity of the obstacle problem for a fractional power of the Laplace operator*, Comm. Pure Appl. Math. **60** (2007), no. 1, 67–112, DOI 10.1002/cpa.20153. MR2270163

[20] Pablo Raúl Stinga and José Luis Torrea, *Extension problem and Harnack's inequality for some fractional operators*, Comm. Partial Differential Equations **35** (2010), no. 11, 2092–2122, DOI 10.1080/03605301003735680. MR2754080

[21] Pablo Raúl Stinga and Chao Zhang, *Harnack's inequality for fractional nonlocal equations*, Discrete Contin. Dyn. Syst. **33** (2013), no. 7, 3153–3170, DOI 10.3934/dcds.2013.33.3153. MR3007742

[22] G. M. Zaslavsky, *Chaos, fractional kinetics, and anomalous transport*, Phys. Rep. **371** (2002), no. 6, 461–580, DOI 10.1016/S0370-1573(02)00331-9. MR1937584

IMAS (UBA-CONICET); and Departamento de Matemática, Facultad de Ciencias Exactas y Naturales, Universidad de Buenos Aires, Ciudad Universitaria, 1428 Buenos Aires, Argentina

Email address: `pdenapo@dm.uba.ar`

Department of Mathematics, Iowa State University, 396 Carver Hall, Ames, Iowa 50011, United States of America

Email address: `stinga@iastate.edu`

Contemporary Mathematics
Volume **723**, 2019
https://doi.org/10.1090/conm/723/14570

Obstacle problems for nonlocal operators

Donatella Danielli, Arshak Petrosyan, and Camelia A. Pop

ABSTRACT. We prove existence, uniqueness, and regularity of viscosity solutions to the stationary and evolution obstacle problems defined by a class of nonlocal operators that are not stable-like and may have supercritical drift. We give sufficient conditions on the coefficients of the operator to obtain Hölder and Lipschitz continuous solutions. The class of nonlocal operators that we consider include non-Gaussian asset price models widely used in mathematical finance, such as Variance Gamma Processes and Regular Lévy Processes of Exponential type. In this context, the viscosity solutions that we analyze coincide with the prices of perpetual and finite expiry American options.

1. Introduction

The nonlocal operators that we consider are the infinitesimal generators of strong Markov processes that are solutions to stochastic equations of the form:

$$dX(t) = b(X(t-))\,dt + \int_{\mathbb{R}^n\setminus\{O\}} F((X(t-),y)\,\widetilde{N}(dt,dy), \quad \forall\, t>0, \tag{1.1}$$

where O denotes the origin in $\mathbb{R}^n$, $\widetilde{N}(dt,dy)$ is a compensated Poisson random measure with Lévy measure $\nu(dy)$, and the coefficients $b(x)$ and $F(x,y)$ appearing in identity (1.1) are assumed to satisfy:

ASSUMPTION 1.1 (Coefficients). There is a positive constant K such that:

1. For all $x_1, x_2 \in \mathbb{R}^n$, we have that

$$\int_{\mathbb{R}^n\setminus\{O\}} |F(x_1,y)-F(x_2,y)|^2\,d\nu(y) \leq K|x_1-x_2|^2, \tag{1.2}$$

$$\sup_{z\in B_{|y|}} |F(x,z)| \leq \rho(y), \quad \forall\, x,y\in\mathbb{R}^n, \quad \text{and} \quad \int_{\mathbb{R}^n\setminus\{O\}} (|y|\vee\rho(y))^2\,\nu(dy) \leq K, \tag{1.3}$$

   where $\rho:\mathbb{R}^n\to[0,\infty)$ is a measurable function.
2. The coefficient $b:\mathbb{R}^n\to\mathbb{R}^n$ is bounded and Lipschitz continuous, i.e., $b\in C^{0,1}(\mathbb{R}^n)$.

2010 *Mathematics Subject Classification.* Primary 35R35; Secondary 60G51, 91G80.
Key words and phrases. Obstacle problem, nonlocal operators, Lévy processes, American options, viscosity solutions, existence and uniqueness.

Throughout our paper, for all $r > 0$ and $x \in \mathbb{R}^n$, we denote by $B_r(x)$ the Euclidean ball of radius r centered at x. When $x = O$, we denote for brevity $B_r(O)$ by B_r. We also use the notation:

$$a \wedge b := \min\{a, b\} \quad \text{and} \quad a \vee b := \max\{a, b\}, \quad \forall\, a, b \in \mathbb{R}.$$

Let $(\Omega, \mathcal{F}, \{\mathcal{F}_t\}_{t\geq 0}, \mathbb{P})$ be a filtered probability space that satisfies the usual hypotheses, [**11**, § I.1], and that supports a Poisson random measure, $N(dt, dy)$, with Lévy measure, $\nu(dy)$. Assumption 1.1 and [**11**, Theorem V.3.6] ensure that, for any initial condition $X^x(0) = x \in \mathbb{R}^n$, the stochastic equation (1.1) admits a unique strong solution $\{X^x(t)\}_{t\geq 0}$ with right-continuous and left-limits (RCLL) paths a.s.. Moreover, from [**11**, Theorem V.6.32]* it follows that the process $\{X^x(t)\}_{t\geq 0}$ satisfies the strong Markov property. Thus, the Markov process $\{X^x(t)\}_{t\geq 0}$ is completely characterized by its infinitesimal generator, which is given by the nonlocal operator:

(1.4)

$$Lu(x) := b(x)\cdot\nabla u(x) + \int_{\mathbb{R}^n\setminus\{O\}} \left(u(x + F(x, y)) - u(x) - \nabla u(x)\cdot F(x, y)\right) \nu(dy),$$

for all $u \in C^2(\mathbb{R}^n)$, where we let $C^2(\mathbb{R}^n)$ denote the space of functions with bounded and continuous derivatives up to and including order 2. Our goal is to study the existence, uniqueness, and regularity properties of viscosity solutions to the stationary and evolution obstacle problems associated to the operator L.

1.1. Stationary obstacle problem. In this section we state our results related to the existence, uniqueness, and regularity of viscosity solutions to the stationary obstacle problem defined by the nonlocal operator L,

$$\min\{-Lv + cv - f, v - \varphi\} = 0, \quad \text{on } \mathbb{R}^n, \tag{1.5}$$

where $c : \mathbb{R}^n \to \mathbb{R}$ is the zeroth order term, $f : \mathbb{R}^n \to \mathbb{R}$ is the source function, and $\varphi : \mathbb{R}^n \to \mathbb{R}$ is the obstacle function. For the existence results in the stationary case we will also assume that the jump size $F(x, y)$ does not depend on the state variable, that is,

$$F(x, y) = F(y), \quad \forall\, x, y \in \mathbb{R}^n. \tag{1.6}$$

Let $\mathcal{T}$ denote the set of $\mathbb{P}$-a.s. finite stopping times adapted to the filtration $\{\mathcal{F}_t\}_{t\geq 0}$. Solutions to the obstacle problem (1.5) are constructed using the stochastic representation formula:

$$v(x) := \sup\{v(x; \tau) : \tau \in \mathcal{T}\}, \tag{1.7}$$

where we denote

$$v(x; \tau) := \mathbb{E}\left[e^{-\int_0^\tau c(X^x(s))\,ds}\varphi(X^x(\tau)) + \int_0^\tau e^{-\int_0^t c(X^x(s))\,ds} f(X^x(t))\,dt\right], \tag{1.8}$$

where $\{X^x(t)\}_{t\geq 0}$ is the unique solution to the stochastic equation (1.1) with initial condition $X^x(0) = x$, for all $x \in \mathbb{R}^n$. We denote by $C(\mathbb{R}^n)$ the space of continuous functions $u : \mathbb{R}^n \to \mathbb{R}$ such that

$$\|u\|_{C(\mathbb{R}^n)} := \sup_{x\in\mathbb{R}^n} |u(x)| < \infty.$$

*[**11**, Theorem V.6.32] applies to the case when $b \equiv 0$ in the stochastic equation (1.1), but it is not difficult to see that the proof immediately extends to the case of Lipschitz continuous drift coefficients $b(x)$, as we suppose in Assumption 1.1.

For all $\alpha \in (0,1]$, a function $u : \mathbb{R}^n \to \mathbb{R}$ belongs to $C^{0,\alpha}(\mathbb{R}^n)$ if

$$\|u\|_{C^{0,\alpha}(\mathbb{R}^n)} := \|u\|_{C(\mathbb{R}^n)} + [u]_{C^{0,\alpha}(\mathbb{R}^n)} < \infty,$$

where, as usual, we define

$$[u]_{C^{0,\alpha}(\mathbb{R}^n)} := \sup_{x_1,x_2\in\mathbb{R}^n, x_1\neq x_2} \frac{|u(x_1)-u(x_2)|}{|x_1-x_2|^\alpha}.$$

When $\alpha \in (0,1)$, we denote for brevity $C^\alpha(\mathbb{R}^n) := C^{0,\alpha}(\mathbb{R}^n)$. We first state:

PROPOSITION 1.2 (Regularity of the value function). *Suppose that Assumption 1.1 and condition (1.6) hold. Let $c, \varphi, f : \mathbb{R}^n \to \mathbb{R}$ be bounded Lipschitz continuous functions, and assume that there is a positive constant, c_0, with the property that*

$$c(x) \geq c_0 > 0, \quad \forall x \in \mathbb{R}^n. \tag{1.9}$$

Then the following hold:

(i) (Hölder continuity) *There is a constant, $\alpha = \alpha([b]_{C^{0,1}(\mathbb{R}^n)}, c_0) \in (0,1)$, such that the value function v defined in (1.7) belongs to $C^\alpha(\mathbb{R}^n)$.*

(ii) (Lipschitz continuity) *If in addition we have that*

$$c_0 \geq [b]_{C^{0,1}(\mathbb{R}^n)}, \tag{1.10}$$

then the value function v in (1.7) belongs to $C^{0,1}(\mathbb{R}^n)$.

DEFINITION 1.3 (Viscosity solutions). Let $v \in C(\mathbb{R}^n)$. We say that v is a viscosity subsolution (supersolution) to the stationary obstacle problem (1.5) if, for all $u \in C^2(\mathbb{R}^n)$ such that $v-u$ has a global max (min) at $x_0 \in \mathbb{R}^n$ and $u(x_0) = v(x_0)$, then

$$\min\{-Lu(x_0) + c(x_0)u(x_0) - f(x_0), u(x_0) - \varphi(x_0)\} \leq (\geq)\, 0. \tag{1.11}$$

We say that v is a viscosity solution to equation (1.5) if it is both a sub- and supersolution.

THEOREM 1.4 (Existence of viscosity solution). *Suppose that the hypotheses of Proposition 1.2 hold, and that*

$$\int_{\mathbb{R}^n\setminus\{O\}} |F(y)|^{2\alpha}\, \nu(dy) < \infty \tag{1.12}$$

where $\alpha \in (0,1)$ is the constant appearing in Proposition 1.2 (i). Then the value function v defined in (1.7) is a viscosity solution to the stationary obstacle problem (1.5).

THEOREM 1.5 (Uniqueness of viscosity solution). *Suppose that Assumption 1.1 holds, that c, f, φ belong to $C(\mathbb{R}^n)$, c satisfies condition (1.9), and*

$$\lim_{y\to O} F(x,y) = 0, \quad \forall x \in \mathbb{R}^n. \tag{1.13}$$

If the stationary obstacle problem (1.5) has a viscosity solution, then it is unique.

REMARK 1.6 (Condition (1.9) on the zeroth order term $c(x)$). Condition (1.9) in Theorem 1.5 can be replaced by the less restrictive assumption that $c(x)$ is a positive function on $\mathbb{R}^n$ and

$$\limsup_{|x|\to\infty} \frac{1}{|x|c(x)} = 0. \tag{1.14}$$

1.2. Evolution obstacle problem. We next consider the questions of existence, uniqueness, and regularity of viscosity solutions to the evolution obstacle problem defined by the nonlocal operator L,

$$\begin{cases} \min\{-v_t - Lv + cv - f, v - \varphi\} & = 0, & \text{on } [0,T) \times \mathbb{R}^n, \\ v(T,\cdot) & = g, & \text{on } \mathbb{R}^n, \end{cases} \tag{1.15}$$

where we assume the compatibility assumption

$$g \geq \varphi(T,\cdot) \quad \text{on} \quad \mathbb{R}^n. \tag{1.16}$$

Let $\mathcal{T}_t$ denote the set of stopping times $\tau \in \mathcal{T}$ bounded by t, for all $t \geq 0$. Solutions to problem (1.15) are constructed using the stochastic representation formula,

$$v(t,x) := \sup\{v(t,x;\tau) : \tau \in \mathcal{T}_{T-t}\}, \tag{1.17}$$

where we define

$$\begin{aligned} v(t,x;\tau) := {} & \mathbb{E}\left[e^{-\int_0^\tau c(t+s,X^x(s))\,ds}\varphi(t+\tau, X^x(\tau))\mathbf{1}_{\{\tau<T-t\}}\right] \\ & + \mathbb{E}\left[e^{-\int_0^\tau c(t+s,X^x(s))\,ds}g(X^x(T-t))\mathbf{1}_{\{\tau=T-t\}}\right] \\ & + \mathbb{E}\left[\int_0^\tau e^{-\int_0^s c(t+r,X^x(r))\,dr} f(t+s, X^x(s))\,ds\right], \end{aligned} \tag{1.18}$$

for all $(t,x) \in [0,T]\times \in \mathbb{R}^n$.

For all $T > 0$, we denote by $C_t^{\frac{1}{2}}C_x^{0,1}([0,T]\times\mathbb{R}^n)$ the space of functions $u : [0,T]\times\mathbb{R}^n \to \mathbb{R}$ such that

$$\|u\|_{C_t^{\frac{1}{2}}C_x^{0,1}([0,T]\times\mathbb{R}^n)} := \|u\|_{C([0,T]\times\mathbb{R}^n)} + \sup_{\substack{t_1,t_2\in[0,T],t_1\neq t_2\\ x_1,x_2\in\mathbb{R}^n, x_1\neq x_2}} \frac{|u(t_1,x_1)-u(t_2,x_2)|}{|t_1-t_2|^{\frac{1}{2}} + |x_1-x_2|} < \infty,$$

and we let $C_t^1C_x^2([0,T]\times\mathbb{R}^n)$ denote the space of functions $u : [0,T]\times\mathbb{R}^n \to \mathbb{R}$ such that the first order derivative in the time variable and the second order derivatives in the spatial variables are continuous and bounded.

The value function $v(t,x)$ satisfies:

PROPOSITION 1.7 (Regularity of the value function). *In addition to Assumption* 1.1 *suppose that* c, φ, f *belong to* $C^{0,1}([0,T]\times\mathbb{R}^n)$, *the final condition* g *is in* $C^{0,1}(\mathbb{R}^n)$, *and the compatibility condition* (1.16) *holds. Then the value function* v *defined in* (1.17) *belongs to* $C_t^{\frac{1}{2}}C_x^{0,1}([0,T]\times\mathbb{R}^n)$.

We next define a notion of viscosity solution for the evolution obstacle problem (1.15) extending that of its stationary analogue for equation (1.5) similarly to the ideas described in [**7**, § 8]:

DEFINITION 1.8 (Viscosity solutions). Let $v \in C(\mathbb{R}^n)$. We say that v is a viscosity subsolution (supersolution) to the evolution obstacle problem (1.15) if

$$v(T,\cdot) \leq (\geq) g, \tag{1.19}$$

and, for all $u \in C_t^1C_x^2([0,T]\times\mathbb{R}^n)$ such that $v - u$ has a global max (min) at $(t_0,x_0) \in [0,T)\times\mathbb{R}^n$ and $u(t_0,x_0) = v(t_0,x_0)$, we have that

$$\begin{aligned} \min\{-u_t(t_0,x_0) - Lu(t_0,x_0) + c(t_0,x_0)u(t_0,x_0) - f(t_0,x_0), \\ u(t_0,x_0) - \varphi(t_0,x_0)\} \leq (\geq)\, 0. \end{aligned} \tag{1.20}$$

We say that v is a viscosity solution to equation (1.15) if it is both a sub- and supersolution.

THEOREM 1.9 (Existence of viscosity solution). *Suppose that the hypotheses of Proposition* 1.7 *hold. Then the value function* v *defined in* (1.17) *is a viscosity solution to the evolution obstacle problem* (1.15).

REMARK 1.10 (Assumptions in evolution vs stationary cases). We note that in the case of the evolution obstacle problem we allow the jump size $F(x, y)$ to depend on the spatial variable x, in contrast to assumption (1.6) in the case of the stationary obstacle problem.

We also note that we do not require condition (1.12) to hold in the statement of Theorem 1.9. We are able to remove this condition because in the evolution case Proposition 1.7 shows that the value function is Lipschitz continuous in the spatial variable, as opposed to the stationary case where we prove in Proposition 1.2 that the value function is α-Hölder continuous.

THEOREM 1.11 (Uniqueness of viscosity solution). *Suppose that Assumption* 1.1 *is satisfied,* g *belongs to* $C(\mathbb{R}^n)$*,* c, f, φ *are in* $C([0,T] \times \mathbb{R}^n)$*, the compatibility condition* (1.16) *holds, and* F *satisfies* (1.13). *If the obstacle problem* (1.15) *has a solution, then it is unique.*

1.3. Applications to mathematical finance. In mathematical finance stochastic representations of the form (1.7) and (1.17) have the meaning of the prices of American perpetual and finite expiry options, respectively. To make this correspondence, in the evolution obstacle problem (1.15), we set $g \equiv \varphi$, $f \equiv 0$, and we choose the obstacle function φ to depend only on the spatial variable x, and to coincide with the payoff of the American option. In addition, we assume that $n = 1$, the zeroth order term $c \equiv r > 0$, where r is the risk-free interest rate, and the asset price process can be written in the form $S(t) = e^{X(t)}$, where $\{X(t)\}_{t\geq 0}$ solves the stochastic equation (1.1). Most importantly, we need to ensure that the discounted asset price process $\{e^{-rt}S(t)\}_{t\geq 0}$ is a martingale in order to obtain an arbitrage-free market. Because the markets containing asset prices driven by discontinuous Lévy processes that are not Poisson processes are incomplete, the motivation to choose to price options using the risk-free probability measure given by the distribution of the asset price requires careful thought. However, we do not address this problem in our paper, but see [**2**, § 1.3.4] and [**6**, Chapter 9] for more discussions on this problem.

Assume that $\{X(t)\}_{t\geq 0}$ is a one-dimensional Lévy process that satisfies the stochastic equation:

$$dX(t) = b\,dt + \int_{\mathbb{R}^n} y\,\widetilde{N}(dt, dy), \quad \forall\, t > 0, \tag{1.21}$$

where b is a real constant and $\widetilde{N}(dt, dy)$ is a compensated Poisson random measure with Lévy measure $\nu(dy)$. Using [**1**, Theorem 5.2.4 and Corollary 5.2.2] a sufficient condition that guarantees that the discounted asset price process $\{e^{-rt+X(t)}\}_{t\geq 0}$ is a martingale is:

$$\begin{aligned} &\int_{|x|\geq 1} e^x\,\nu(dx) < \infty, \\ &-r + \psi(-i) = 0, \end{aligned} \tag{1.22}$$

where $\psi(\xi)$ denotes the characteristic exponent of the Lévy process $\{X(t)\}_{t\geq 0}$, that is,

$$\psi(\xi) = ib\cdot\xi + \int_{\mathbb{R}\setminus\{0\}} (e^{ix\xi} - 1 - ix\xi)\,\nu(dx). \tag{1.23}$$

Examples in mathematical finance to which our results apply include the Variance Gamma Process [**9**] and Regular Lévy Processes of Exponential type (RLPE) [**2**].

When the jump-part of the nonlocal operator L corresponding to the integral term in the characteristic exponent (1.23) has sublinear growth as $|\xi| \to \infty$, we say that the drift term $b\cdot\nabla$ corresponding to $ib\cdot\xi$ in the characteristic exponent (1.23) is supercritical. An example of a nonlocal operator with supercritical drift is the Variance Gamma Process described below in §1.3.1.

1.3.1. *Variance Gamma Process.* Following [**5**, Identity (6)], the Variance Gamma Process $\{X(t)\}_{t\geq 0}$ with parameters ν, σ, and θ has Lévy measure given by

$$\nu(dx) = \frac{1}{\nu|x|}\left(e^{-\frac{|x|}{\eta_p}}\mathbf{1}_{\{x>0\}} + e^{-\frac{|x|}{\eta_n}}\mathbf{1}_{\{x<0\}}\right)dx,$$

where $\eta_p > \eta_n$ are the roots of the equation $x^2 - \theta\nu x - \sigma^2\nu/2 = 0$, and ν, σ, θ are positive constants. From [**5**, Identity (4)], we have that the characteristic exponent of the Variance Gamma Process with constant drift $b \in \mathbb{R}$, $\{X(t) + bt\}_{t\geq 0}$, has the expression:

$$\psi_{\mathrm{VG}}(\xi) = \frac{1}{\nu}\ln\left(1 - i\theta\nu\xi + \frac{1}{2}\sigma^2\nu\xi^2\right) + ib\xi, \quad \forall\,\xi \in \mathbb{C},$$

and so the infinitesimal generator of $\{X(t) + bt\}_{t\geq 0}$ is given by

$$L = \frac{1}{\nu}\ln(1 - \theta\nu\nabla - \frac{1}{2}\sigma^2\nu\Delta) + b\cdot\nabla,$$

which is a sum of a pseudo-differential operator of order less than any $s > 0$ and one of order 1. When $\eta_p < 1$ and $r = \psi_{VG}(-i)^\dagger$, condition (1.22) is satisfied and the discounted asset price process $\{e^{-rt+X(t)}\}_{t\geq 0}$ is a martingale. Thus, applying the results in § 1.1 and § 1.2 to the Variance Gamma Process $\{X(t)\}_{t\geq 0}$ with constant drift b, we obtain that the prices of perpetual and finite expiry American options with bounded and Lipschitz payoffs are Lipschitz functions in the spatial variable. Given that the nonlocal component of the infinitesimal generator L has order less than any $s > 0$, this may be the optimal regularity of solutions that we can expect.

1.3.2. *Regular Lévy Processes of Exponential type.* Following [**2**, Chapter 3], for parameters $\lambda_- < 0 < \lambda_+$, a Lévy process is said to be of exponential type $[\lambda_-, \lambda_+]$ if it has a Lévy measure $\nu(dx)$ such that

$$\int_{-\infty}^{-1} e^{-\lambda_+ x}\nu(dx) + \int_1^\infty e^{-\lambda_- x}\nu(dx) < \infty.$$

Regular Lévy Processes of Exponential type $[\lambda_-, \lambda_+]$ and order ν are non-Gaussian Lévy processes of exponential type $[\lambda_-, \lambda_+]$ such that, in a neighborhood of zero, the Lévy measure can be represented as $\nu(dx) = f(x)\,dx$, where $f(x)$ satisfies the property that

$$|f(x) - c|x|^{-\nu-1}| \leq C|x|^{-\nu'-1}, \quad \forall\,|x| \leq 1,$$

---

†The fact that $r > 0$ and $r = \psi_{VG}(-i)$ implies that $1 - \theta\nu - \frac{1}{2}\sigma^2\nu$ is a positive constant, and the drift b satisfies the inequality $b > -\frac{1}{\nu}\ln\left(1 - \theta\nu - \frac{1}{2}\sigma^2\nu\right)$.

for constants $\nu' < \nu$, $c > 0$, and $C > 0$. Our results apply to RLPE type $[\lambda_-, \lambda_+]$, when we choose the parameters $\lambda_- \leq -1$ and $\lambda_+ \geq 1$[‡].

The class of RLPE include the CGMY/KoBoL processes introduced in [**5**]. Following [**5**, Equation (7)], CGMY/KoBoL processes are characterized by a Lévy measure of the form

$$\nu(dx) = \frac{C}{|x|^{1+Y}} \left(e^{-G|x|} \mathbf{1}_{\{x<0\}} + e^{-M|x|} \mathbf{1}_{\{x>0\}} \right) dx,$$

where the parameters $C > 0$, $G, M \geq 0$, and $Y < 2$. Our results apply to CGMY/KoBoL processes, when we choose the parameter $M > 1$ and $Y < 2$, or $M = 1$ and $0 < Y < 2$[‡].

We remark that a sufficient condition on the Lévy measure to ensure that perpetual American put option prices are Lipschitz continuous, but not continuously differentiable, is provided in [**2**, Theorem 5.4, p. 133]. However, the condition is in terms of the Wiener-Hopf factorization for the characteristic exponent of the Lévy process, and it is difficult to find a concrete example for which it holds.

1.4. Comparison with previous research. In [**10**], the author establishes closed-form formulas for prices of perpetual American call and put options on a stock driven by a general Lévy process, in terms of the distribution of the supremum and the infimum of the process, respectively. In [**2**, **3**], in the framework of Regular Lévy Processes of Exponential type, the authors obtain closed-form formulas for prices of perpetual American call and put options via the Wiener-Hopf factorization method distribution of the supremum and the infimum of the process. Compared with [**2**, **3**, **10**], in our work we allow more general payoff functions for which we study both the perpetual and the finite expiry American options, together with the regularity properties of the option prices. Our results apply to multi-dimensional Markov processes that may not be Lévy processes, and when restricted to the class of Lévy processes, we cover a smaller family than in [**10**], but a more general family of processes than the one analyzed in [**2**, **3**].

The nonlocal operators most often studied in the context of obstacle problems are stable-like [**4**]. However, the nonlocal operators often arising in applications in mathematical finance are not of this form, and in our work we include operators relevant in this field as we described in § 1.3. Their analytic properties appear to be quite different, as the case of Variance Gamma Processes in § 1.3.1 shows, and we prove regularity properties of solutions using probabilistic and viscosity solutions arguments. The Lipschitz regularity of solutions that we establish in Theorem 1.4 is optimal for a subclass of nonlocal operators, as [**2**, Theorem 5.4, p. 133] proves.

1.5. Structure of the paper. We prove the main results stated in the introduction in § 2 and § 3, respectively. In addition to these results, we first prove in the stationary case a Dynamic Programming Principle and a Comparison Principle in Lemma 2.1 and Theorem 2.2, respectively. The Dynamic Programming Principle is used in the proof of Theorem 1.4 where we establish the existence of viscosity solutions, while the Comparison Principle is used in the proof of Theorem 1.5 to establish the uniqueness of solutions to the stationary obstacle problem (1.5). Analogous results are obtained for the evolution case in Lemma 3.2 and Theorem 3.3, respectively.

[‡]See the first identity in condition (1.22).

2. Stationary obstacle problem

In this section, we give the proofs of Proposition 1.2, and Theorems 1.4 and 1.5. In addition, we prove a Dynamical Programming Principle in Lemma 2.1 and a comparison principle in Theorem 2.2. We begin with:

PROOF OF PROPOSITION 1.2. We denote by $\{X^x(t)\}_{t\geq 0}$ the unique strong solution to the stochastic equation (1.1) with initial condition $X(0) = x \in \mathbb{R}^n$. Because the functions φ and f are bounded and the zeroth order term c satisfies property (1.9), it is clear that the value function v defined in (1.7) is bounded. To prove the Hölder continuity of v, we use the fact that

$$|v(x_1) - v(x_2)| \leq \sup_{\tau\in\mathcal{T}} |v(x_1;\tau) - v(x_2;\tau)|, \quad \forall x_1, x_2 \in \mathbb{R}^n, \tag{2.1}$$

and we can assume without loss of generality that $|x_1 - x_2| < 1$ since v is bounded. Let $T > 0$ be a constant. Using definition (1.8) of the function $v(x;\tau)$ and condition (1.9), we see that

$$\begin{aligned}
&|v(x_1;\tau) - v(x_2;\tau)| \\
&\quad \leq \mathbb{E}\left[e^{-\int_0^\tau c(X^{x_1}(s))\,ds} |\varphi(X^{x_1}(\tau)) - \varphi(X^{x_2}(\tau))| \mathbf{1}_{\{\tau\leq T\}}\right] \\
&\quad + \mathbb{E}\left[\left|e^{-\int_0^\tau c(X^{x_1}(s))\,ds} - e^{-\int_0^\tau c(X^{x_2}(s))\,ds}\right| |\varphi(X^{x_2}(\tau))| \mathbf{1}_{\{\tau\leq T\}}\right] \\
&\quad + \mathbb{E}\left[\int_0^{\tau\wedge T} e^{-\int_0^t c(X^{x_1}(s))\,ds} |f(X^{x_1}(t)) - f(X^{x_2}(t)|\,dt\right] \\
&\quad + \mathbb{E}\left[\int_0^{\tau\wedge T} \left|e^{-\int_0^t c(X^{x_1}(s))\,ds} - e^{-\int_0^t c(X^{x_2}(s))\,ds}\right| |f(X^{x_2}(t))|\,dt\right] \\
&\quad + 2\left(\|\varphi\|_{C(\mathbb{R}^n)} + \|f\|_{C(\mathbb{R}^n)}\right) e^{-c_0 T}.
\end{aligned}$$

Property (1.9) and the fact that the functions c, f, and φ are Lipschitz continuous give us that there is a positive constant, $C{=}C(c_0, [c]_{C^{0,1}(\mathbb{R}^n)}, \|\varphi\|_{C^{0,1}(\mathbb{R}^n)}, \|f\|_{C^{0,1}(\mathbb{R}^n)})$, such that

$$\begin{aligned}
&|v(x_1;\tau) - v(x_2;\tau)| \\
&\quad \leq C\mathbb{E}\left[e^{-c_0(\tau\wedge T)}|X^{x_1}(\tau\wedge T) - X^{x_2}(\tau\wedge T)|\right] \\
&\quad + C\mathbb{E}\left[\int_0^{\tau\wedge T} e^{-c_0 t}|X^{x_1}(t) - X^{x_2}(t)|\,dt\right] \\
&\quad + C\mathbb{E}\left[\int_0^{\tau\wedge T} e^{-c_0 t}\int_0^t |X^{x_1}(s) - X^{x_2}(s)|\,ds\,dt\right] + Ce^{-c_0 T}.
\end{aligned} \tag{2.2}$$

Using assumption (1.6) in the stochastic equation (1.1), it follows that

$$X^{x_i}(t) = x_i + \int_0^t b(X^{x_i}(s-))\,ds + \int_0^t \int_{\mathbb{R}^n\setminus\{O\}} F(y)\,\widetilde{N}(ds,dy), \quad \text{for} i = 1, 2, \quad \forall t > 0,$$

which gives us that

$$X^{x_1}(t) - X^{x_2}(t) = x_1 - x_2 + \int_0^t (b(X^{x_1}(s-)) - b(X^{x_2}(s-)))\,ds, \quad \forall t > 0,$$

and, using the fact that the drift coefficients $b(x)$ are Lipschitz continuous functions, we obtain

$$|X^{x_1}(t) - X^{x_2}(t)| \leq |x_1 - x_2| + [b]_{C^{0,1}(\mathbb{R}^n)} \int_0^t |X^{x_1}(s-) - X^{x_2}(s-)|\, ds, \quad \forall t > 0.$$

Gronwall's inequality now gives us that

$$|X^{x_1}(t) - X^{x_2}(t)| \leq |x_1 - x_2| e^{\beta t}, \quad \forall t > 0,$$

where we denote for brevity $\beta := [b]_{C^{0,1}(\mathbb{R}^n)}$. The preceding inequality together with (2.2) imply

$$|v(x_1;\tau) - v(x_2;\tau)| \leq C\left(|x_1 - x_2|(e^{(\beta - c_0)T} + 1) + e^{-c_0 T}\right), \quad \forall \tau \in \mathcal{T}. \tag{2.3}$$

Letting $\gamma := c_0/\beta$ and choosing $T > 0$ large enough such that

$$e^{-c_0 T} = |x_1 - x_2|^{\gamma}, \tag{2.4}$$

we have that

$$|x_1 - x_2| e^{(\beta - c_0)T} = |x_1 - x_2|^{1+\gamma-\gamma\beta/c_0} = |x_1 - x_2|^{\gamma}. \tag{2.5}$$

Letting now $\alpha := 1 \wedge \gamma$, it follows from estimates (2.3), (2.4), and (2.5) that

$$|v(x_1;\tau) - v(x_2;\tau)| \leq C|x_1 - x_2|^{\alpha}, \quad \forall x_1, x_2 \in \mathbb{R}^n, \quad \forall \tau \in \mathcal{T}. \tag{2.6}$$

Thus, using identity (2.1) we obtain that the value function v belongs to $C^{\alpha}(\mathbb{R}^n)$. When inequality (1.10) holds, the fact that the value function v belongs to $C^{0,1}(\mathbb{R}^n)$ is an immediate consequence of the fact that $\gamma \geq 1$, and so $\alpha = 1$. This completes the proof. □

For all $r > 0$ and $x \in \mathbb{R}^n$, we let

$$\tau_r := \inf\{t \geq 0 : X^x(t) \notin B_r(x)\}, \tag{2.7}$$

where $\{X^x(t)\}_{t\geq 0}$ is the unique solution to equation (1.1) with initial condition $X^x(0) = x$. We prove Theorem 1.4 with the aid of the following Dynamic Programming Principle.

LEMMA 2.1 (Dynamic Programming Principle). *Suppose that the hypotheses of Proposition 1.2 hold. Then the value function $v(x)$ defined in (1.7) satisfies:*

$$v(x) = \sup\{v(x;r,\tau) : \tau \leq \tau_r\}, \quad \forall r > 0, \tag{2.8}$$

where we define

$$\begin{aligned} v(x;r,\tau) := {}& \mathbb{E}\left[e^{-\int_0^\tau c(X^x(s))\,ds}\left(\varphi(X^x(\tau))\mathbf{1}_{\{\tau<\tau_r\}} + v(X^x(\tau))\mathbf{1}_{\{\tau=\tau_r\}}\right)\right] \\ & + \mathbb{E}\left[\int_0^{\tau\wedge\tau_r} e^{-\int_0^t c(X^x(s))\,ds} f(X^x(t))\,dt\right]. \end{aligned} \tag{2.9}$$

PROOF. We denote by $w(x)$ the right hand-side of identity (2.8), and we divide the proof into two steps.

STEP 1 (Proof of inequality $v(x) \leq w(x)$). Let $\tau \in \mathcal{T}$. From definition (1.8) of the function $v(x;\tau)$, conditioning on the σ-algebra $\mathcal{F}_{\tau_r}$, we have that
(2.10)

$$\begin{aligned} v(x;\tau) &= \mathbb{E}\left[e^{-\int_0^\tau c(X^x(s))\,ds}\varphi(X^x(\tau))\mathbf{1}_{\{\tau<\tau_r\}} + \int_0^{\tau\wedge\tau_r} e^{-\int_0^t c(X^x(s))\,ds} f(X^x(t))\,dt\right] \\ &\quad + \mathbb{E}\left[\mathbf{1}_{\{\tau\geq\tau_r\}} e^{-\int_0^{\tau_r} c(X^x(s))\,ds}\right. \\ &\quad \left.\times\, \mathbb{E}\left[e^{-\int_{\tau_r}^{\tau} c(X^x(s))\,ds}\varphi(X^x(\tau)) + \int_{\tau_r}^{\tau} e^{-\int_{\tau_r}^t c(X^x(s))\,ds} f(X^x(t))\,dt \Big| \mathcal{F}_{\tau_r}\right]\right]. \end{aligned}$$

Let $\theta := (\tau - \tau_r) \vee 0$. We next prove that the last term in the preceding identity can be written in the form
(2.11)

$$\begin{aligned} v(X^x(\tau_r);\theta) &= \mathbb{E}\left[e^{-\int_{\tau_r}^{\tau} c(X^x(s))\,ds}\varphi(X^x(\tau)) + \int_{\tau_r}^{\tau} e^{-\int_{\tau_r}^t c(X^x(s))\,ds} f(X^x(t))\,dt \Big| \mathcal{F}_{\tau_r}\right], \\ &\qquad \mathbb{P}\text{-a.s. on } \{\tau \geq \tau_r\}. \end{aligned}$$

For this purpose we need to show that there is a regular conditional probability distribution of $\mathbb{P}$ given $\mathcal{F}_{\tau_r}$, and that θ is a $\{\mathcal{F}_{\tau_r+t}\}_{t\geq 0}$-stopping time.

To construct regular conditional probability distributions, we can assume without loss of generality that the sample space Ω of the Poisson random measure $N(dt,dy)$ with Lévy measure $\nu(dy)$ is the space of functions $\omega : [0,\infty) \to \mathbb{R}^n$ that are right-continuous and have left-limits (RCLL). We endow the space Ω with the Borel σ-algebra generated by the Skorohod topology, [**8**, Chapter 3.5], which we denote by $\mathcal{G}$. For all $t \geq 0$, we let $\mathcal{G}_t$ denote the σ-algebra generated by the Skorohod topology on $\Omega_t := \{\omega : \omega : [0,t] \to \mathbb{R}^n \text{ RCLL}\}$. Letting $\mathcal{G}'(\mathcal{G}'_t)$ be the σ-algebra generated by $\mathcal{G}$ ($\mathcal{G}_t$) and the $\mathbb{P}$-null sets of $\mathcal{G}$, and setting $\mathcal{F} := \mathcal{G}'$ and $\mathcal{F}_t = \cap_{s>t}\mathcal{G}'_s$, we obtain that the filtered probability space $(\Omega, \mathcal{F}, \{\mathcal{F}_t\}_{t\geq 0}, \mathbb{P})$ satisfies the usual hypotheses in [**11**, § I.1]. Moreover, applying [**8**, Theorem 3.5.6] we obtain that Ω endowed with the Skorohod topology is a complete separable metric space, and so [**12**, Theorems 1.1.6 and 1.1.8] give us that there is a regular conditional probability distribution of $\mathbb{P}$ given $\mathcal{F}_{\tau_r}$, which we denote by $\mathbb{P}_{\tau_r}$. To conclude the proof of Step 1, we next prove

CLAIM 1. The random time $\theta := (\tau - \tau_r) \vee 0$ is a $\{\mathcal{F}_{\tau_r+t}\}_{t\geq 0}$-stopping time.

PROOF OF CLAIM 1. We will show that for all $t \geq 0$ we have that $\{\theta \leq t\} \in \mathcal{F}_{\tau_r+t}$, which is equivalent to proving that for all $t < s$ we have that the event $S := \{\tau \leq \tau_r + t\} \cap \{\tau_r + t \leq s\}$ belongs to $\mathcal{F}_s$. Because the filtration $\{\mathcal{F}_t\}_{t\geq 0}$ is right-continuous, the preceding inclusion can be replaced with the condition

$$S := \{\tau < \tau_r + t\} \cap \{\tau_r + t < s\} \in \mathcal{F}_s, \quad \forall 0 \leq t < s. \tag{2.12}$$

The event S can be written in the form

$$S = \cup_{n\in\mathbb{N}} \cup_{q\in[0,s-t)\cap\mathbb{Q}} \left(\{q - 1/n < \tau_r < q\} \cap \{\tau < q - 1/n + t\}\right),$$

which clearly implies that they belong to the σ-algebra $\mathcal{F}_s$. Thus property (2.12) is satisfied, which concludes the proof of Claim 1. □

Applying Claim 1 and the strong Markov property of the process $\{X^x(t)\}_{t\geq 0}$, we obtain that

$$\mathbb{E}\left[e^{-\int_{\tau_r}^{\tau} c(X^x(s))\,ds}\varphi(X^x(\tau)) + \int_{\tau_r}^{\tau} e^{-\int_{\tau_r}^{t} c(X^x(s))\,ds} f(X^x(t))\,dt \Big| \mathcal{F}_{\tau_r}\right]$$
$$= \mathbb{E}_{\mathbb{P}_{\tau_r}}\left[e^{-\int_0^{\theta} c(X^x(s))\,ds}\varphi(X^x(\theta)) + \int_0^{\theta} e^{-\int_0^{t} c(X^x(s))\,ds} f(X^x(t))\,dt\right],$$

$\mathbb{P}$-a.s. on $\{\tau \geq \tau_r\}$,

which together with definition (1.8) give us that identity (2.11) holds. Identities (2.10) and (2.11) imply that

$$\begin{aligned} v(x;\tau) &\leq \mathbb{E}\left[e^{-\int_0^{\tau} c(X^x(s))\,ds}\varphi(X^x(\tau))\mathbf{1}_{\{\tau<\tau_r\}} + \int_0^{\tau\wedge\tau_r} e^{-\int_0^{t} c(X^x(s))\,ds} f(X^x(t))\,dt\right] \\ &\quad + \mathbb{E}\left[\mathbf{1}_{\{\tau\geq\tau_r\}} e^{-\int_0^{\tau_r} c(X^x(s))\,ds} v(X^x(\tau_r))\right] \\ &\qquad \text{(because } v(X^x(\tau_r;\theta)) \leq v(X^x(\tau_r)) \text{ by (1.7))} \\ &= v(x; r, \tau\wedge\tau_r) \quad \text{(by (2.9))}. \end{aligned}$$

Inequality $v(x) \leq w(x)$ now follows from the preceding expression by taking the supremum over all stopping times $\tau \in \mathcal{T}$.

STEP 2 (Proof of inequality $w(x) \leq v(x)$). We fix $\varepsilon \in (0,1)$, and we choose a stopping time $\tau = \tau(\varepsilon) \leq \tau_r$ with the property that

$$w(x) < v(x; r, \tau) + \varepsilon. \tag{2.13}$$

We choose a countable family of Borel measurable sets, $\{A_k\}_{k\in\mathbb{N}}$, that partitions $\mathbb{R}^n$, and a sequence of points, $\{x_k\}_{k\in\mathbb{N}}$, such that $x_k \in A_k \subseteq B_\varepsilon(x_k)$. For all $k \in \mathbb{N}$, we choose a stopping time, $\theta_k \in \mathcal{T}$, such that

$$v(x_k) < v(x_k; \theta_k) + \varepsilon. \tag{2.14}$$

Let $\alpha \in (0,1)$ be the Hölder exponent appearing in the statement of Proposition 1.2. For all $\omega \in \{\tau = \tau_r \text{ and } X^x(\tau_r) \in A_k\}$, we have that

$$\begin{aligned} v(X^x(\tau_r)) &\leq v(x_k) + C\varepsilon^\alpha \quad \text{(by (2.6) and the fact that } A_k \subseteq B_\varepsilon(x_k)) \\ &\leq v(x_k;\theta_k) + \varepsilon + C\varepsilon^\alpha \quad \text{(by our choice of } \theta_k \text{ in (2.14))} \\ &\leq v(X^x(\tau_r);\theta_k) + (v(x_k;\theta_k) - v(X^x(\tau_r);\theta_k)) + \varepsilon + C\varepsilon^\alpha \\ &\leq v(X^x(\tau_r);\theta_k) + \varepsilon + 2C\varepsilon^\alpha \quad \text{(by (2.6) and the fact that } A_k \subseteq B_\varepsilon(x_k)). \end{aligned}$$

The preceding construction together with definition (2.9) of $v(x; r, \tau)$ and inequality (2.13) yield

$$\begin{aligned} w(x) &< \mathbb{E}\left[e^{-\int_0^{\tau} c(X^x(s))\,ds}\varphi(X^x(\tau))\mathbf{1}_{\{\tau<\tau_r\}} + \int_0^{\tau\wedge\tau_r} e^{-\int_0^{t} c(X^x(s))\,ds} f(X^x(t))\,dt\right] \\ &\quad + \mathbb{E}\left[e^{-\int_0^{\tau} c(X^x(s))\,ds} \sum_{k\in\mathbb{N}} v(X^x(\tau_r);\theta_k)\mathbf{1}_{\{\tau=\tau_r\}}\mathbf{1}_{\{X^x(\tau_r)\in A_k\}}\right] + C\varepsilon^\alpha. \end{aligned} \tag{2.15}$$

We notice that, by letting $S_{\tau_r}(\omega)(t) := \omega(\tau_r + t)$ denote the shift operator, we have that

CLAIM 2. The random time defined by

$$\bar{\tau} := \tau_r + \sum_{k\in\mathbb{N}} \theta_k(S_{\tau_r})\mathbf{1}_{\{\tau=\tau_r\}}\mathbf{1}_{\{X^x(\tau_r)\in A_k\}} \tag{2.16}$$

is a $\{\mathcal{F}_t\}_{t\geq 0}$-stopping time.

PROOF OF CLAIM 2. Because τ_r is a $\{\mathcal{F}_t\}_{t\geq 0}$-stopping time, the conclusion of Claim 2 follows once we prove that the random time $\bar{\theta} := \bar{\tau} - \tau_r$ is a $\{\mathcal{F}_{\tau_r+t}\}_{t\geq 0}$-stopping time. For this purpose, it is sufficient to prove that the set $E := \{\bar{\theta} \leq t\}$ is $\mathcal{F}_{\tau_r+t}$-measurable, for all $t \geq 0$. From the definitions of $\bar{\theta}$ and $\bar{\tau}$, using the fact that the family of sets $\{A_k\}_{k\in\mathbb{N}}$ partitions $\mathbb{R}^n$, we can write E as a disjoint union of sets $\{E_k\}_{k\in\mathbb{N}}$, where

$$E_k := \{X^x(\tau_r) \in A_k\} \cap \{\tau = \tau_r\} \cap \{\theta_k(S_{\tau_r}) \leq t\}, \quad \forall\, k \in \mathbb{N}.$$

The proof is concluded once we prove that the set E_k is $\mathcal{F}_{\tau_r+t}$-measurable, for all $k \in \mathbb{N}$. Using the fact that the shift operator $S_{\tau_r} : (\Omega, \mathcal{F}_{\tau_r+t}) \to (\Omega, \mathcal{F})$ is a measurable function, and θ_k is a $\{\mathcal{F}_t\}_{t\geq 0}$-stopping time, we obtain that the event $\{\theta_k(S_{\tau_r}) \leq t\}$ is $\mathcal{F}_{\tau_r+t}$-measurable. Combining this property with the fact that $\{X^x(S_{\tau_r}) \in A_k\}$ and $\{\tau = \tau_r\}$ are $\mathcal{F}_{\tau_r}$-measurable sets, we deduce that E_k is $\mathcal{F}_{\tau_r+t}$-measurable. Hence, the set E is $\mathcal{F}_{\tau_r+t}$-measurable, for all $t \geq 0$, which completes the proof of Claim 2. □

Using definition (1.8), we have that

$$\begin{aligned}
&\sum_{k\in\mathbb{N}} v(X^x(\tau_r);\theta_k)\mathbf{1}_{\{\tau=\tau_r\}}\mathbf{1}_{\{X^x(\tau_r)\in A_k\}}\\
&\quad = \mathbf{1}_{\{\tau=\tau_r\}}\mathbb{E}_{\mathbb{P}_{\tau_r}}\left[e^{-\int_{\tau_r}^{\bar{\tau}} c(X^x(s))\,ds}\varphi(X^x(\bar{\tau})) + \int_{\tau_r}^{\bar{\tau}} e^{-\int_0^t c(X^x(s))\,ds} f(X^x(t))\,dt\right].
\end{aligned}$$

Recalling that $\mathbb{P}_{\tau_r}$ is a regular condition probability distribution given $\mathcal{F}_{\tau_r}$, it follows from the preceding identity and (2.15) that

$$\begin{aligned}
w(x) < \mathbb{E}&\left[e^{-\int_0^\tau c(X^x(s))\,ds}\varphi(X^x(\tau))\mathbf{1}_{\{\tau<\tau_r\}} + \int_0^{\tau\wedge\tau_r} e^{-\int_0^t c(X^x(s))\,ds} f(X^x(t))\,dt\right]\\
&+ \mathbb{E}\left[e^{-\int_0^\tau c(X^x(s-))\,ds}\mathbf{1}_{\{\tau=\tau_r\}}\right.\\
&\qquad \left.\times\, \mathbb{E}\left[e^{-\int_{\tau_r}^{\bar{\tau}} c(X^x(s))\,ds}\varphi(X^x(\bar{\tau})) + \int_{\tau_r}^{\bar{\tau}} e^{-\int_{\tau_r}^t c(X^x(s))\,ds} f(X^x(t))\,dt\Big|\mathcal{F}_{\tau_r}\right]\right]\\
&+ C\varepsilon^\alpha.
\end{aligned}$$

Claim 2 yields that $\bar{\tau}$ belongs to $\mathcal{T}$, and using definition (1.8) we can write the sum of the first two terms on the right-hand side of the preceding inequality as $v(x;\bar{\tau})$. Hence, we deduce that $w(x) < v(x;\bar{\tau}) + c\varepsilon^\alpha$, from which it follows that $w(x) < v(x) + C\varepsilon^\alpha$ by identity (1.7). We obtain the desired inequality $w(x) \leq v(x)$ by letting ε tend to zero.

Steps 1 and 2 combined complete the proof. □

We next apply Lemma 2.1 to give the

PROOF OF THEOREM 1.4. We divide the proof into two steps.

STEP 1 (Verification that v is a subsolution). To prove that v is a subsolution to equation (1.5), let $u \in C^2(\mathbb{R}^n)$ be such that $u(x_0) = v(x_0)$ and $u(x) \geq v(x)$, for all $x \in \mathbb{R}^n$. We need to show that

$$-Lu(x_0) + c(x_0)u(x_0) \leq f(x_0) \quad \text{or} \quad u(x_0) \leq \varphi(x_0).$$

We assume without loss of generality that $u(x_0) > \varphi(x_0)$, otherwise the conclusion that v is a subsolution is obtained. For all stopping times $\tau \in \mathcal{T}$, Itô's rule applied to u and the unique strong solution, $\{X^{x_0}(t)\}_{t\geq 0}$, to equation (1.1) with initial condition $X^{x_0}(0) = x_0$, gives us

$$\begin{aligned} & e^{-\int_0^\tau c(X^{x_0}(s))\,ds} u(X^{x_0}(\tau)) \\ &\quad = u(x_0) + \int_0^\tau e^{-\int_0^t c(X^{x_0}(s))\,ds} (L - c(X^{x_0}(s)))u(X^{x_0}(t))\,dt + M(\tau), \end{aligned}$$

where we define the process

$$M(t) := \int_0^t \int_{\mathbb{R}^n\setminus\{O\}} e^{-\int_0^s c(X^{x_0}(r))\,dr} \left(u(X^{x_0}(s-) + F(y)) - u(X^{x_0}(s-))\right) \widetilde{N}(ds, dy),$$

for all $t \geq 0$. It follows by Proposition 1.2, condition (1.12), and [**1**, Theorem 4.2.3] that the process $\{M(t)\}_{t\geq 0}$ is a martingale, which gives us that

$$\begin{aligned} u(x_0) &= \mathbb{E}\left[e^{-\int_0^\tau c(X^{x_0}(s))\,ds} u(X^{x_0}(\tau))\right] \\ &\quad + \mathbb{E}\left[\int_0^\tau e^{-\int_0^t c(X^{x_0}(s))\,ds} (-L + c(X^{x_0}(t)))u(X^{x_0}(t))\,dt\right]. \end{aligned}$$

Assume by contradiction that there are positive constants, ε and r, such that $(-L + c)u(x) \geq f(x) + \varepsilon$, for all $x \in B_r(x_0)$. Using the fact that $v(x_0) = u(x_0)$, and replacing the stopping times τ in the preceding identity for $u(x_0)$ by $\tau \wedge \tau_r$, it follows that

$$\begin{aligned} v(x_0) &\geq \mathbb{E}\left[e^{-\int_0^{\tau\wedge\tau_r} c(X^{x_0}(s))\,ds} u(X^{x_0}(\tau \wedge \tau_r))\right] \\ &\quad + \mathbb{E}\left[\int_0^{\tau\wedge\tau_r} e^{-\int_0^t c(X^{x_0}(s))\,ds} \left(f(X^{x_0}(t)) + \varepsilon\right)\,dt\right]. \end{aligned}$$

Letting $M := \|c\|_{C(\mathbb{R}^n)}$ and using the fact that

$$\begin{aligned} \mathbb{E}\left[\int_0^{\tau\wedge\tau_r} e^{-\int_0^t c(X^{x_0}(s))\,ds}\,dt\right] &\geq \mathbb{E}\left[\int_0^{\tau\wedge\tau_r} e^{-Mt}\,dt\right] \\ &= \mathbb{E}\left[\frac{1}{M}\left(1 - e^{-M\tau\wedge\tau_r}\right)\right] \\ &\geq \mathbb{E}\left[\frac{1}{2M}\mathbf{1}_{\{e^{-M\tau\wedge\tau_r}\leq 1/2\}} + \frac{1}{2}\tau\wedge\tau_r\mathbf{1}_{\{e^{-M\tau\wedge\tau_r}>1/2\}}\right] \\ &\geq \frac{1}{2}\mathbb{E}\left[\tau\wedge\tau_r\wedge\frac{1}{M}\right], \end{aligned}$$

it follows that

(2.17)
$$\begin{aligned} v(x_0) &\geq \mathbb{E}\left[e^{-\int_0^{\tau\wedge\tau_r} c(X^{x_0}(s))\,ds} u(X^{x_0}(\tau\wedge\tau_r)) + \int_0^{\tau\wedge\tau_r} e^{-\int_0^t c(X^{x_0}(s))\,ds} f(X^{x_0}(t))\,dt\right] \\ &\quad + \frac{\varepsilon}{2}\mathbb{E}\left[\tau\wedge\tau_r\wedge\frac{1}{M}\right]. \end{aligned}$$

Recall that $u(x_0) > \varphi(x_0)$, $u(x_0) = v(x_0)$, and $u \geq v$ on $\mathbb{R}^n$, and so we can assume without loss of generality that r is chosen small enough so that $u(x) \geq v(x) > \varphi(x)$, for all $x \in B_r(x_0)$. This implies that

$$\begin{aligned}
&\sup_{\tau\in\mathcal{T}} \mathbb{E}\left[e^{-\int_0^{\tau\wedge\tau_r} c(X^{x_0}(s))\,ds} u(X^{x_0}(\tau\wedge\tau_r)) + \int_0^{\tau\wedge\tau_r} e^{-\int_0^t c(X^{x_0}(s))\,ds} f(X^{x_0}(t))\,dt\right]\\
&\quad\geq \sup_{\tau\in\mathcal{T}} \mathbb{E}\left[e^{-\int_0^{\tau\wedge\tau_r} c(X^{x_0}(s))\,ds}\left(\varphi(X^{x_0}(\tau\wedge\tau_r))\mathbf{1}_{\{\tau<\tau_r\}} + v(X^{x_0}(\tau\wedge\tau_r))\mathbf{1}_{\{\tau\geq\tau_r\}}\right)\right]\\
&\qquad + \mathbb{E}\left[\int_0^{\tau\wedge\tau_r} e^{-\int_0^t c(X^{x_0}(s))\,ds} f(X^{x_0}(t))\,dt\right]\\
&\quad = v(x_0),
\end{aligned}$$

where in the last line we applied Lemma 2.1. Thus, for all $\delta > 0$, we can choose a stopping time $\tau^\delta \in \mathcal{T}$ with the property that

(2.18)

$$\begin{aligned}
&\mathbb{E}\left[e^{-\int_0^{\tau^\delta\wedge\tau_r} c(X^{x_0}(s))\,ds}\left(\varphi(X^{x_0}(\tau^\delta\wedge\tau_r))\mathbf{1}_{\{\tau^\delta<\tau_r\}} + v(X^{x_0}(\tau^\delta\wedge\tau_r))\mathbf{1}_{\{\tau^\delta\geq\tau_r\}}\right)\right]\\
&\quad + \mathbb{E}\left[\int_0^{\tau^\delta\wedge\tau_r} e^{-\int_0^t c(X^{x_0}(s))\,ds} f(X^{x_0}(t))\,dt\right] > v(x_0) - \delta.
\end{aligned}$$

Letting $\tau = \tau^\delta$ in inequality (2.17) and using (2.18), it follows that, for all $\delta > 0$, we have the inequality

$$v(x_0) \geq v(x_0) - \delta + \frac{\varepsilon}{2}\mathbb{E}\left[\tau^\delta \wedge \tau_r \wedge \frac{1}{M}\right]. \tag{2.19}$$

We next claim that

$$\liminf_{\delta\to 0} \mathbb{E}\left[\tau^\delta \wedge \tau_r \wedge \frac{1}{M}\right] > 0. \tag{2.20}$$

When property (2.20) does not hold, there is a sequence $\{\delta_k\}_{k\in\mathbb{N}} \subset (0,\infty)$ decreasing to zero such that $\{\mathbb{E}\left[\tau^{\delta_k} \wedge \tau_r \wedge \frac{1}{M}\right]\}_{k\in\mathbb{N}}$ converges to zero, and so $\{\tau^{\delta_k} \wedge \tau_r\}_{k\in\mathbb{N}}$ also converges to zero $\mathbb{P}$-a.s.. Replacing δ by δ^k in inequality (2.18), letting k tend to ∞, and applying the previous property together with the Dominated Convergence Theorem, we obtain

$$\varphi(x_0)\mathbb{P}_{x_0}(\tau_r > 0) + v(x_0)\mathbb{P}_{x_0}(\tau_r = 0) \geq v(x_0).$$

This implies that $\varphi(x_0)\mathbb{P}_{x_0}(\tau_r > 0) \geq v(x_0)\mathbb{P}_{x_0}(\tau_r > 0)$, and using the fact that the event $\tau_r > 0$ has positive probability, we obtain the inequality $\varphi(x_0) \geq v(x_0)$, which contradicts our assumption that $u(x_0) = v(x_0) > \varphi(x_0)$. Thus, property (2.20) holds. Combining now (2.20) with inequality (2.19), where we let δ tend to zero, we obtain the contradiction $v(x_0) > v(x_0)$. This implies that $-Lu(x_0) + c(x_0)u(x_0) \leq f(x_0)$, which guarantees that

$$\min\{-Lu(x_0) + c(x_0)u(x_0) - f(x_0), u(x_0) - \varphi(x_0)\} \leq 0,$$

and so, indeed v defined in (1.7) is a viscosity subsolution to the obstacle problem (1.5). This concludes the proof of Step 1.

STEP 2 (Verification that v is a supersolution). Let $u \in C^2(\mathbb{R}^n)$ be such that $u(x_0) = v(x_0)$ and $u(x) \leq v(x)$, for all $x \in \mathbb{R}^n$. Our goal is to prove that

$$\min\{-Lu(x_0) + c(x_0)u(x_0) - f(x_0), u(x_0) - \varphi(x_0)\} \geq 0,$$

and so conclude that the value function v is a supersolution to the obstacle problem (1.5). Definition (1.7) implies that $v(x) \geq \varphi(x)$, for all $x \in \mathbb{R}^n$, and in particular that $u(x_0) \geq \varphi(x_0)$. Thus, we need to show that

$$-Lu(x_0) + c(x_0)u(x_0) \geq f(x_0). \tag{2.21}$$

Assuming by contradiction that the preceding inequality does not hold, there are positive constants, ε and r, such that

$$-Lu(x) + c(x)u(x) \leq f(x) - \varepsilon, \quad \forall x \in B_r(x_0). \tag{2.22}$$

Using the fact that $u(x_0) = v(x_0)$, and applying Itô's rule to u and the unique strong solution, $\{X^{x_0}(t)\}_{t\geq 0}$, to equation (1.1) with initial condition $X^{x_0}(0) = x_0$, we obtain similarly to Step 1 that

$$\begin{aligned} v(x_0) &= \mathbb{E}\Big[e^{-\int_0^{\tau_r} c(X^{x_0}(s))\,ds} u(X^{x_0}(\tau_r)) \\ &\qquad + \int_0^{\tau_r} e^{-\int_0^t c(X^{x_0}(s))\,ds}(-L + c(X^{x_0}(t-)))u(X^{x_0}(t))\,dt\Big] \\ &\leq \mathbb{E}\left[e^{-\int_0^{\tau_r} c(X^{x_0}(s))\,ds} v(X^{x_0}(\tau_r)) + \int_0^{\tau_r} e^{-\int_0^t c(X^{x_0}(s))\,ds} f(X^{x_0}(t))\,dt\right] \\ &\quad - \varepsilon\mathbb{E}\left[\int_0^{\tau_r} e^{-\int_0^t c(X^{x_0}(s))\,ds}\,dt\right], \end{aligned}$$

where in the second inequality we used the fact that $u(x) \leq v(x)$, for all $x \in \mathbb{R}^n$, together with assumption (2.22). Notice that the first term in the second inequality above is $v(x_0; r, \tau_r)$ by identity (2.9). Applying Lemma 2.1 we have that $v(x_0; r, \tau_r) \leq v(x_0)$, which gives us that

$$v(x_0) \leq v(x_0) - \varepsilon\mathbb{E}_{x_0}\left[\int_0^{\tau_r} e^{-\int_0^t c(X(s))\,ds}\,dt\right].$$

Because $\mathbb{P}_{x_0}(\tau_r > 0)$ is positive, we obtain the contradiction $v(x_0) < v(x_0)$. This implies that inequality (2.21) holds, which shows that v is a supersolution to equation (1.5).

Steps 1 and 2 complete the proof that the value function defined in (1.7) is a viscosity solution to the obstacle problem (1.5). □

We prove Theorem 1.5 with the aid of the following comparison principle:

THEOREM 2.2 (Comparison principle). *Suppose that Assumption* 1.1 *holds, the coefficient* $c \in C(\mathbb{R}^n)$ *satisfies condition* (1.9), $f \in C(\mathbb{R}^n)$, *and condition* (1.13) *holds. If* u *and* v *are a viscosity subsolution and supersolution to the obstacle problem* (1.5), *respectively, then* $u \leq v$.

PROOF. As usual in comparison arguments for viscosity solutions on unbounded domains, [**7**, Proof of Theorem 5.1], we let α and ε be positive constants and we define:

$$M_{\alpha,\varepsilon} := \sup_{x,y\in\mathbb{R}^n} \left\{u(x) - v(y) - \frac{\alpha}{2}|x-y|^2 - \varepsilon(|x|^2 + |y|^2)\right\}. \tag{2.23}$$

Letting $x_{\alpha,\varepsilon}$ and $y_{\alpha,\varepsilon}$ be points where the supremum of $M_{\alpha,\varepsilon}$ is attained, it follows by [**7**, Lemma 3.1] that

$$\alpha|x_{\alpha,\varepsilon} - y_{\alpha,\varepsilon}|^2 \to 0, \quad \text{as } \alpha \to \infty, \tag{2.24}$$

and so, we can assume without loss of generality that

$$x_{\alpha,\varepsilon}, y_{\alpha,\varepsilon} \to x_\varepsilon, \quad \text{as } \alpha \to \infty, \tag{2.25}$$

and applying again [**7**, Lemma 3.1], we have that

$$u(x_\varepsilon) - v(x_\varepsilon) - 2\varepsilon|x_\varepsilon|^2 = \sup_{x\in\mathbb{R}^n} \{u(x) - v(x) - 2\varepsilon|x|^2\}. \tag{2.26}$$

For all $\delta > 0$, condition (1.13) ensures that there is $r = r(\alpha, \delta, \varepsilon) \in (0, \delta)$ such that

$$B_r \subseteq \{y \in \mathbb{R}^n : |F(x_{\alpha,\varepsilon}, y)| < \delta \text{ and } |F(y_{\alpha,\varepsilon}, y)| < \delta\}.$$

We consider the the auxiliary function

$$\hat{u}(x) := u(x_{\alpha,\varepsilon}) + \frac{\alpha}{2}\left(|x - y_{\alpha,\varepsilon}|^2 - |x_{\alpha,\varepsilon} - y_{\alpha,\varepsilon}|^2\right) + \varepsilon(|x|^2 - |x_{\alpha,\varepsilon}|^2),$$

$$\bar{u}(x) := \begin{cases} \hat{u}(x), & \text{if } x \in B_r(x_{\alpha,\varepsilon}), \\ u(x), & \text{if } x \in (B_r(x_{\alpha,\varepsilon}))^c, \end{cases}$$

for all $x \in \mathbb{R}^n$. Above we denoted by $(B_r(x_{\alpha,\varepsilon}))^c$ the complement of $B_r(x_{\alpha,\varepsilon})$. From definition (2.23) of $M_{\alpha,\varepsilon}$ and from the choice of the points $x_{\alpha,\varepsilon}$ and $y_{\alpha,\varepsilon}$, we see that $u \leq \bar{u}$ on $\mathbb{R}^n$ and $x_{\alpha,\varepsilon}$ is a point at which $u - \bar{u}$ attains its maximum. Similarly, we define

$$\hat{v}(y) := v(y_{\alpha,\varepsilon}) + \frac{\alpha}{2}\left(|x_{\alpha,\varepsilon} - y_{\alpha,\varepsilon}|^2 - |x_{\alpha,\varepsilon} - y|^2\right) - \varepsilon(|y|^2 - |y_{\alpha,\varepsilon}|^2),$$

$$\bar{v}(y) := \begin{cases} \hat{v}(y), & \text{if } y \in B_r(y_{\alpha,\varepsilon}), \\ v(y), & \text{if } y \in (B_r(y_{\alpha,\varepsilon}))^c, \end{cases}$$

for all $y \in \mathbb{R}^n$, and we see that $v \geq \bar{v}$ on $\mathbb{R}^n$ and $y_{\alpha,\varepsilon}$ is a point at which $v - \bar{v}$ attains its minimum. By construction, the auxiliary functions $\bar{u}$ and $\bar{v}$ are C^2 in a neighborhood of $x_{\alpha,\varepsilon}$ and $y_{\alpha,\varepsilon}$, respectively, and are continuous functions on $\mathbb{R}^n \setminus \partial B_r(x_{\alpha,\varepsilon})$ and $\mathbb{R}^n \setminus \partial B_r(y_{\alpha,\varepsilon})$, respectively. A mollification argument applied to $\bar{u}$ and $\bar{v}$ shows that even though $\bar{u}$ and $\bar{v}$ are not C^2 functions on $\mathbb{R}^n$, we can still apply Definition 1.3 to obtain that

$$\begin{aligned} \min\{-L\bar{u}(x_{\alpha,\varepsilon}) + c(x_{\alpha,\varepsilon})\bar{u}(x_{\alpha,\varepsilon}) - f(x_{\alpha,\varepsilon}), \bar{u}(x_{\alpha,\varepsilon}) - \varphi(x_{\alpha,\varepsilon})\} &\leq 0, \\ \min\{-L\bar{v}(y_{\alpha,\varepsilon}) + c(y_{\alpha,\varepsilon})\bar{v}(y_{\alpha,\varepsilon}) - f(y_{\alpha,\varepsilon}), \bar{v}(y_{\alpha,\varepsilon}) - \varphi(y_{\alpha,\varepsilon})\} &\geq 0. \end{aligned} \tag{2.27}$$

For $\varepsilon > 0$ fixed, if there is a sequence $\{\alpha_k\}_{k\in\mathbb{N}}$ converging to infinity such that $\bar{u}(x_{\alpha_k,\varepsilon}) \leq \varphi(x_{\alpha_k,\varepsilon})$, then it follows by (2.25) that $\bar{u}(x_\varepsilon) \leq \varphi(x_\varepsilon)$. The second inequality in (2.27) shows that $\bar{v}(y_{\alpha,\varepsilon}) \geq \varphi(y_{\alpha,\varepsilon})$, for all $\alpha > 0$, and so we have that

$$v(x_\varepsilon) \geq \varphi(x_\varepsilon) \geq u(x_\varepsilon). \tag{2.28}$$

If we cannot find a sequence $\{\alpha_k\}_{k\in\mathbb{N}}$ satisfying the preceding property, then the inequalities in (2.27) hold for the first terms on the left-hand side, which by subtraction imply that

$$\begin{aligned} c(x_{\alpha,\varepsilon})u(x_{\alpha,\varepsilon}) - c(y_{\alpha,\varepsilon})v(y_{\alpha,\varepsilon}) &= c(x_{\alpha,\varepsilon})\bar{u}(x_{\alpha,\varepsilon}) - c(y_{\alpha,\varepsilon})\bar{v}(y_{\alpha,\varepsilon}) \\ &\leq L\bar{u}(x_{\alpha,\varepsilon}) - L\bar{v}(y_{\alpha,\varepsilon}) + f(x_{\alpha,\varepsilon}) - f(y_{\alpha,\varepsilon}). \end{aligned} \tag{2.29}$$

We write the last term in the preceding inequality as a sum $I_1+I_2+I_3$, where each term is defined by

$$
\begin{aligned}
I_1 &:= b(x_{\alpha,\varepsilon})\cdot(\alpha z_{\alpha,\varepsilon}+2\varepsilon x_{\alpha,\varepsilon}) - b(y_{\alpha,\varepsilon})\cdot(\alpha z_{\alpha,\varepsilon}-2\varepsilon y_{\alpha,\varepsilon}) + f(x_{\alpha,\varepsilon}) - f(y_{\alpha,\varepsilon}),\\
I_2 &:= \int_{B_r\setminus\{O\}} (\bar u(x_{\alpha,\varepsilon}+F(x_{\alpha,\varepsilon},y)) - \bar u(x_{\alpha,\varepsilon}) - \nabla \bar u(x_{\alpha,\varepsilon})\cdot F(x_{\alpha,\varepsilon},y))\,\nu(dy)\\
&\quad - \int_{B_r\setminus\{O\}} (\bar v(y_{\alpha,\varepsilon}+F(y_{\alpha,\varepsilon},y)) - \bar v(y_{\alpha,\varepsilon}) - \nabla \bar v(y_{\alpha,\varepsilon})\cdot F(y_{\alpha,\varepsilon},y))\,\nu(dy),\\
I_3 &:= \int_{B_r^c} (\bar u(x_{\alpha,\varepsilon}+F(x_{\alpha,\varepsilon},y)) - \bar u(x_{\alpha,\varepsilon}) - \nabla \bar u(x_{\alpha,\varepsilon})\cdot F(x_{\alpha,\varepsilon},y))\,\nu(dy)\\
&\quad - \int_{B_r^c} (\bar v(y_{\alpha,\varepsilon}+F(y_{\alpha,\varepsilon},y)) - \bar v(y_{\alpha,\varepsilon}) - \nabla \bar v(y_{\alpha,\varepsilon})\cdot F(y_{\alpha,\varepsilon},y))\,\nu(dy),
\end{aligned}
$$

where we denote $z_{\alpha,\varepsilon} := x_{\alpha,\varepsilon} - y_{\alpha,\varepsilon}$. Rearranging terms and using the definitions of the functions $\bar u$ and $\bar v$, we obtain by direct computations:

$$
\begin{aligned}
I_1 &= \alpha\,(b(x_{\alpha,\varepsilon}) - b(y_{\alpha,\varepsilon}))\cdot z_{\alpha,\varepsilon} + 2\varepsilon\,(b(x_{\alpha,\varepsilon})\cdot x_{\alpha,\varepsilon} + b(y_{\alpha,\varepsilon})\cdot y_{\alpha,\varepsilon}) + f(x_{\alpha,\varepsilon}) - f(y_{\alpha,\varepsilon}),\\
I_2 &= \left(\frac{\alpha}{2}+\varepsilon\right)\int_{B_r\setminus\{O\}} \left(|F(x_{\alpha,\varepsilon},y)|^2 + |F(y_{\alpha,\varepsilon},y)|^2\right)\,\nu(dy),\\
I_3 &= \int_{B_r^c} (u(x_{\alpha,\varepsilon}+F(x_{\alpha,\varepsilon},y)) - u(x_{\alpha,\varepsilon}) - v(y_{\alpha,\varepsilon}+F(y_{\alpha,\varepsilon},y)) + v(y_{\alpha,\varepsilon}))\,\nu(dy)\\
&\quad - \alpha\int_{B_r^c} z_{\alpha,\varepsilon}\cdot(F(x_{\alpha,\varepsilon},y) - F(y_{\alpha,\varepsilon},y))\,d\nu(dy)\\
&\quad - 2\varepsilon\int_{B_r^c} (x_{\alpha,\varepsilon}\cdot F(x_{\alpha,\varepsilon},y) + y_{\alpha,\varepsilon}\cdot F(y_{\alpha,\varepsilon},y))\cdot\nu(dy).
\end{aligned}
$$

Using definition (2.23) of $M_{\alpha,\varepsilon}$, it follows that

$$
\begin{aligned}
&u(x_{\alpha,\varepsilon}+F(x_{\alpha,\varepsilon},y)) - u(x_{\alpha,\varepsilon}) - v(y_{\alpha,\varepsilon}+F(y_{\alpha,\varepsilon},y)) + v(y_{\alpha,\varepsilon})\\
&\qquad\le \alpha z_{\alpha,\varepsilon}\cdot(F(x_{\alpha,\varepsilon},y) - F(y_{\alpha,\varepsilon},y))\\
&\qquad\quad + \frac{\alpha}{2}\left|F(x_{\alpha,\varepsilon},y) - F(y_{\alpha,\varepsilon},y)\right|^2 + 2\varepsilon x_{\alpha,\varepsilon}\cdot F(x_{\alpha,\varepsilon},y) + 2\varepsilon y_{\alpha,\varepsilon}\cdot F(y_{\alpha,\varepsilon},y)\\
&\qquad\quad + \varepsilon\left(|F(x_{\alpha,\varepsilon},y)|^2 + |F(y_{\alpha,\varepsilon},y)|^2\right),
\end{aligned}
$$

which implies that

$$
\begin{aligned}
I_3 &\le \frac{\alpha}{2}\int_{B_r^c} |F(x_{\alpha,\varepsilon},y) - F(y_{\alpha,\varepsilon},y)|^2\,\nu(dy) + \varepsilon\int_{B_r^c} |\left(F(x_{\alpha,\varepsilon},y)|^2 + |F(y_{\alpha,\varepsilon},y)|^2\right)\,\nu(dy)\\
&\le K\left(\frac{\alpha}{2}|x_{\alpha,\varepsilon} - y_{\alpha,\varepsilon}|^2 + 2\varepsilon\right),
\end{aligned}
$$

where in the last line we used conditions (1.2) and (1.3). Property (1.3) implies that I_2 converges to zero, as $r\downarrow 0$. Applying the preceding observations to the right-hand side of identity (2.29) and letting r tend to zero, we obtain

$$
\begin{aligned}
&c(x_{\alpha,\varepsilon})u(x_{\alpha,\varepsilon}) - c(y_{\alpha,\varepsilon})v(y_{\alpha,\varepsilon})\\
&\qquad\le \alpha\,(b(x_{\alpha,\varepsilon}) - b(y_{\alpha,\varepsilon}))\cdot z_{\alpha,\varepsilon} + 2\varepsilon\,(b(x_{\alpha,\varepsilon})\cdot x_{\alpha,\varepsilon} + b(y_{\alpha,\varepsilon})\cdot y_{\alpha,\varepsilon})\\
&\qquad\quad + f(x_{\alpha,\varepsilon}) - f(y_{\alpha,\varepsilon}) + K\left(\frac{\alpha}{2}|z_{\alpha,\varepsilon}|^2 + 2\varepsilon\right).
\end{aligned}
$$

Because b is a bounded, Lipschitz continuous function, we can find a positive constant, C, such that

$$c(x_{\alpha,\varepsilon})u(x_{\alpha,\varepsilon}) - c(y_{\alpha,\varepsilon})v(y_{\alpha,\varepsilon}) \leq C\left(\alpha|z_{\alpha,\varepsilon}|^2 + \varepsilon + \varepsilon(|x_{\alpha,\varepsilon}| + |y_{\alpha,\varepsilon}|)\right) + |f(x_{\alpha,\varepsilon}) - f(y_{\alpha,\varepsilon})|.$$

Letting now α tend to infinity, and using properties (2.24), (2.25), and the fact that $f \in C(\mathbb{R}^n)$, it follows that

$$c(x_\varepsilon)\left(u(x_\varepsilon) - v(x_\varepsilon)\right) \leq C\left(\varepsilon + \varepsilon|x_\varepsilon|\right), \tag{2.30}$$

and using condition (1.9) satisfied by the coefficient $c(x)$, we have

$$u(x_\varepsilon) - v(x_\varepsilon) \leq C\left(\varepsilon + \varepsilon|x_\varepsilon|\right).$$

It is clear from identity (2.26) that $\varepsilon|x_\varepsilon|^2$ is bounded in ε, and that

$$\lim_{\varepsilon\downarrow 0}(u(x_\varepsilon) - v(x_\varepsilon)) = \sup_{x\in\mathbb{R}^n}\{u(x) - v(x)\}.$$

Letting now ε tend to zero in the preceding two properties, we obtain that $u(x) - v(x) \leq 0$, for all $x \in \mathbb{R}^n$. Together with inequality (2.28), this completes the proof. □

REMARK 2.3. Examining the proof of Theorem 2.2, we see from the inequality (2.30) that

$$u(x_\varepsilon) - v(x_\varepsilon) \leq \frac{C}{c(x_\varepsilon)}\left(\varepsilon + \varepsilon|x_\varepsilon|\right),$$

where the right-hand side converges to zero when we assume that $c(x)$ is a positive function on $\mathbb{R}^n$ and satisfies property (1.14). Thus, as indicated in Remark 1.14, we again obtain that $u(x)-v(x) \leq 0$, for all $x \in \mathbb{R}^n$, and the conclusion of Theorem 2.2 holds.

PROOF OF THEOREM 1.5. It is an obvious consequence of Theorem 2.2. □

3. Evolution obstacle problem

In this section, we outline the proofs of Proposition 1.7, and Theorems 1.9 and 1.11, and of a Dynamical Programming Principle in Lemma 3.2 and a comparison principle in Theorem 3.3. Because the proofs are very similar to those for the stationary obstacle problem, we only point out the main changes that need to be done to the proofs in § 2.

We begin with an auxiliary lemma which we use to prove Proposition 1.7:

LEMMA 3.1 (Continuity properties of $\{X(t)\}_{t\geq 0}$). *Suppose that Assumption* 1.1 *is satisfied. Then there is a positive constant,* $C = C(\|b\|_{C^{0,1}(\mathbb{R}^n)}, K)$ *such that*

$$\mathbb{E}\left[\max_{s\in[0,t]}|X^{x_1}(s) - X^{x_2}(s)|^2\right] \leq C|x_1 - x_2|^2 e^{Ct}, \quad \forall\, x_1, x_2 \in \mathbb{R}^n,\ t \geq 0, \tag{3.1}$$

$$\mathbb{E}\left[\max_{r\in[s,t]}|X^{x}(r) - X^{x}(s)|^2\right] \leq C|t-s| \vee |t-s|^2, \quad \forall\, x \in \mathbb{R}^n,\ 0 \leq s < t. \tag{3.2}$$

PROOF. To prove inequality (3.1), using the stochastic equation (1.1), we have that

$$X^{x_1}(t) - X^{x_2}(t) = x_1 - x_2 + \int_0^t (b(X^{x_1}(s-)) - b(X^{x_2}(s-)))\,ds + M(t), \quad \forall\, t \geq 0, \tag{3.3}$$

where we denote by $\{M(t)\}_{t\geq 0}$ the square-integrable martingale:

$$M(t) := \int_0^t \int_{\mathbb{R}^n\setminus\{O\}} \left(F(X^{x_1}(s-),y) - F(X^{x_2}(s-),y)\right) \widetilde{N}(ds,dy).$$

Applying Doob's martingale inequality [**1**, Theorem 2.1.5] and [**1**, Lemma 4.2.2] to $\{M(t)\}_{t\geq 0}$, and using property (1.2), it follows that

$$\begin{aligned}
\mathbb{E}\left[\sup_{s\in[0,t]} |M(s)|^2\right] &\leq 4\mathbb{E}\left[|M(t)|^2\right]\\
&\leq \mathbb{E}\left[\int_0^t \int_{\mathbb{R}^n\setminus\{O\}} \left|F(X^{x_1}(s-),y) - F(X^{x_2}(s-),y)\right|^2 \nu(dy)\, ds\right]\\
&\leq K\mathbb{E}\left[\int_0^t |X^{x_1}(s) - X^{x_2}(s)|^2\, ds\right].
\end{aligned}$$

The preceding inequality, identity (3.3), and the Lipschitz continuity of the drift coefficient $b(x)$, together with Gronwall's inequality, imply estimate (3.1). To prove inequality (3.2), we again use the stochastic equation (1.1) and we obtain that

$$X^x(t) - X^x(s) = \int_s^t b(X^x(r))\, dr + N(t), \quad \forall\, 0\leq s<t, \tag{3.4}$$

where we denote by $\{N(t)\}_{t\geq 0}$ the square-integrable martingale:

$$N(t) := \int_s^t \int_{\mathbb{R}^n\setminus\{O\}} F(X^x(r-),y)\widetilde{N}(dr,dy), \quad \forall\, t>s.$$

Applying Doob's martingale inequality [**1**, Theorem 2.1.5] and [**1**, Lemma 4.2.2] to $\{N(t)\}_{t\geq 0}$, and using the boundedness of the coefficient $b(x)$, it follows from identity (3.4) that

$$\begin{aligned}
&\mathbb{E}\left[\sup_{r\in[s,t]} |X^x(r) - X^x(s)|^2\right]\\
&\quad\leq C\left(|t-s|^2 + \mathbb{E}\left[\int_s^t \int_{\mathbb{R}^n\setminus\{O\}} |F(X^x(r-),y)|^2\, \nu(dy)\, ds\right]\right)\\
&\quad\leq C\left(|t-s|^2 + |t-s|\int_{\mathbb{R}^n\setminus\{O\}} |\rho(y)|^2\, \nu(dy)\right) \qquad \text{(by condition (1.3))}\\
&\quad\leq C|t-s| \vee |t-s|^2.
\end{aligned}$$

This concludes the proof of inequality (3.2) and of Lemma 3.1. □

PROOF OF PROPOSITION 1.7. We divide the proof into two steps.

STEP 1 (Lipschitz regularity in the spatial variable). Using the expression of the value function (1.17), we see that $|v(t,x_1)-v(t,x_2)| \leq \sup\{|v(t,x_1;\tau)-v(t,x_2;\tau)| : \tau\in\mathcal{T}_{T-t}\}$, and so our goal is to prove that there is a positive constant C such that

$$|v(t,x_1;\tau) - v(t,x_2;\tau)| \leq C|x_1-x_2|, \quad \forall\, t\in[0,T], \quad \forall\, x_1,x_2\in\mathbb{R}^n, \quad \forall\, \tau\in\mathcal{T}_{T-t}. \tag{3.5}$$

Similarly to the proof of Proposition 1.2, using the Lipschitz continuity of c, f, φ, and g, and the boundedness of the zeroth order term c and of the stopping time

$\tau \in [0,T]$, we obtain that

$$|v(t,x_1;\tau) - v(t,x_2;\tau)| \leq C\mathbb{E}\left[\max_{0\leq s\leq T-t} |X^{x_1}(s) - X^{x_2}(s)|\right].$$

Applying Hölder's inequality and estimate (3.1) to the right-hand side of the preceding inequality, we obtain that (3.5) holds, and so we have that

$$\sup\{\|v(t,\cdot)\|_{C^{0,1}(\mathbb{R}^n)} : t \in [0,T]\} < \infty. \tag{3.6}$$

STEP 2 ($C^{\frac{1}{2}}$-regularity in the time variable). We assume without loss of generality that $0 \leq t_1 < t_2 \leq T$, and using the fact that $\mathcal{T}_{T-t_2} \subset \mathcal{T}_{T-t_1}$, we have that the following inequalities hold:

$$v(t_2,x) - v(t_1,x) \leq \sup_{\tau\in\mathcal{T}_{T-t_2}} (v(t_2,x;\tau) - v(t_1,x;\tau')),$$
$$v(t_1,x) - v(t_2,x) \leq \sup_{\tau\in\mathcal{T}_{T-t_1}} (v(t_1,x;\tau) - v(t_2,x;\tau'')),$$

where we used the notation

$$\tau' := \tau\mathbf{1}_{\{\tau<T-t_2\}} + (T-t_1)\mathbf{1}_{\{\tau=T-t_2\}} \quad \text{and} \quad \tau'' := \tau\wedge(T-t_2) \in \mathcal{T}_{T-t_2}. \tag{3.7}$$

Our goal is to prove that there is a positive constant C such that

$$v(t_2,x;\tau) - v(t_1,x;\tau') \leq C|t_1-t_2|, \quad \forall \tau\in\mathcal{T}_{T-t_2}, \tag{3.8}$$

$$v(t_1,x;\tau) - v(t_2,x;\tau'') \leq C|t_1-t_2|^{\frac{1}{2}}, \quad \forall \tau\in\mathcal{T}_{T-t_1}. \tag{3.9}$$

The preceding four inequalities will imply that

$$\sup\{\|v(\cdot,x)\|_{C^{\frac{1}{2}}([0,T])} : x\in\mathbb{R}^n\} < \infty. \tag{3.10}$$

We split the proofs of inequalities (3.8) and (3.9) into two cases.

CASE 1 (Proof of inequality (3.8)). We denote for simplicity

$$E(t,s,x) := e^{-\int_0^s c(t+r,X^x(r))\,dr}, \quad \forall t\in[0,T],\ s\in[0,T-t],\ x\in\mathbb{R}^n,$$

and using the choice of the stopping times τ' in (3.7), we obtain the decomposition:

$$\begin{aligned}
&v(t_2,x;\tau) - v(t_1,x;\tau')\\
&\quad= \mathbb{E}\left[\mathbf{1}_{\{\tau<T-t_2\}}\left(E(t_2,\tau,x) - E(t_1,\tau,x)\right)\varphi(t_2+\tau,X^x(\tau))\right]\\
&\qquad+ \mathbb{E}\left[\mathbf{1}_{\{\tau<T-t_2\}}E(t_1,\tau,x)\left(\varphi(t_2+\tau,X^x(\tau)) - \varphi(t_1+\tau,X^x(\tau))\right)\right]\\
&\qquad+ \mathbb{E}\left[\mathbf{1}_{\{\tau=T-t_2\}}\left(E(t_2,\tau,x) - E(t_1,\tau,x)\right)g(X^x(\tau))\right]\\
&\qquad\quad(\tau' = T-t_1 \text{ when } \tau = T-t_2 \text{ by (3.7)})\\
&\qquad+ \mathbb{E}\left[\int_0^\tau \left(E(t_2,r,x) - E(t_1,r,x)\right)f(t_2+r,X^x(r))\,dr\right]\\
&\qquad+ \mathbb{E}\left[\int_0^\tau E(t_1,r,x)\left(f(t_2+r,X^x(r) - f(t_1+r,X^x(r))\right)\,dr\right]\\
&\qquad- \mathbb{E}\left[\mathbf{1}_{\{\tau=T-t_2\}}\int_{T-t_2}^{T-t_1} E(t_1,r,x)f(t_1+r,X^x(r))\,dr\right].
\end{aligned}$$

Inequality (3.8) is an immediate consequence of the preceding expression, the boundedness of the zeroth order term c, and of the Lipschitz continuity of the functions c, f, φ and g.

CASE 2 (Proof of inequality (3.9)). Similarly to the proof of Case 1, by direct computation we can write the difference as a sum of three terms corresponding to the cases $\tau < T - t_2$, $T - t_2 \leq \tau < T - t_1$, and $\tau = T - t_1$:

$$v(t_1, x; \tau) - v(t_2, x; \tau'') = I_1 + I_2 + I_3,$$

where we denote:

$$\begin{aligned}
I_1 &:= \mathbb{E}\left[\mathbf{1}_{\{\tau<T-t_2\}}\left(E(t_1,\tau,x)\varphi(t_1+\tau,X^x(\tau)) - E(t_2,\tau,x)\varphi(t_2+\tau,X^x(\tau))\right)\right]\\
&\quad + \mathbb{E}\left[\int_0^{\tau\wedge(T-t_2)} \left(E(t_1,r,x)f(t_1+r,X^x(r-)) - E(t_2,r,x)f(t_2+r,X^x(r))\right) dr\right],\\
I_2 &:= \mathbb{E}\left[\mathbf{1}_{\{T-t_2\leq\tau<T-t_1\}}\left(E(t_1,\tau,x)\varphi(t_1+\tau,X^x(\tau)) - E(t_2,T-t_2,x)g(X^x(T-t_2))\right)\right]\\
&\quad + \mathbb{E}\left[\mathbf{1}_{\{T-t_2\leq\tau<T-t_1\}}\int_{T-t_2}^{\tau} E(t_1,r,x)f(t_1+r,X^x(r)\,dr\right],\\
I_3 &:= \mathbb{E}\left[\mathbf{1}_{\{\tau=T-t_1\}}\left(E(t_1,\tau,x)g(X^x(T-t_1)) - E(t_2,\tau,x)g(X^x(T-t_2))\right)\right].
\end{aligned}$$

To prove inequality (3.9) we further expand each term as follows:

$$\begin{aligned}
I_1 &= \mathbb{E}\left[\mathbf{1}_{\{\tau<T-t_2\}}\left(E(t_1,\tau,x) - E(t_2,\tau,x)\right)\varphi(t_1+\tau,X^x(\tau))\right]\\
&\quad + \mathbb{E}\left[\mathbf{1}_{\{\tau<T-t_2\}}E(t_2,\tau,x)\left(\varphi(t_1+\tau,X^x(\tau)) - \varphi(t_2+\tau,X^x(\tau))\right)\right]\\
&\quad + \mathbb{E}\left[\int_0^{\tau\wedge(T-t_2)} \left(E(t_1,r,x) - E(t_2,r,x)\right) f(t_2+r,X^x(r))\,dr\right]\\
&\quad + \mathbb{E}\left[\int_0^{\tau\wedge(T-t_2)} E(t_2,r,x)\left(f(t_1+r,X^x(r-)) - f(t_2+r,X^x(r))\right) dr\right],\\
I_2 &= \mathbb{E}\left[\mathbf{1}_{\{T-t_2\leq\tau<T-t_1\}}\left(E(t_1,\tau,x) - E(t_2,\tau,x)\right)\varphi(t_1+\tau,X^x(\tau))\right]\\
&\quad + \mathbb{E}\left[\mathbf{1}_{\{T-t_2\leq\tau<T-t_1\}}\left(E(t_2,\tau,x) - E(t_2,T-t_2,x)\right)\varphi(t_1+\tau,X^x(\tau))\right]\\
&\quad + \mathbb{E}\left[\mathbf{1}_{\{T-t_2\leq\tau<T-t_1\}}E(t_2,T-t_2,x)\left(\varphi(t_1+\tau,X^x(\tau)) - \varphi(T,X^x(\tau))\right)\right]\\
&\quad + \mathbb{E}\left[\mathbf{1}_{\{T-t_2\leq\tau<T-t_1\}}E(t_2,T-t_2,x)\left(\varphi(T,X^x(\tau)) - g(X^x(\tau))\right)\right]\\
&\quad + \mathbb{E}\left[\mathbf{1}_{\{T-t_2\leq\tau<T-t_1\}}E(t_2,T-t_2,x)\left(g(X^x(\tau)) - g(X^x(T-t_2))\right)\right]\\
&\quad + \mathbb{E}\left[\mathbf{1}_{\{T-t_2\leq\tau<T-t_1\}}\int_{T-t_2}^{\tau} E(t_1,r,x)f(t_1+r,X^x(r)\,dr\right],\\
I_3 &= \mathbb{E}\left[\mathbf{1}_{\{\tau=T-t_1\}}\left(E(t_1,\tau,x) - E(t_2,\tau,x)\right)g(X^x(T-t_1))\right]\\
&\quad + \mathbb{E}\left[\mathbf{1}_{\{\tau=T-t_1\}}\left(E(t_2,\tau,x) - E(t_2,T-t_2,x)\right)g(X^x(T-t_1))\right]\\
&\quad + \mathbb{E}\left[\mathbf{1}_{\{\tau=T-t_1\}}E(t_2,T-t_2,x)\left(g(X^x(T-t_1)) - g(X^x(T-t_2))\right)\right].
\end{aligned}$$

The compatibility condition (1.16) gives us that the fourth term in the expression of I_2 is non-positive. Using the fact that c is a bounded function and g is Lipschitz continuous, we bound the fifth term in the expression of I_2 and the last term in the expression of I_3 by

$$\begin{aligned}
&\left|\mathbb{E}\left[\mathbf{1}_{\{T-t_2\leq\tau<T-t_1\}}E(t_2,T-t_2,x)\left(g(X^x(\tau)) - g(X^x(T-t_2))\right)\right]\right|\\
&\quad + \left|\mathbb{E}\left[\mathbf{1}_{\{\tau=T-t_1\}}E(t_2,T-t_2,x)\left(g(X^x(T-t_1)) - g(X^x(T-t_2))\right)\right]\right|\\
&\quad \leq C\mathbb{E}\left[\sup_{T-t_2\leq s<T-t_1} |X^x(s) - X^x(T-t_2)|\right],
\end{aligned}$$

for a positive constant C. Using again the fact that the zeroth order term c is bounded and c, f, φ and g are Lipschitz functions, the remaining terms in the expression of the difference $v(t_1, x; \tau) - v(t_2, x; \tau'')$ can be bounded by $C|t_1 - t_2|$, for a positive constant C. Hence, it follows that

$$v(t_1, x; \tau) - v(t_2, x; \tau'') \leq C|t_1 - t_2| + C\mathbb{E}\left[\sup_{T-t_2 \leq s < T-t_1} |X^x(s) - X^x(T - t_2)|\right].$$

Applying Hölder's inequality and estimate (3.2) to the second term on the right-hand side, it follows that property (3.9) holds.

The proofs of inequalities (3.8) and (3.9) imply estimate (3.10).

Properties (3.6) and (3.10) complete the proof. □

We have the following analogue in the evolution case of the Dynamic Programming Principle in the stationary case proved in Lemma 2.1:

LEMMA 3.2 (Dynamic Programming Principle). *Suppose that the hypotheses of Proposition* 1.7 *hold. Then, the value function* $v(t, x)$ *defined in* (1.17) *satisfies:*

$$v(t, x) = \sup\{v(t, x; r, \tau) : \tau \leq \tau_r \wedge (T - t)\}, \quad \forall (t, x) \in [0, T) \times \mathbb{R}^n, \tag{3.11}$$

where we denote

$$\begin{aligned} v(t, x; r, \tau) := &\ \mathbb{E}\left[e^{-\int_0^\tau c(t+s, X^x(s))\,ds} \varphi(t + \tau, X^x(\tau)) \mathbf{1}_{\{\tau < \tau_r \wedge (T-t)\}}\right] \\ &+ \mathbb{E}\left[e^{-\int_0^\tau c(t+s, X^x(s))\,ds} v(t + \tau, X^x(\tau)) \mathbf{1}_{\{\tau = \tau_r, \tau < T-t\}}\right] \\ &+ \mathbb{E}\left[e^{-\int_0^\tau c(t+s, X^x(s))\,ds} g(X^x(\tau)) \mathbf{1}_{\{\tau < \tau_r, \tau = T-t\}}\right] \\ &+ \mathbb{E}\left[\int_0^\tau e^{-\int_0^\rho c(t+s, X^x(s))\,ds} f(t + \rho, X^x(\rho))\,d\rho\right]. \end{aligned} \tag{3.12}$$

PROOF. To prove Lemma 3.2 we can employ the argument used to establish Lemma 2.1 with the following changes. We choose the stopping times τ in $\mathcal{T}_{T-t}$ instead of $\mathcal{T}$, and we replace the use of the function $v(x; r, \tau)$ defined in (2.9) by that of $v(t, x; r, \tau)$ in (3.12). In Step 1 of the proof of Lemma 2.1 we condition on the σ-algebra $\mathcal{F}_{\tau_r \wedge (T-t)}$ instead of $\mathcal{F}_{\tau_r}$, and in Step 2 of Lemma 2.1, we choose the family of sets $\{A_k\}_{k \in \mathbb{N}}$ such that it partitions $[0, T] \times \mathbb{R}^n$ instead of $\mathbb{R}^n$, and we replace the application of Proposition 1.2 by that of Proposition 1.7. We omit the remaining details of the proof for brevity. □

We can now use Lemma 3.2 to prove the existence of viscosity solutions to the evolution obstacle problem (1.15).

PROOF OF THEOREM 1.9. The proof is very similar to that of Theorem 1.4 with the observation that we replace the application of Lemma 2.1 with that of Lemma 3.2. We omit the detailed proof. □

THEOREM 3.3 (Comparison principle). *Suppose that Assumption* 1.1 *is satisfied,* g *belongs to* $C(\mathbb{R}^n)$*,* c, f, φ *are in* $C([0, T] \times \mathbb{R}^n)$*, and conditions* (1.13) *and* (1.16) *hold. If* u *and* v *are a viscosity subsolution and supersolution to the evolution obstacle problem* (1.15)*, respectively, then* $u \leq v$.

PROOF. The proof is identical to that of Theorem 2.2 except for the following modifications. We replace the definition of $M_{\alpha,\varepsilon}$ in (2.23) by

(3.13)

$$M_{\alpha,\varepsilon} := \sup_{(t,x),(s,y)\in[0,T]\times\mathbb{R}^n} \left\{ u(t,x) - v(s,y) - \frac{\alpha}{2}\left(|x-y|^2 + |t-s|^2\right) - \varepsilon(|x|^2+|y|^2) \right\}.$$

In the proof of Theorem 2.2 we used the fact that the zeroth order coefficient $c(x)$ is positive, which is not necessarily true under the hypotheses of Theorem 3.3. We notice that because $c(x)$ is bounded, there is a positive constant λ such that $\lambda + c(x) > 0$. From Definition 1.8 it is clear that if v is a viscosity sub- or supersolution to the evolution obstacle problem (1.15), then $e^{\lambda(T-t)}v$ is a viscosity sub- or supersolution to equation (1.15), where we replace $c(x)$ by $c(x) + \lambda > 0$, $f(t,x)$ by $e^{\lambda(T-t)}f(t,x)$, and $\varphi(t,x)$ by $e^{\lambda(T-t)}\varphi(t,x)$. We omit the remainder of the proof for brevity. □

PROOF OF THEOREM 1.11. The conclusion of Theorem 1.11 is an immediate consequence of Theorem 3.3. □

References

[1] David Applebaum, *Lévy processes and stochastic calculus*, 2nd ed., Cambridge Studies in Advanced Mathematics, vol. 116, Cambridge University Press, Cambridge, 2009. MR2512800

[2] S. I. Boyarchenko and S. Z. Levendorskiĭ, *Non-Gaussian Merton-Black-Scholes theory*, Advanced Series on Statistical Science & Applied Probability, vol. 9, World Scientific Publishing Co., Inc., River Edge, NJ, 2002.

[3] S. I. Boyarchenko and S. Z. Levendorskiĭ, *Perpetual American options under Lévy processes*, SIAM J. Control Optim. **40** (2002), no. 6, 1663–1696, DOI 10.1137/S0363012900373987. MR1897191

[4] Luis Caffarelli, Xavier Ros-Oton, and Joaquim Serra, *Obstacle problems for integro-differential operators: regularity of solutions and free boundaries*, Invent. Math. **208** (2017), no. 3, 1155–1211, DOI 10.1007/s00222-016-0703-3. MR3648978

[5] P. Carr, H. Geman, D. B. Madan, and M. Yor, *The fine structure of asset returns: An empirical investigation*, The Journal of Business **75** (2002), no. 2, 305–332.

[6] Rama Cont and Peter Tankov, *Financial modelling with jump processes*, Chapman & Hall/CRC Financial Mathematics Series, Chapman & Hall/CRC, Boca Raton, FL, 2004. MR2042661

[7] Michael G. Crandall, Hitoshi Ishii, and Pierre-Louis Lions, *User's guide to viscosity solutions of second order partial differential equations*, Bull. Amer. Math. Soc. (N.S.) **27** (1992), no. 1, 1–67, DOI 10.1090/S0273-0979-1992-00266-5. MR1118699

[8] Stewart N. Ethier and Thomas G. Kurtz, *Markov processes*, Wiley Series in Probability and Mathematical Statistics: Probability and Mathematical Statistics, John Wiley & Sons, Inc., New York, 1986. Characterization and convergence. MR838085

[9] D. B. Madan and E. Seneta, *The Variance Gamma (V.G.) model for share market returns*, The Journal of Business **63** (1990), 511–524.

[10] Ernesto Mordecki, *Optimal stopping and perpetual options for Lévy processes*, Finance Stoch. **6** (2002), no. 4, 473–493, DOI 10.1007/s007800200070. MR1932381

[11] Philip E. Protter, *Stochastic integration and differential equations*, Stochastic Modelling and Applied Probability, vol. 21, Springer-Verlag, Berlin, 2005. Second edition. Version 2.1; Corrected third printing. MR2273672

[12] Daniel W. Stroock and S. R. Srinivasa Varadhan, *Multidimensional diffusion processes*, Grundlehren der Mathematischen Wissenschaften [Fundamental Principles of Mathematical Sciences], vol. 233, Springer-Verlag, Berlin-New York, 1979. MR532498

Department of Mathematics, Purdue University, West Lafayette, IN 47907
Email address: danielli@math.purdue.edu

Department of Mathematics, Purdue University, West Lafayette, IN 47907
Email address: arshak@math.purdue.edu

School of Mathematics, University of Minnesota, Minneapolis, MN 55455
Email address: capop@umn.edu

Selected Published Titles in This Series

723 **Donatella Danielli, Arshak Petrosyan, and Camelia A. Pop, Editors,** New Developments in the Analysis of Nonlocal Operators, 2019

720 **Alexandre Girouard, Editor,** Spectral Theory and Applications, 2018

719 **Florian Sobieczky, Editor,** Unimodularity in Randomly Generated Graphs, 2018

718 **David Ayala, Daniel S. Freed, and Ryan E. Grady, Editors,** Topology and Quantum Theory in Interaction, 2018

717 **Federico Bonetto, David Borthwick, Evans Harrell, and Michael Loss, Editors,** Mathematical Problems in Quantum Physics, 2018

716 **Alex Martsinkovsky, Kiyoshi Igusa, and Gordana Todorov, Editors,** Surveys in Representation Theory of Algebras, 2018

715 **Sergio R. López-Permouth, Jae Keol Park, S. Tariq Rizvi, and Cosmin S. Roman, Editors,** Advances in Rings and Modules, 2018

714 **Jens Gerlach Christensen, Susanna Dann, and Matthew Dawson, Editors,** Representation Theory and Harmonic Analysis on Symmetric Spaces, 2018

713 **Naihuan Jing and Kailash C. Misra, Editors,** Representations of Lie Algebras, Quantum Groups and Related Topics, 2018

712 **Nero Budur, Tommaso de Fernex, Roi Docampo, and Kevin Tucker, Editors,** Local and Global Methods in Algebraic Geometry, 2018

711 **Thomas Creutzig and Andrew R. Linshaw, Editors,** Vertex Algebras and Geometry, 2018

710 **Raphaël Danchin, Reinhard Farwig, Jiří Neustupa, and Patrick Penel, Editors,** Mathematical Analysis in Fluid Mechanics, 2018

709 **Fernando Galaz-García, Juan Carlos Pardo Millán, and Pedro Solórzano, Editors,** Contributions of Mexican Mathematicians Abroad in Pure and Applied Mathematics, 2018

708 **Christian Ausoni, Kathryn Hess, Brenda Johnson, Ieke Moerdijk, and Jérôme Scherer, Editors,** An Alpine Bouquet of Algebraic Topology, 2018

707 **Nitya Kitchloo, Mona Merling, Jack Morava, Emily Riehl, and W. Stephen Wilson, Editors,** New Directions in Homotopy Theory, 2018

706 **Yeonhyang Kim, Sivaram K. Narayan, Gabriel Picioroaga, and Eric S. Weber, Editors,** Frames and Harmonic Analysis, 2018

705 **Graham J. Leuschke, Frauke Bleher, Ralf Schiffler, and Dan Zacharia, Editors,** Representations of Algebras, 2018

704 **Alain Escassut, Cristina Perez-Garcia, and Khodr Shamseddine, Editors,** Advances in Ultrametric Analysis, 2018

703 **Andreas Malmendier and Tony Shaska, Editors,** Higher Genus Curves in Mathematical Physics and Arithmetic Geometry, 2018

702 **Mark Grant, Gregory Lupton, and Lucile Vandembroucq, Editors,** Topological Complexity and Related Topics, 2018

701 **Joan-Carles Lario and V. Kumar Murty, Editors,** Number Theory Related to Modular Curves, 2018

700 **Alexandre Girouard, Dmitry Jakobson, Michael Levitin, Nilima Nigam, Iosif Polterovich, and Frédéric Rochon, Editors,** Geometric and Computational Spectral Theory, 2017

699 **Mark L. Agranovsky, Matania Ben-Artzi, Catherine Bénéteau, Lavi Karp, Dmitry Khavinson, Simeon Reich, David Shoikhet, Gilbert Weinstein, and Lawrence Zalcman, Editors,** Complex Analysis and Dynamical Systems VII, 2017

For a complete list of titles in this series, visit the
AMS Bookstore at **www.ams.org/bookstore/conmseries/**.

